Sun Kwok

Organic Matter in the Universe

Related Titles

Dvorak, R. (ed.)

Extrasolar Planets
Formation, Detection and Dynamics

2008
ISBN: 978-3-527-40671-5

Horneck, G., Rettberg, P. (eds)

Complete Course in Astrobiology

2007
ISBN: 978-3-527-40660-9

Shaw, A. M.

Astrochemistry
From Astronomy to Astrobiology

2006
ISBN: 978-0-470-09137-1

Stahler, S. W., Palla, F.

The Formation of Stars

2004
ISBN: 978-3-527-40559-6

Spitzer, L.

Physical Processes in the Interstellar Medium

1998
ISBN: 978-0-471-29335-4

Sun Kwok

Organic Matter in the Universe

WILEY-VCH

WILEY-VCH Verlag GmbH & Co. KGaA

The Author

Prof. Sun Kwok
The University of Hong Kong
Faculty of Science
Pukfulam Road
Hong Kong

Library of Congress Card No.: applied for

British Library Cataloguing-in-Publication Data:
A catalogue record for this book is available from the British Library.

Bibliographic information published by the Deutsche Nationalbibliothek
The Deutsche Nationalbibliothek lists this publication in the Deutsche Nationalbibliografie; detailed bibliographic data are available on the Internet at http://dnb.d-nb.de.

© 2012 WILEY-VCH Verlag GmbH & Co. KGaA, Boschstr. 12, 69469 Weinheim, Germany

Typesetting   le-tex publishing services GmbH, Leipzig
Printing and Binding   Fabulous Printers Pte Ltd

Cover Design   Grafik-Design Schulz, Fußgönheim

Printed in Singapore
Printed on acid-free paper

ISBN Print 978-3-527-40986-0

ISBN oBook 978-3-527-63703-4
ISBN eBook 978-3-527-63705-8
ISBN ePub 978-3-527-41119-1
ISBN Mobi 978-3-527-63706-5

Contents

Preface

Since the 1970s, millimeter and submillimeter observations have detected rotational transitions of over 160 molecules, including hydrocarbons, alcohols, acids, aldehydes, ketones, amines, ethers, and other organic molecules. Infrared ground based, airborne, and space-based spectroscopic observations have found evidence of complex carbonaceous compounds with aromatic and aliphatic structures in circumstellar and interstellar media. These infrared emission features are seen in many distant galaxies, suggesting organic synthesis had taken place even during the early days of the Universe. At the same time, complex organics are being found in meteorites, comets, and interplanetary dust particles. The *Cassini* mission and the *Huygens* probe have returned new results regarding the chemical composition of planetary and satellite atmospheres. There is an increasing recognition that organic compounds are major constituents of the atmosphere and surface of Titan. The sample return from the *STARDUST* mission is currently providing us with a great opportunity to examine the content of stellar material in the Solar System.

Laboratory isotopic analysis of meteorites and interplanetary dust collected in the upper atmosphere, and now cometary materials have revealed the presence of presolar grains similar to those formed in evolved stars. There also exists isotopic evidence that some Solar System organics have an interstellar chemical heritage. The direct link between star dust and the Solar System therefore suggests that the early Solar System was chemically enriched by both stellar ejecta and the products of interstellar processing.

Although this may sound surprising to many scientists, organic compounds are prevalent in the Universe, from our local backyard of the Solar System to distant galaxies formed billions of years ago. How stars and galaxies manage to produce such a large amount of complex organics is still unknown. The solution to this question and some long-standing mysteries such as the diffuse interstellar bands, the 220 nm extinction feature, the extended red emission, and the family of "unidentified infrared emission" bands will depend on the close working relationships between astronomers and laboratory spectroscopists.

The main purpose of this book is to bring awareness to the scientific community that organic matter is indeed commonly present everywhere in the Universe. The question that we pose, and would like to seek answers to, is whether organic matter in diverse locations are related or have a common origin.

Since the study of organic matter in space requires collaborative efforts between researchers in the astronomical, Solar System, and chemical/biological laboratory communities, it is desirable to have one volume that summarizes the current state of knowledge that can serve as a starting point for researchers who wish to embark on research in this field. The book would also be of interest to the nonspecialist scientist to gain insight into the scientific basis of how our perception of the origin of life has changed in the last 10 years.

Although the book is mainly focused on astronomical observations, I have tried to make it comprehensible to readers from diverse backgrounds. With introduction and background materials, an experienced scientist should be able to grasp an understanding of the issues involved. With extensive discussions on techniques, results, and references, a researcher can use it as a starting point for further research. A casual reader can glance over the technical details and still get a message that organic matter is everywhere in the Universe and there is a likely link between the stars and us.

One of the most enjoyable aspects of working in this interdisciplinary area of research is the opportunity to meet scientists in many different fields. Other than astronomers and physicists, I have met and learned from many chemists, biochemists, biologists, geologists, and planetary scientists. Being present in meetings and conferences, and listening to people whose expertise is completely outside of my own gives one a sense of humility. It is this quality of humility that drives me to continue to learn.

In spite of the amount of work and effort it has taken to write this book, I must say that it has been a very rewarding experience for me. It allows me to summarize my own thoughts on the subject and to express them in a coherent manner. Since my working hours are mostly occupied by administrative duties, writing this book was confined to evenings, weekends, and holidays. Over time, I found solace in writing and came to believe that this book preserved my sanity.

I must first express my gratitude to my colleague Anisia Tang who has skillfully drawn many of the figures and checked the references and the proofs with diligence. Her expert knowledge of LaTex has also saved me from many difficult situations. Many scientists have provided me with advice and expert opinions on the various subjects covered in this book. These individuals include Peter Bernath, Dale P. Cruikshank, Richard Grieve, George Jacoby, David Kring, Stephen Ridgway, Farid Salama, and Scott Sandford. I am also grateful to the scientists who kindly gave me permission to use their figures in the book. I would like to thank my HKU colleagues Jesse Chan, Patrick Toy, and Aixin Yan who provided me with comments based on expert knowledge in their respective fields. Especially, I would like to thank my science writer daughter Roberta, who carefully read the manuscript and provided many useful comments, suggestions, and editorial corrections.

During the early parts of my career, I learned from the examples of Lawrence Aller and Gehard Herzberg. They showed me their dedication to science, and their influences have remained with me to this day.

I am grateful to my wife Emily who has tolerated me as I spent precious family time on this project. The understanding and support of my daughters Roberta and Kelly have always been a strong motivation factor for me.

Hong Kong, 2011   *Sun Kwok*

Abbreviations

The profusive use of acronyms in scientific literature makes it increasingly difficult for people from other disciplines to read the literature. Since this book covers elements of astronomy, biology, chemistry, geology, and planetary science, a list of the acronyms used within the text is given below. Please consult this list first before looking up a term in the subject index.

AAA	aminoadipic acid
ABA	aminobutyric acid
AFGL	Air Force Geophysical Laboratory
AGB	asymptotic giant branch stars
AIB	aromatic infrared bands
ALMA	Atacama LargeMillimeter Array
ARO	Arizona Radio Observatory
ATP	adeninosine triphosphates
AU	astronomical unit
BIMA	Berkeley-Illinois-Maryland Association Array
CCD	charged coupled device
CIRS	composite infrared spectrometer
CNP	carbon nanoparticles
CSO	Caltech submillimeter Observatory
CVD	chemical vapor deposition
DAE	distillate aromatic extract
DAH	diaminohexanoic acid
DAMN	diaminomaleonitrile
DAP	diaminopentanoic acid
DHA	1,3-dihydroxyacetone
DIB	diffuse interstellar bands
DNA	deoxyribonucleic acid
ERE	Extended Red Emission
ESA	European Space Agency
ESO	European Southern Observatory
GCMS	gas chromatography-mass spectrometry
GTP	guanosine triphosphates

HAC	hydrogenated amorphous carbon
HETG-ACIS	High-Energy Transmission Grating-Advanced CCD Imaging Spectrometer
HMT	hexamethylenetetramine
HRTEM	high resolution transmission electron microscopy
HST	Hubble Space Telescope
IDP	Interplanetary dust particles
INMS	ion neutral mass spectrometer
IOM	Insoluble Organic Matter
IR	infrared
IRAC	infrared array camera
IRAM	Institut de Radioastronomie Millimétrique
IRAS	Infrared Astronomical Satellite
IRIS	infrared radiometer interferometer and spectrometer
IRS	Infrared Spectrograph
IRTF	Infrared Telescope Facility
IRTS	Infrared Telescope in Space
ISM	interstellar medium
ISO	Infrared Space Observatory
ISOCAM	Infrared Space Observatory infrared camera
ISOPHO	Infrared Space Observatory photo-polarimeter
IUE	International Ultraviolet Explorer
JCMT	James–Clerk–Maxwell Telescope
KAO	Kuiper Airborne Observatory
KBO	Kuiper Belt Objects
LHB	late heavy bombardment
LRS	Low Resolution Spectrometer
LTE	local thermodynamical equilibrium
LUCA	Last Universal Common Ancestor
LWS	Long Wavelength Spectrometer
m-FTIR	microscope-based Fourier transform infrared
MIRS	mid-infrared spectral
NASA	National Aeronautics and Space Administration
NIRS/NIRSPEC	Near Infrared Spectrometer
NMR	nuclear magnetic resonance
NRAO	National Radio Astronomy Observatory
PAH	polycyclic aromatic hydrocarbon
PDR	photodissociatino regions
POM	polyoxymethylenes
QCC	quenched carbonaceous composites
RAE	residual aromatic extract
RNA	ribonucleic acid
SCM	small carbonaceous molecules
SED	spectral energy distributions
SEST	Swedish-ESO Submillimeter Telescope

SIMS	secondary ion mass spectroscopy
SpeX	medium resolution spectrograph
SSB	spinnig side bands
STXM	scanning transmission X-ray microscope
SWAS	Submillimeter Wave Astronomy Satellite
SWS	Short Wavelength Spectrometer
TNO	Trans-Neptunian Objects
UIE	unidentified infrared emission
UKIRT	United Kingdom Infrared Telescope
UV	ultraviolet
VIMS	visible-infrared mapping spectrometer
VLA	Very Large Array
VLT	The Very Large Telescope array
WFPC	wide field planetary camera
XANES	X-ray absorption near-edge structure

Color Plates

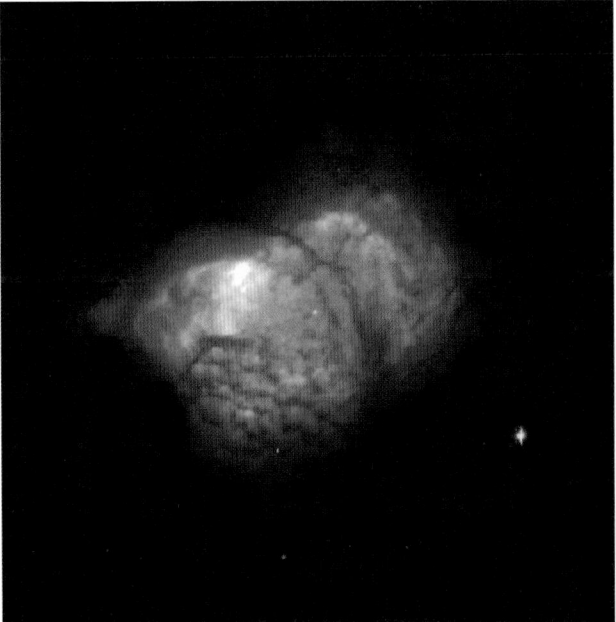

Figure C 1.4 Color composite image of HST WFPC 2 observations of the young planetary nebula NGC 7027, the first astronomical object to be found to have AIB emission.

Figure C 1.5 HST WFPC2 false color image of HD 44179, the Red Rectangle. The Red Rectangle is the stellar object with the strongest AIB emissions. Credit: ESA, NASA, H. Van Winckel, and M. Cohen.

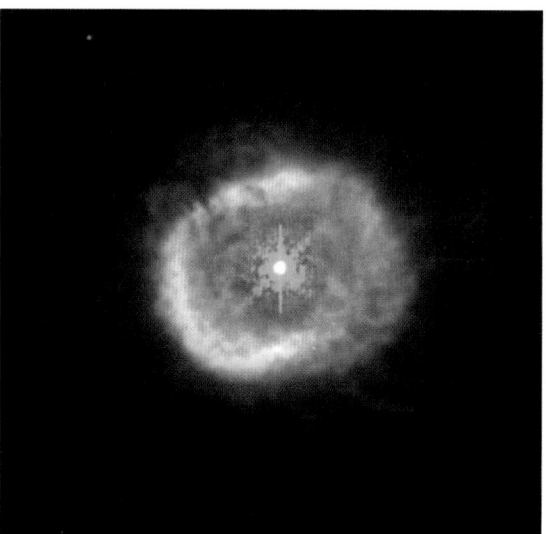

Figure C 6.2 Color composite image of HST WFPC 2 observations of BD+30° 3639, another planetary nebula with rich organic content.

Figure C 6.3 Color composite image of HST WFPC 3 observations of the planetary nebula NGC 6302. The bright color lobes are regions of photoionized gas and the molecular and solid materials are in the dark regions near the waist of the bipolar nebulosity. Credit: NASA, ESA, and the Hubble SM4 ERO Team.

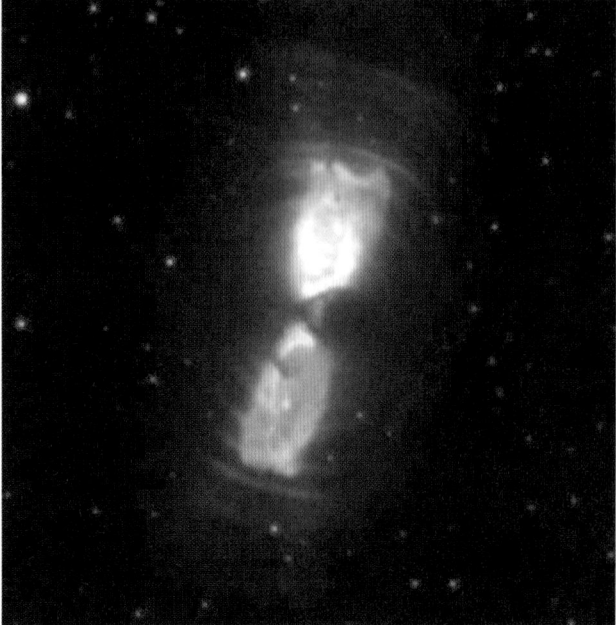

Figure C 6.5 Color composite image of HST WFPC 2 observations of the protoplanetary nebula IRAS 17150-3224 (Cotton Candy Nebula) [551].

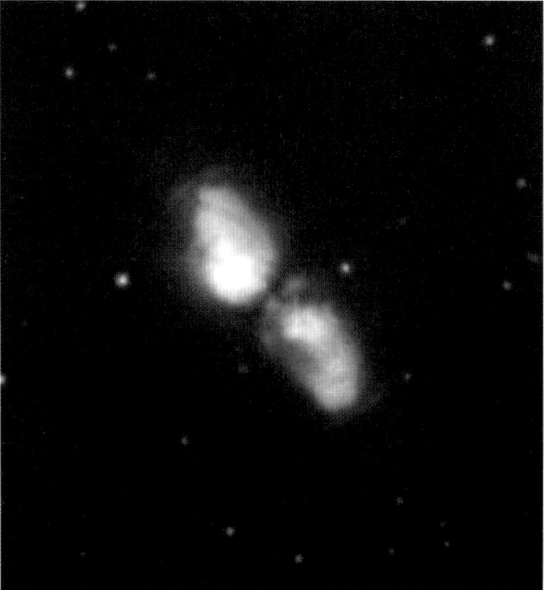

Figure C 6.6 Color composite image of HST WFPC observations of the protoplanetary nebula IRAS 17441-2411 (the Silkworm Nebula) [552].

Figure C 6.7 Color composite image of HST WFPC 2 I- and V-band observations of IRAS 16594-4656 (Water Lily Nebula) [553], a protoplanetary nebula that shows AIB emissions [554].

Plate C 1 Optical image of the starburst galaxy M 82. M 82 is the galaxy with the strongest AIB emission, showing that organic matter is widespread in the Universe. Credit: N.A. Sharp, AURA/NOAO/NSF.

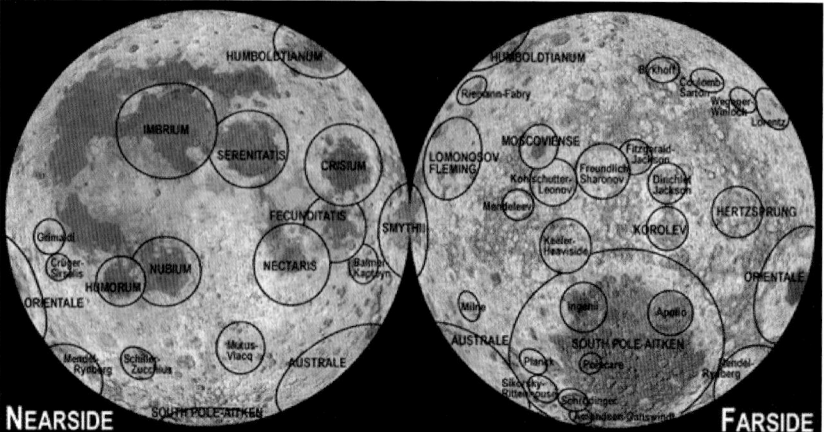

Plate C 2 A schematic map showing the most energetic impact basins on the Moon with diameters >300 km. The map of the nearside of the Moon is on the left and the farside of the Moon is on the right. The colors represent the topography of the lunar surface from deep blue, indicating −8 km below the mean (color white) to +8 km shown in red. Image Credit: LPI (Paul D. Spudis and David A. Kring).

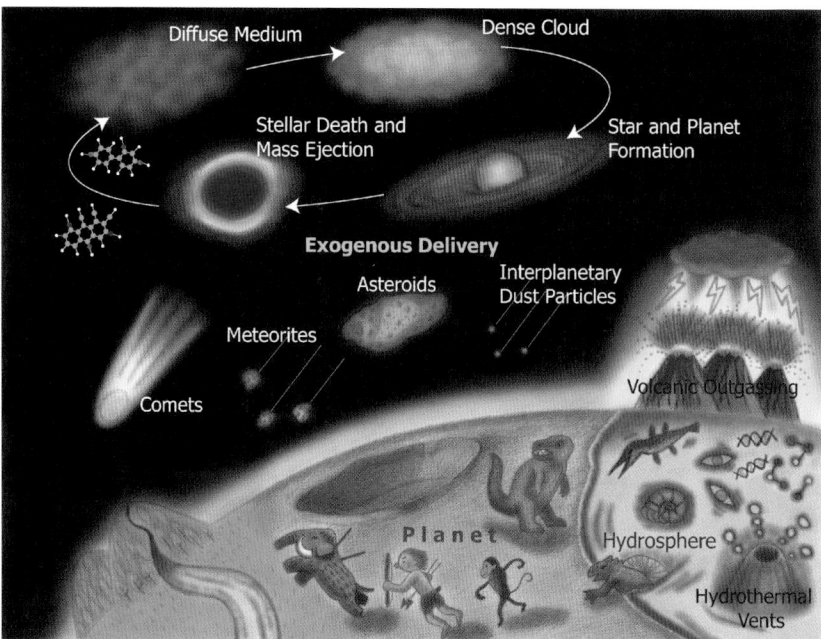

Plate C 3 A cartoon illustrating the manufacturing of organic compounds in old stars and their ejection into interstellar space. The primordial Solar System formed from interstellar clouds inherited these stellar materials which rained onto the early Earth, leading to the creation of life.

1
History and Introduction

The term organic matter was originally created to refer to compounds derived from natural living things which are fundamentally different from those derived from nonliving substances (inorganic matter). It was believed that living things posses a "vital force" which is absent in nonliving things. By the early nineteenth century, advances in chemical techniques had led to the isolation and discovery of an increasing number of organic molecules from living biological organisms. These included amino acids such as asparagine (isolated from asparagus in 1806), cysteine (extracted in 1810 from urinary calculi), leucine (1819, from fermenting cheese), and glycine (1820, from gelatin), fatty acids (1823), proteins (1838), DNA (1869, from yeast nuclei), and the nucleic acid bases guanine (1882), thymine (1883), adenine (1886), cytosine (1894), and uracil (1900) as well as deoxyribose (1909). These and other organic molecules represent the building blocks of life.

At the same time, it was commonly and firmly believed by many chemists that these molecules could only be produced by living organisms. While inorganic matter could be produced in the laboratory by chemical means, scientists thought organic matter could not be synthesized from inorganic matter because it lacked the "vital force". Although the form of this "vital force" was never precisely described or defined, it was believed to be electrical in nature and involved in the rearrangement of molecular structures. In 1823, Friedrich Wöhler (1800–1882) heated an inorganic salt ammonium cyanate (NH_4NCO) and turned it into urea [$(NH_2)_2CO$], an organic compound isolated from urine. Although ammonium cyanate and urea are made up of the same atoms, their molecular structures are different. This experiment suggested that it was possible to convert an inorganic molecule into an organic one by artificial means, without the magic of "vitalism". This was the beginning of the disappearance of the concept of "vital force" from the scientific arena.

This pioneering work on abiotic synthesis was followed by the laboratory synthesis of the amino acid alanine (from a mixture of acetaldehyde, ammonia, and hydrogen cyanide) by Adolph Strecker (1822–1871) in 1850 and the synthesis of sugars (from formaldehyde) by Aleksandr Mikhailovich Butlerov (1828–1886) in 1861. However, it was not until the 1960s that the first nucleobase adenine ($C_5H_5N_5$) was synthesized abiotically (from HCN and NH_3) [1]. This was followed by the synthesis of guanine [2] and cytosine [3].

Organic Matter in the Universe, First Edition. Sun Kwok.
© 2012 WILEY-VCH Verlag GmbH & Co. KGaA. Published 2012 by WILEY-VCH Verlag GmbH & Co. KGaA.

Biochemistry developed as the systematic study of biological forms and functions in terms of chemical structures and reactions. In the mid-nineteenth century, it was thought that "vitalism" from living yeast cells was the key to the fermentation of sugar into alcohol. In 1897, Eduard Büchner (1860–1917) discovered that yeast extracts could ferment sugar into alcohol and living cells were not necessary. This marked the beginning of the realization that biomolecules (which we now call enzymes), not the "vital force", are responsible for fermentation. Enzymes have since been shown to be the catalysts that accelerate chemical reactions in biological systems.

In 1926, James Sumner found that urease, an enzyme that catalyzes the hydrolysis of urea into CO_2 and NH_3, is a protein. Shortly after, it was found that several other crystallized digestive enzymes are also proteins. The basis of one of the key elements of life – enhancing the rates of chemical reactions efficiently and selectively – was reduced to the study of the structures and functions of protein molecules.

Now, our definition of organic matter has evolved from something that possesses a special nonphysical element such as the "vital force" to a group of molecules and compounds based on the chemical element carbon (C). The element carbon is unique in its versatility for forming different chemical bonds. Not only is carbon able to connect with other C atoms to form different structures (see Chapter 2), it can also combine with other elements such as hydrogen (H), oxygen (O), nitrogen (N), sulfur (S), and phosphorus (P) to form a great variety of molecular forms. This group of molecules forms the basis of living organisms.

We note that four elements – hydrogen, oxygen, carbon, and nitrogen – make up more than 99% of the mass of most cells. These four elements are also the first, third, fourth, and fifth most abundant elements in the Universe. The existence of biomolecules is therefore built upon chemical elements that are abundantly available. The first step in molecular synthesis depends on the creation and distribution of chemical elements in the Universe.

1.1
Origin of Chemical Elements

Studies on stellar nucleosynthesis in the 1950s have led to the current realization that most of the chemical elements are synthesized in stars [4]. Helium is made by fusing hydrogen into helium in the core during the main sequence and in a shell above the core in the red giant phase. The element carbon is created by helium-burning[1] (the triple-α process), first through core burning and later through shell burning above an electron-degenerate carbon-oxygen core. For massive (> 10 $M_\odot$) stars, direct nuclear burning continues with the production of oxygen, neon, magnesium, silicon, and so on, culminating in the synthesis of iron, the heaviest el-

---

1) Although the term "burning" has the connotation of a chemical reaction, it is used in astronomical literature to mean nuclear fusion.

ement that can be formed through direct nuclear burning. The other heavy elements, from yttrium and zirconium to uranium and beyond, are produced by neutron capture followed by β decay [5].

For the majority of stars ($\sim$ 95%, corresponding to stars with initial masses less than $\sim$ 8 $M_\odot$), direct nuclear burning does not proceed beyond helium, and carbon is never ignited. Most of the nucleosynthesis occurs through slow neutron capture (the s process) during the asymptotic giant branch (AGB), a brief phase ($\sim 10^6$ yr) of stellar evolution where hydrogen and helium burn alternately in a shell. These newly synthesized elements are raised to the surface through periodic "dredge-up" episodes, and the observation of short-lived isotopes in stellar atmospheres provides direct evidence that nucleosynthesis is occurring in AGB stars [6].

Other than the light elements H, He, D, and Li, which were produced in significant quantities during the Big Bang, all the other natural chemical elements are made in stellar furnaces. They are made in or near the core of stars, brought to the surface by convection, and ejected into interstellar space by stellar winds and supernovae explosions. With spectroscopic observations of distant galaxies (which means also looking back in time), we can detect the same elements through their atomic transitions. The element hydrogen (H) can be detected through its recombination lines (e.g., Lyman α, $n = 2-1$) to a distance of redshift (z) of 7 [7], corresponding to more than 10 billion years back in time.[2] The fine-structure lines of oxygen have been detected in galaxies with $z = 3.9$ [8]. In the far infrared, fine-structure lines of C, N, and O and rotational lines of simple molecules such as CH, OH, and H_2O can be seen in emission and absorption respectively in the spectra of galaxies (Figure 1.1). With modern radio telescopes equipped with sensitive receivers, the molecule carbon monoxide (CO) has been detected in very distant quasars and galaxies [9, 10]. The most distant detection of molecular gas is in the quasar J1148+5251 at the redshift of 6.42 [11] (Figure 1.2). Photometric observations at the infrared and submillimeter wavelengths have detected excess infrared emission from galaxies and quasars at even similar distances [12]. This infrared excess cannot be due to starlight and is generally interpreted to be due to reemission by interstellar solid-state particles heated by starlight. Since these solids must be made of heavy elements,[3] we know that the synthesis of chemical elements occurred soon ($< 10^9$ yr) after the Big Bang.

This tells us that the laws of physics and chemistry are spatially universal over the Universe and also through time. Chemical elements were made in the first generation of stars, and the heavy elements created by these stars are used as raw materials to form the next generation of stars.

---

2) Redshift (z) is a measure of how much the observed wavelength of a spectral line has changed as a result of Doppler effect due to the expansion of the Universe. Given a cosmological model, the value of redshift can be used to infer the distance of the galaxy from which the line is emitted and the time it took for the light to reach us (see Appendix C).

3) Astronomers refer to all elements heavier than H and He as heavy elements. They are also referred to as "metals". This usage is different from definitions used in physics or chemistry.

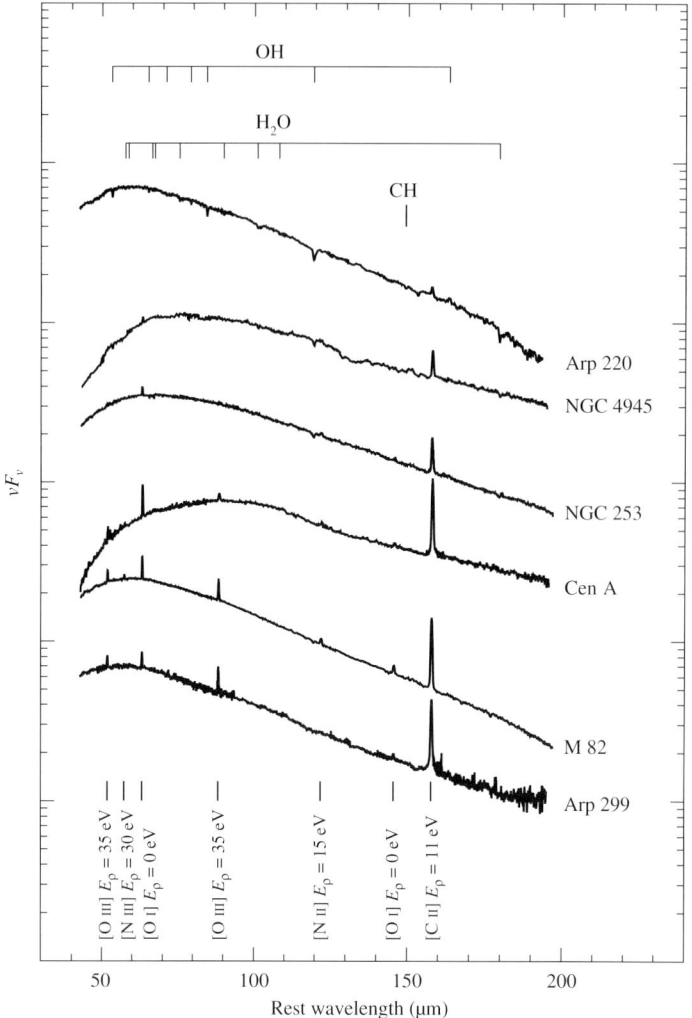

Figure 1.1 ISO LWS spectra of six infrared bright galaxies showing that external galaxies have the same atomic and molecular spectra as in the terrestrial laboratory. Emission lines due to ions of C, O, and N can be seen in the spectrum. Absorption features due to simple molecules such as OH, CH, and H_2O can also be seen. All spectra have been shifted in wavelength to rest wavelengths. Figure adapted from [13].

We can derive the relative abundances of chemical elements from the strengths of atomic lines in the photospheric spectra of stars by making use of atomic parameters from laboratory measurements and models of stellar atmospheres [14]. Assuming hydrostatic and thermodynamic equilibrium and using the equations of radiation transfer, the observed fluxes of atomic lines can be translated into column densities, from which the relative abundances are obtained. The most abundant element is H (71% by mass), followed by He (27%), O (1%), C (0.3%), and N (0.1%).

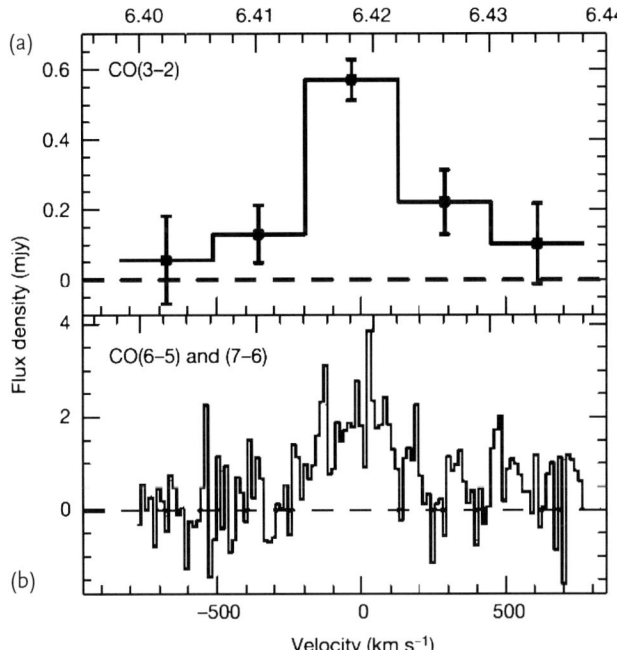

Figure 1.2 The redshifted CO rotational transitions $J = 4-3$ and $J = 5-4$ from the quasar J1148+5251 at the redshift of 6.42, showing that molecular synthesis was already happening over 10 billion years ago. (a) The spectrum for CO $J = 3-2$ line was obtained at the Very Large Array (VLA) and (b) is an average of the CO $J = 6-5$ and $J = 7-6$ line obtained at Plateau de Bure Interferometer (PdBI). Figure adapted from [11].

The derived chemical abundances (primarily from solar observations) are called cosmic abundances.[4] These values are different from terrestrial chemical abundances as the Earth's atmosphere has been unable to keep lighter elements within the grasp of the Earth's gravity. The observed cosmic abundances of chemical elements are generally consistent with present models of primordial (Big Bang) and stellar nucleosynthesis.

In the interstellar medium (ISM), atomic nuclei are not only in atomic or ionic forms, but can also be part of molecules and solids. The molecular abundances relative to molecular hydrogen (H_2) can be derived from spectroscopic observations of the rotational or vibrational lines (Chapter 3). The observed fluxes of the lines are analyzed with excitation models based on our knowledge of the temperature and density of the region, together with solutions to the equation of transfer. The abundances of solids can also be derived from the strengths of vibrational bands, but the results are less accurate than those of molecular abundances.

Given the availability of the elements, it is therefore no surprise that the four most common elements in the Universe (H, O, C, N) comprise 99% of the mass

4) Strictly speaking, these are solar abundances, which can be different from the elemental abundances determined from meteorites or from values derived in the interstellar medium.

of living matter. Because of its chemical inertness, He is not an essential element in the chemistry of life. Among molecules, water (H_2O) is the most abundant in living organisms, often accounting for the majority of the mass. Living organisms also contain highly complex molecules, in particular, polymers such as proteins, carbohydrates, and nucleic acids. These polymers are built from monomers of about three dozen or so of amino acids, sugars, purines, and pyrimidines. These biomolecules in turn are probably synthesized from simpler molecules such as H_2O, CH_4, CO_2, and NH_3.

The Universe consists of not just atoms and ions, but also complex molecules and solids. At this time, we do not know of the upper limits for the complexity of these compounds in the Universe. We do know that more than 160 molecules and a variety of complex solids are made naturally by stars and interstellar clouds, and they are spread over the entire Galaxy.

1.2
Extraterrestrial Organics

While it has been demonstrated that biomolecules can be artificially produced in the laboratory from inorganic matter, the question remains as to whether this process can occur naturally. Although carbon is the fourth most abundant element in the Universe, the possibility of the widespread presence of organic matter in space was not seriously contemplated at first because space density was thought to be too low for the synthesis of complex molecules. The first evidence for the existence of molecules in space can be traced to 1937, when Theodore Dunham and Walter Sydney Adams found that the spectra of early-type stars obtained at Mount Wilson Observatory contained a number of unidentified absorption lines in addition to known atomic lines [15]. One of the lines at 430.03 nm was identified by Pol Swings and Leon Rosenfeld [16] as due to the $A^2\Delta - X^2\Pi$ band of CH molecule by simply comparing the astronomical spectra to the spectra of CH in a flame of an ordinary Bunsen burner. In 1940, lines from the other excited electronic states of $B^2\Sigma^-$ and $C^2\Sigma^+$ to the ground $X^2\Pi$ at 388.9 and 314.5 nm were measured in the laboratory [17]. These lines of CH as well as the $B^2\Sigma^+ - X^2\Sigma^+$ transition of CN all have interstellar counterparts in the spectra of Dunham and Adams. Figure 1.3 is a modern spectrum showing the $R(1)$, $R(0)$, and $P(1)$ lines of CN in absorption against the stellar continuum background. Another unidentified interstellar line at 423.26 nm was identified as the $A^1\Pi - X^1\Sigma^+$ transition of CH^+ [18].

After these early discoveries, it took another 26 years before the next molecule, OH, was discovered by radio spectroscopy. The λ 18 cm Λ-doublet transition of OH was observed in absorption against a background radio source in 1963 [20]. This was followed by the detection of the λ 1.3 cm inversion transition of ammonia (NH_3) in 1968 [21], the λ 1.4 cm $5_{23}-6_{16}$ maser rotational transition of water (H_2O) in 1969 [22], and the λ 6.2 cm $1_{10}-1_{11}$ of formaldehyde (H_2CO) in 1969 [23]. The development of millimeter-wave receivers led to the detection of $J = 1-0$ rotational transitions of CO, HCN, and HC_3N in 1970.

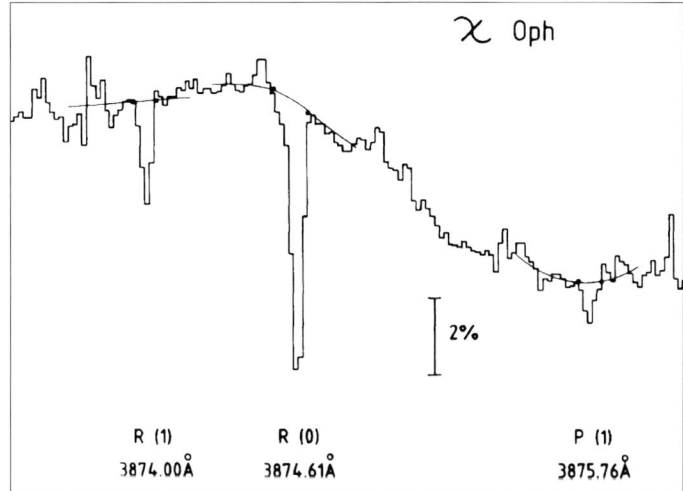

Figure 1.3 Interstellar CN lines appearing in absorption against the stellar continuum of the star χ Oph. These electronic transitions are the first indication of the existence of molecules in the diffuse interstellar medium. Figure adapted from [19].

The explosion in the number of detections of gas-phase molecules in the interstellar medium led to the recognition of astrochemistry as a new scientific discipline. The detection of molecules in the outflow of evolved stars also led to the realization that molecules can form in the low-density environment of stellar winds, soon after the element carbon is synthesized by nuclear reactions in the stellar core, dredged up to the surface, and released from the atmosphere [24].

While the millimeter-wave technique is capable of detecting molecules consisting of more than a dozen atoms, the high degree of complexity of interstellar organics was not appreciated until the development of astronomical infrared spectroscopy. A family of strong infrared emission bands at 3.3, 6.2, 7.7, 8.6, 11.3 and 12.7 μm were first detected by the Kuiper Airborne Observatory (KAO) in the young carbon-rich planetary nebula NGC 7027 [25] (Figure 1.4) and reflection nebula HD 44179 [26] (Figure 1.5). Since the initial discovery, these features are now widely observed in H$_{II}$ regions, reflection nebulae, planetary nebulae, proto-planetary nebulae, and the diffuse ISM of our own and other galaxies.

The widths of the features are much broader than molecular linewidths broadened by Doppler effects or turbulence and are therefore designated as emission bands. The energy emitted in these bands can be a significant fraction of the total infrared continuum energy output of galaxies, and the identification of the carriers of these features is therefore important for understanding the chemical makeup of the ISM and galaxies.

The fact that the strengths of these infrared emission features correlate with the C/O ratio of planetary nebulae suggests a carbon-based carrier. Comparison with laboratory infrared spectroscopy of organic compounds has led to the identification that these features arise from the stretching and bending modes of various CH

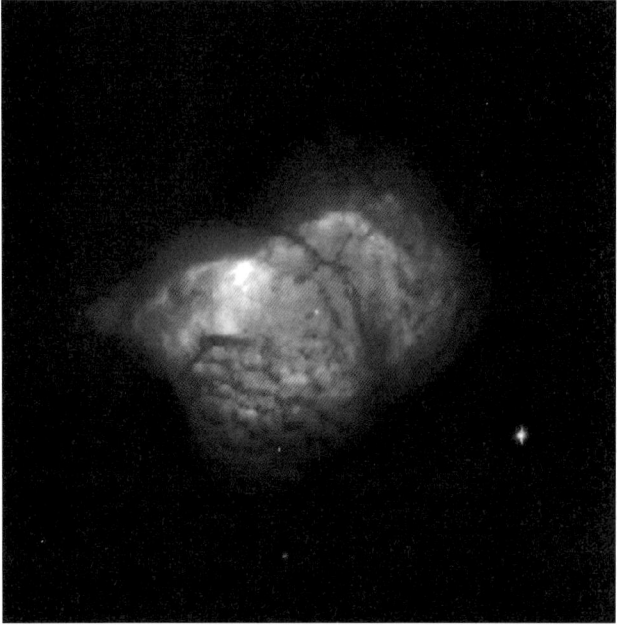

Figure 1.4 Color composite image of HST WFPC 2 observations of the young planetary nebula NGC 7027, the first astronomical object to be found to have AIB emission. For a color version of this figure, please see the Color Plates at the beginning of the book.

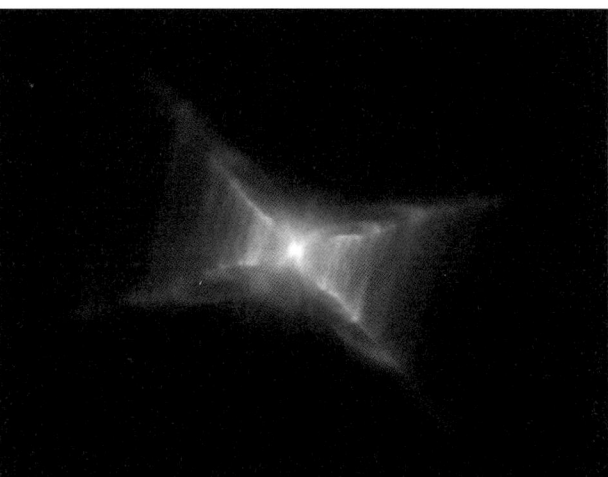

Figure 1.5 HST WFPC2 false color image of HD 44179, the Red Rectangle. The Red Rectangle is the stellar object with the strongest AIB emissions. Credit: ESA, NASA, H. Van Winckel, and M. Cohen. For a color version of this figure, please see the Color Plates at the beginning of the book.

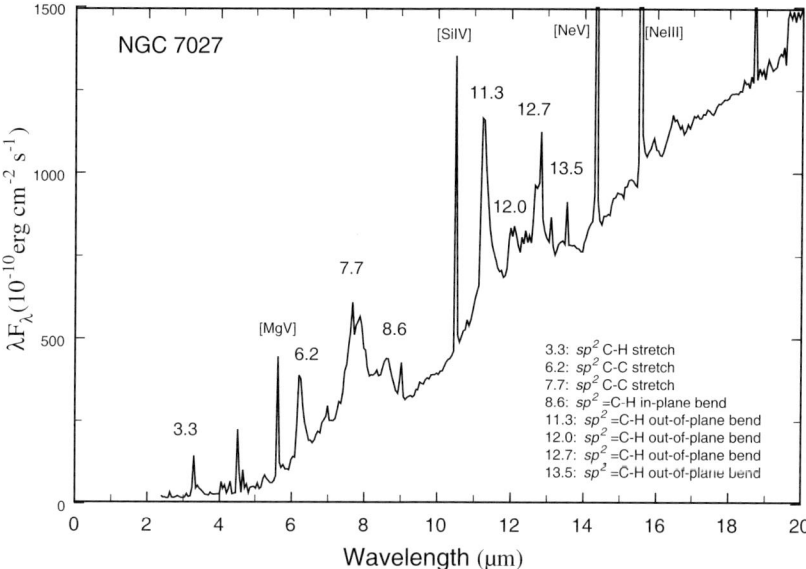

Figure 1.6 The ISO SWS spectrum of the planetary nebula NGC 7027. The narrow features are atomic emission lines due to fine-structure transitions of heavy elements (e.g., the [N iii] line at 15.6 μm). The broad features marked by wavelengths are due to stretching and bending modes of aromatic compounds. Their identifications are listed in the legend.

and CC bonds in aromatic hydrocarbons [27]. For this reason, these strong infrared emission features are now known as the aromatic infrared bands (AIB) (Figure 1.6).

In this book, we summarize the evidence for the presence of organic matter in the Universe. The organics include molecules in the gaseous form and large, bulk solids consisting of hundreds or thousands of atoms. Methods of detecting organic compounds range from remote spectroscopic observations with telescopes to laboratory analysis of samples collected from space or on Earth. With the technique of infrared and millimeter-wave spectroscopy, specific molecules can be identified through their vibrational and rotational transitions and examples of groups of organic molecules are discussed in Chapter 3. We discuss the diverse environments where organic compounds are found, including clouds in the diffuse ISM (Chapter 4), distant galaxies (Chapter 5), the ejecta of stars (Chapter 6), and various objects in the Solar System (Chapter 7). Unexplained astronomical phenomena that may be due to organic compounds are discussed in Chapter 8. The chemical structures and possible laboratory analogs of extraterrestrial organics are presented in Chapter 9. Near the end of the book, some speculations on the origin of these compounds and the possible links between stars and the Solar System are offered.

2
The Chemistry of Organic Matter

Carbon (C) is the fourth most abundant element in the Universe after H, He, and O. In the ISM, the total carbon abundance is estimated to be 225 ± 50 atoms per 10^6 H atoms. In the chemistry of living organisms, carbon plays a dominant role, accounting for more than half of the dry weight of cells. This can be attributed to the versatility of the C atom, which can form single, double, and triple bonds with other C atoms. It can also form single and double bonds with oxygen (O) and nitrogen (N) atoms, giving rise to functional groups such as alcohol, ether, amine, carbonyl, aldehyde, and ketone.

The carbon atom has a ground-state electron configuration of $1s^2 2s^2 2p^2$ and therefore four valence electrons (2s and 2p). The atomic wave functions from the 2s and 2p electrons can combine to form hybrid orbits for bonding. This mixing of atomic orbitals is known as hybridization and is responsible for the richness of carbon chemistry. For example, methane (CH_4) has four identical C–H bonds in the form of a tetrahedral structure with each H–C–H forming an angle of 109.5°. This hybrid orbital is a mixture of one s orbital and three p orbitals, and is labeled as sp^3. sp^3 is the most common form of hybridization for carbon.

If the 2s orbital combines with only two of the three available 2p orbitals, the result is a planar structure with an H–C–H angle of 120°. This is known as sp^2 hybridization. When two sp^2 hybridized carbons approach one another, they form a σ bond by sp^2–sp^2 overlap, whereas the unhybridized p orbitals on each carbon form π bonds on the side. The combination of the σ and π bonds results in the sharing of four electrons and the formation of a C=C double bond. An example is ethylene ($H_2C=CH_2$) which is a planar (flat) molecule with the two carbon atoms connected by a double bond. A molecule in the sp^2 hybridization of particular interest is benzene (C_6H_6). The six C atoms of benzene form a symmetric ring structure, each with an H atom attached.

A third form of hybridization for carbon occurs when the 2s orbital combines with a single 2p orbital, which forms a linear structure with the remaining two 2p orbitals unchanged. This is called sp hybridization and acetylene (C_2H_2) is an example of such a structure. The two sp-hybridized carbon atoms are joined by one sp–sp σ bond and two p–p π bonds, resulting in a linear molecule with an H–C–C bond angle of 180°. The carbon atoms are joined by a triple bond, and the structure of acetylene can be written as $H-C{\equiv}C-H$.

Organic Matter in the Universe, First Edition. Sun Kwok.
© 2012 WILEY-VCH Verlag GmbH & Co. KGaA. Published 2012 by WILEY-VCH Verlag GmbH & Co. KGaA.

Because of the availability of these hybridizations, the carbon atom is extremely versatile and can form a wide variety of molecules with hydrogen, oxygen, and nitrogen. For comparison, the only stable molecules that oxygen can form with H are water and hydrogen peroxide (H_2O_2). For nitrogen, the only stable molecules are ammonia (NH_3) and hydrazine (N_2H_4). Since H and C are respectively the first and fourth most abundant elements in the universe, organic compounds could be abundantly present in the Universe if physical conditions allow their synthesis.

2.1
Families of Organic Molecules

Organic compounds are grouped into families depending on the functional groups they contain. The family that only contains H and C is called hydrocarbons. Hydrocarbons are divided into aliphatic and aromatic classes, corresponding to their origins in oils and fats and plant extracts with an odor. Aliphatic hydrocarbons include alkanes (containing only C–H and C–C single bonds), alkenes (containing at least one carbon–carbon double bond), and alkynes (containing at least one carbon–carbon triple bond). Since all four electrons of the C atom are paired with the electron from an H atom, alkanes are also called saturated hydrocarbons. The first member of this family is methane, which has one C atom and four H atoms (CH_4). The next in the series is ethane (C_2H_6), and then propane (C_3H_8). The general formula of this family is C_nH_{2n+2}. An example of an alkene is ethylene ($H_2C=CH_2$) and an example of an alkyne is acetylene ($H-C\equiv C-H$). Molecules with ring-like structures are referred to as aromatic compounds. The simplest example of an aromatic hydrocarbon is benzene (C_6H_6) which has the form of a ring. When two or more rings are fused together, they are called polycyclic aromatic molecules.

If we treat an alkane as a combination of a structural unit R (called the alkyl group) and a hydrogen atom, then other families of organic compounds can be viewed as combinations of this hydrocarbon framework and different functional groups. For example, alcohol is represented by ROH, ether as ROR, carboxylic acid as R(C=O)OH, aldehyde as R(C=O)H, and ketone as R(C=O)R. Simple examples of alkyl groups include the methyl group ($-CH_3$), which is the result of removing an H atom from methane. Similarly, the ethyl group ($-CH_2CH_3$) comes from the removal of a H from ethane (CH_3CH_3). From propane ($CH_3CH_2CH_3$), we have the propyl ($CH_3CH_2CH_2-$) and isopropyl (CH_3CHCH_3) groups.

2.2
Different Forms of Carbon

The best known forms of pure carbon are graphite and diamond, which are crystalline structures based on sp^2 and sp^3 hybridizations, respectively. Carbon can also form amorphous structures of different mixed sp^2/sp^3 hybridization ratios.

The mostly widely known example of natural amorphous carbon is coal, which is believed to be of biological origin. Laboratory experimentation has led to the discovery of new forms of carbon, including fullerite solids which are geodesic structures of cage-like spheroids. Other examples of curved structure are nanotubes, which are cylindrical rather than spherical. A fourth crystalline structure of carbon is carbyne, which is a long chain of sp-hybridized carbons. The search for these different forms of carbon in space is of great current interest [28, 29].

2.2.1
Graphite

Graphite is a soft, black and opaque form of pure carbon. It is made up of parallel layers of two-dimensional sheets of benzene-like rings with sp^2 sites. These sheets are referred to as graphene sheets, which are flat monolayers of C atoms [30, 31]. The black color of graphite is attributed to the excitation of the delocalized π electrons, leading to the absorption of lights of all colors. Graphite is formed as the final stage of thermal evolution of kerogen, coal, and oil as these organic solids become more aromatic with time under high temperature and density conditions.

2.2.2
Diamond

Diamond is a hard, colorless and transparent form of pure carbon. It consists of a covalent network solid composed of sp^3-hybridized atoms in a three-dimensional lattice. Diamonds were identified as a form of carbon in 1797, and natural terrestrial diamonds are believed to form deep in the Earth's interior under high pressures and temperatures. Under thermodynamical equilibrium, diamond can only form under high pressure. Diamonds were first synthetically made in the laboratory at pressures of 30 000 bar[1] and temperatures over 700 °C. However, laboratory experiments have shown that diamond-like material can be formed under nonequilibrium conditions, during condensation of carbon vapor on cool substrates. Nanocrystalline diamonds can now be made in the laboratory under low pressure using the chemical vapor deposition (CVD) technique. Once formed, diamond is "metastable" and requires high temperature to convert it to graphite.

Diamondoids are hydrocarbons that have a diamond-like carbon lattice with the dangling carbon bonds terminated with H. The smallest diamondoid is $C_{10}H_{16}$ (adamantane). The higher members have two, three, four, five, ... face-fused cages called diamantane ($C_{14}H_{20}$), triamantane ($C_{18}H_{24}$), tetramantane ($C_{22}H_{28}$), and pentamantanes ($C_{26}H_{32}$), respectively. While tetramantane has four isomers of the same molecular weight, pentamantanes have nine isomers with molecular weight 344 ($C_{26}H_{32}$) and one isomer with molecular weight 330 ($C_{25}H_{30}$). The higher

1) 1 bar = atmospheric pressure

diamondoids (polymantanes) are more varied in molecular geometry and structural complexity. Because of their structural stability, diamondoids are difficult to synthesize, and they are mostly isolated from petroleum.

2.2.3
Fullerenes

In an effort to reproduce the conditions under which long-chain cyanopolyynes are formed in the circumstellar environment, Harry Kroto, Richard Smalley, and Robert Curl accidentally discovered a new form of carbon: a molecule with twelve five-membered and twenty six-membered carbon rings on the surface of a soccer ball-shaped geodesic structure, which they named Buckminsterfullerene [32] (Figure 2.1). The curvature of the molecule is created by the pentagons, none of which shares an edge with another. Each carbon atom is at the vertex of one five-membered and two six-membered rings. Spherical molecules made of fused pentagons and hexagons are now part of the general class of fullerenes, which refer to pure-carbon molecules with a hollow structure. C_{60}, with each pentagonal site surrounded by five hexagonal sites, is the smallest stable fullerene and has no stable isomers. The higher members of the family are C_{70}, C_{74}, C_{76}, C_{80}, C_{82}, C_{84}, and so on.

The Buckminsterfullerene is a remarkably stable molecule as it has no edges and therefore no dangling bonds [33]. Since it is easily made in flames and carbon arcs [34], it is natural to assume that the molecule should be widely present in space. Since hydrogen is abundantly present in the interstellar medium and H atoms can attach to the corners of the C_{60} molecule, hydrides of fullerenes ($C_{60}H_m$, $m = 1, 2, \ldots 60$, also known as polyhydrofullerenes or fulleranes) [35] may also exist in space. Due to the stability of the C_{60} molecule and its assumed presence in interstellar space, it was mentioned in the original discovery paper of Kroto *et al.* [32] that fullerene could be responsible for the diffuse interstellar bands (DIB, Section 8.2). The molecule has also become a popular proposed candidate carrier of a variety of unidentified astronomical features [36].

There is evidence that naturally occurring fullerenes may be present in a carbonaceous Precambrian rock in the Karelia region [37], and researchers have

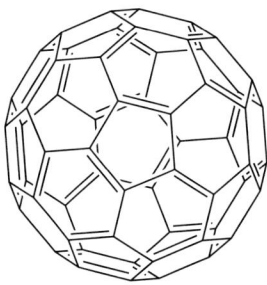

C_{60}

Figure 2.1 A schematic drawing of the structure of the C_{60} molecule.

claimed to find fullerenes (C_{60} to C_{200}) in the Permian–Triassic Boundary corresponding to the massive extinction event of 251 million years ago [38]. However, the origin of geological fullerenes remains unclear. They can either be brought to Earth from extraterrestrial impactors, or formed in high temperature conditions during impact [39].

2.2.4
Nanotubes and Fullerene Onions

Nanotubes are cylindrical fullerenes. They are hollow tubular structures with diameters between 1 and 1000 nm. The first carbon nanotube was created by rolling graphene sheets into long cylinders [40] (Figure 2.2). The ends of the open tubes are closed by half fullerenes.

Fullerene onions are nested fullerenes in the form of concentric shells. This structure can be made by electron beam irradiation of carbon soot in a transmission electron microscope or by heat treatment and electron beam irradiation of nanodiamond.

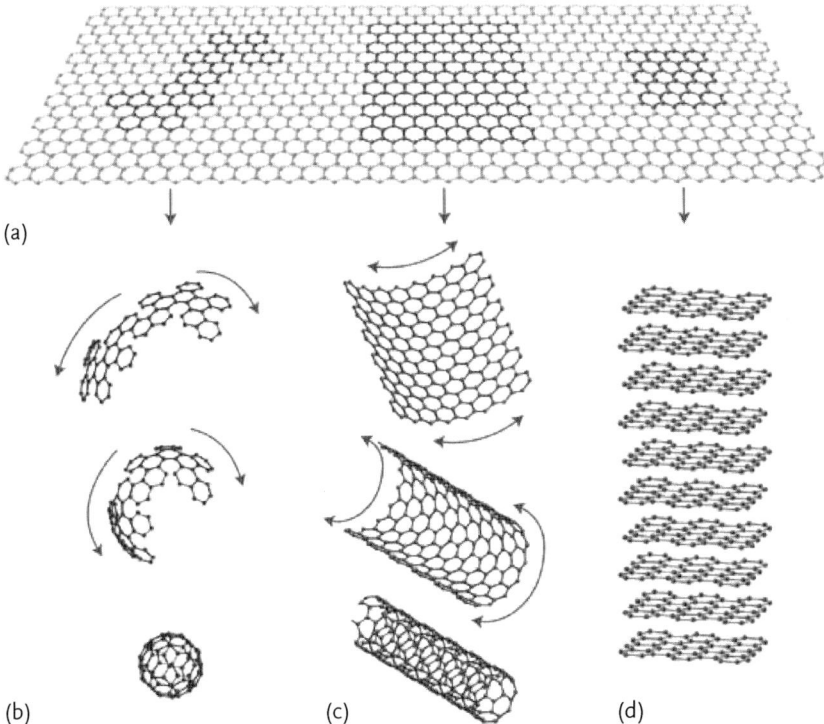

(a)

(b) (c) (d)

Figure 2.2 Graphene sheets (a) can be wrapped into fullerenes (b), nanotubes (c), or graphite (d). Figure adapted from [31].

2.2.5
Carbynes

Carbynes are long linear chains of sp and sp^2-hybridized carbon atoms. They can be divided into the subclasses of polyynes with alternating single and triple bonds ($\cdot$C≡C–C≡C...C≡C–C≡C$\cdot$, etc.) and cumulenes with successive nearly equivalent length double bonds (:C=C=C...C=C=C:). The acetylenic bonding of polyynes have alternating bond lengths whereas the cumulenes have nearly equal bond lengths. An example of a simple polyyne is diacetylene (H–C≡C–C≡C–H) and an example of cumulene is butatriene (H$_2$C=C=C=CH$_2$). A CN group can be added to the end of the polyynyl radical chains C$_n$H to form a closed-shell linear molecule (HC$_{2n}$CN) called cyanopolyynes. Examples of these structures are shown in Figure 2.3. These molecules have very strong rotational transitions because of their linear or near-linear geometry and their high dipole moments due to the nonbonding electrons [41].

The geometry and electronic structures of carbon cumulene chains C$_n$ depend on their sizes and on n being even or odd. The linear even-numbered chains have $^3\Sigma$ ground electronic states, and the odd-numbered linear chains have $^1\Sigma$ ground electronic states. For $n < 10$, C$_n$ is expected to be linear in its ground states whereas the larger members form monocyclic rings, although cyclic isomers of C$_4$, C$_6$, and C$_8$ have been found to have low-energy structures [43] (Figure 2.4).

Cumulene carbenes (H$_2$C=(C=)$_n$C:) are highly reactive and play important roles in the combustion of hydrocarbons. They are produced in the polymerization of

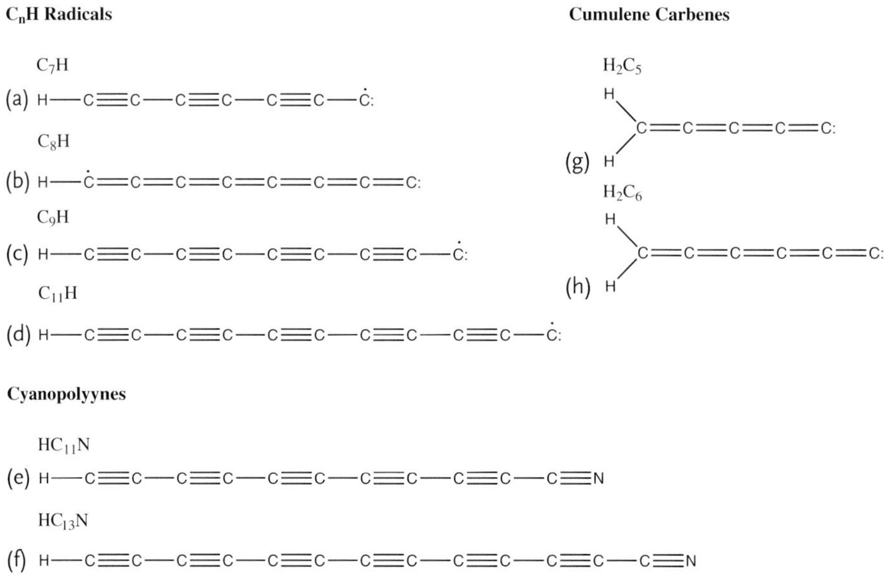

Figure 2.3 Chemical structures of several examples of polyynyl radicals (a)–(d), cyanopolyynes (e) and (f), and cumulene carbenes (g) and (h). Figure adapted from [42].

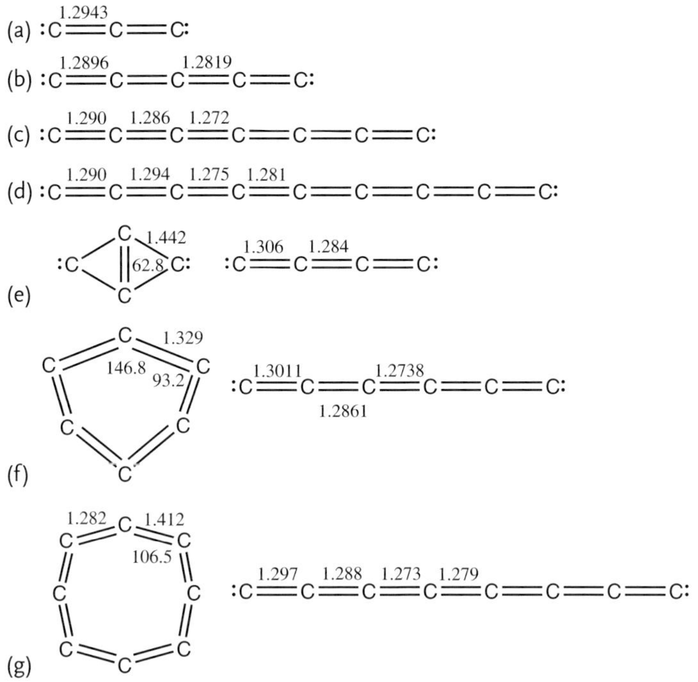

Figure 2.4 Chemical structures of several examples of cumulene chains, for example, (a) C_3, (b) C_5, (c) C_7, (d) C_9, and the ring and chain isomers of (e) C_4, (f) C_6, and (g) C_7. The bond lengths (in Å) and bond angles (in degrees) are marked on the structures. Figure adapted from [43].

acetylene and can lead to the formation of soot [44]. Experiments based on arcing graphite electrodes can produce families of polyynes whose abundances decrease with chain length, as seen in the circumstellar envelopes of carbon stars (Section 6.1). This suggests that the formation mechanisms in stars may be similar to those in carbon arcs [45].

2.2.6
Amorphous Forms of Carbon

Amorphous carbon is characterized by different sp^2/sp^3 hybridization ratios as well as mixed hybridization states. In its pure form, it is made of a random network of sp^2/sp^3 bonds. A common example of amorphous carbon is coal. Coal is formed from fossilized hydrocarbon materials and contains a mixture of sp, sp^2, and sp^3 bonds [46]. The chemical composition and structure of coal have been studied by infrared absorption spectroscopy, magnetic resonance spectroscopy, X-ray diffraction, electron spin resonance, mass spectroscopy, and ultraviolet spectroscopy. In spite of all these efforts, the structure of coal is still not well defined. Generally speaking, coal can be considered to consist of a macromolecular network of aromatic ring clusters linked by bridges, plus a mobile component of lower molecular weight species. Perhaps it is best to consider coal as a complex heterogeneous organic rock which cannot be represented by a single molecular structure [47].

Terrestrial coal is classified by its H and O content relative to C. Coal is believed to have evolved naturally with time through heat and pressure, changing from "low-rank" to "high-rank" through decreasing H and O contents. Lower ranked coal such as peat and lignite have lower carbon contents ($< 50\%$ for peat) than higher ranked coal such as bituminous coal and anthracite. Structurally, it increases its long-range order by stacking aromatic planes to form randomly oriented basic structural units, eventually ending as graphite (which has $> 95\%$ carbon).

Another well known form of amorphous carbon is soot. Soot is familiar to us in the form of smoke from wood-burning fireplaces, diesel engines, industrial furnaces and other forms of combustion. If one places a cool object in a candle flame, it will be covered with black soot. Viewing these soot particles under an electron microscope reveals that they are spheroids of sizes 10–30 nm. In terms of chemical structure, soot particles are C and H atoms in mixed sp^2/sp^3 hybridizations. The exact chemical composition of soot is dependent on the relative mix of the precursor gas and the pressure and temperature of the flame. From this point of view, soot can also be considered as an artificial carbonaceous compound. Since soot is produced rather than intrinsically present on Earth, soot is also characterized by its formation process, which consists of stages of nucleation, growth and coalescence (Section 10.2).

Carbon black is the product of combustion of heavy oil in an oxygen-deficient environment, for example, a furnace. Structurally, it is an aggregate of very small graphene sheets [48, 49]. Although it looks amorphous, it has some degree of order at small scales.

Since there is no one definite structure, the optical properties of amorphous carbon are dependent on the relative number of various bond types and the physical conditions under which they are made [50]. The optical properties of a grain of amorphous carbon are dependent not only on the internal structure, but also on the morphology of the grains. Calculations based on a continuous distribution of ellipsoids show that the emissivity of the grain can change significantly as a function of sp^2 content.

2.3
Molecules of Biological Significance

The idea that biological substances can be divided into different classes based on their chemical compositions has became evident in the mid-nineteenth century. In 1827, William Prout (1785–1850) classified foodstuffs into saccharine, oily, and albuminous groups. Later, the names of these groups evolved to become carbohydrates, lipids, and proteins. Together with nucleic acids, they form the four major classes of organic compounds in living systems. In terms of functions, carbohydrates serve as food for living organisms, fat (a form of lipids) is used to store and transport energy, and lipids form membranes that organize molecules into cells.

Nucleic acids contain the information needed to make proteins, which are responsible for the structure and functions of cells.

2.3.1
Carbohydrates

The term carbohydrate refers to the class of polyhydroxylated aldehydes and ketones. Carbohydrates are separated into the classes of monosaccharides (or simple sugars) such as glucose and fructose which cannot be split into smaller units through hydrolysis, and polysaccharides (or complex carbohydrates) that are made of two or more simple sugars. An example is sucrose (table sugar), which is made of one glucose $(C_6H_{12}O_6)$ molecule linked to a fructose $(C_6H_{12}O_6)$ molecule. Another example of a polysaccharide is cellulose, which is made of thousands of glucose subunits.

Carbohydrates are the most abundant biomolecules on Earth. Animals use carbohydrates to store energy and plants use them to build cell walls. Every year, 10^{11} tons of CO_2 and H_2O in the atmosphere are converted to cellulose through photosynthesis.

2.3.2
Lipids

Unlike carbohydrates and proteins, which are mainly defined by their chemical structures, lipids are characterized by their chemical constituents as well as physical properties such as solubility. Common examples of lipids are fats and waxes. Hydrolysis of fats yields fatty acids, which commonly contain an even number of C atoms. Some common fatty acids are lauric acid $(CH_3(CH_2)_{10}CO_2H$, with 12 C atoms), myristic acid $(CH_3(CH_2)_{12}CO_2H$, with 14 C atoms), palmitic acid $(CH_3(CH_2)_{14}CO_2H$, with 16 C atoms), stearic acid $(CH_3(CH_2)_{16}CO_2H$, with 18 C atoms), and arachidonic acid $(CH_3(CH_2)_{18}CO_2H$, with 20 C atoms).

Lipids are the basic components of cell membranes. Membranes are necessary to keep the molecules synthesized inside the cells away from the environment and to modulate the transport of molecules in and out of cells.

2.3.3
Proteins

Proteins are built from α-amino acids linked together by peptide bonds. Amino acids are molecules that contain an amino $(-NH_2)$ group and a carboxyl $(-COOH)$ group. A peptide bond is formed when the carboxyl group of one amino acid joins the α-amino N of another amino acid (Figure 2.5). Polypeptides are molecules that contain large number of amino acids joined by peptide bonds. Proteins are polypeptides that have some biological function.

Amino acids are classified as α, β, and γ, and so on, based on how far away the amino group is from the carboxylic acid group. If the two groups are connected to

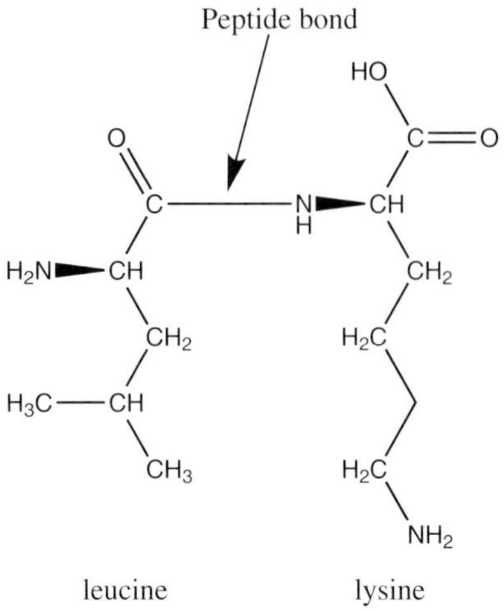

Figure 2.5 A polypeptide is formed when amino acids (leucine and lysine) are joined by a peptide bond.

the same C atom, it is an α amino acid. If they are separated by one C atom, then it is a β-amino acid (e.g., $H_2NCH_2CH_2CO_2H$). One example of a γ amino acid is $H_2NCH_2CH_2CH_2CO_2H$). Glycine (NH_2CH_2COOH) is the smallest amino acid. It is an α amino acid with no side chains. When one of the H atoms connected to the α-carbon of glycine is replaced by a methyl group, it becomes alanine. The 20 amino acids that are normally present in proteins are all α-amino acids. They differ from each other in their side chains. The reason that these 20 amino acids are selected for terrestrial life is not entirely by chance because these molecules represent better choices to serve their specific functions [51]. This however does not preclude the possibility that a larger number of amino acids were used in the early development of life on Earth or that a different set (while probably with some overlapping members) of amino acids is used in life beyond the Earth.

Since an average protein molecule contains on the order of 100 amino acids, there are potentially 20^{100} different kinds of proteins that can be made from the 20 common amino acids. This number is far greater than the $\sim 10^5$ kinds of proteins in living organisms. This suggests that extraterrestrial life could be vastly different from our own, even it obeys the same rules of biochemistry.

2.3.4
Nucleic Acids

DNA (deoxyribonucleic acid) and RNA (ribonucleic acid) are carriers of genetic information. Structurally, DNA and RNA are made of two long strands of carbohydrates each linked by phosphate bridges (PO_4^{2-}), with the two strands held together by heterocyclic aromatic compounds. Two of the heterocyclic bases in DNA are adenine (A) and guanine (G), which are substitute purines (c-$C_5H_4N_4$) (Figure 2.6). The other two are cytosine (C) and thymine (T), which are substitute pyrimidines

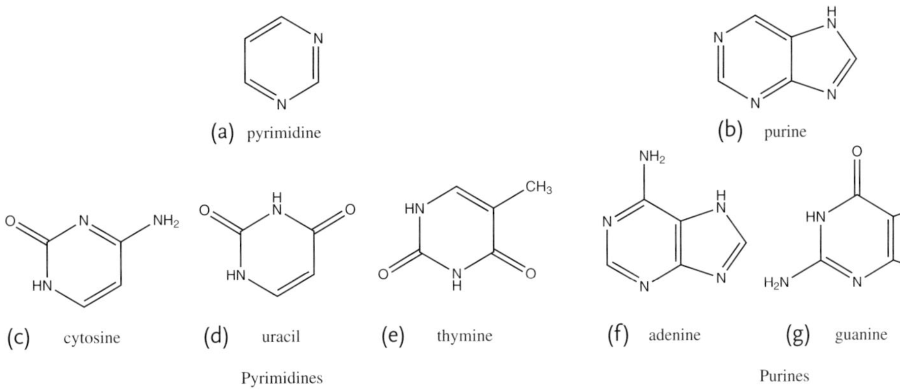

Figure 2.6 The schematic structure of pyrimidine (a) and purine (b) and the schematic structures of the nucleic acid bases cytosine (c), uracil (d), thymine (e), adenine (f), guanine (g).

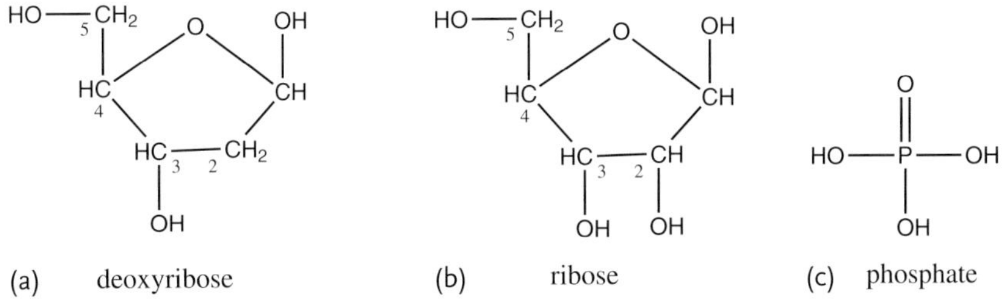

Figure 2.7 The schematic chemical structures of the 5-carbon atom sugar molecules deoxyribose (a) and ribose (b), and phosphate (c). The difference between deoxyribose and ribose lies in whether a H or OH group is covalently linked to the number 2 carbon atom.

(c-$C_4H_4N_2$). In RNA, thymine is replaced by uracil, a different pyrimidine (Figure 2.6). The two strands of DNA backbones are held together by H-bonds between the A-T and C-G base-pairs. Each of the purine and pyrimidine bases represent a letter in the genetic code. Each of the nucleic acid bases is covalently linked to a deoxyribose or ribose to form a nucleoside (Figure 2.7). When a nucleoside is linked with a phosphate through its number five carbon, it is called a nucleotide (Figure 2.8). A nucleotide is joined with the next nucleotide by attaching the phosphate group to the number three carbon atom, thereby creating a linear, unbranched polymeric nucleic acid.

The genetic code is specified by the sequence of three nucleic acid bases called codons. This code dictates the synthesis of specific amino acids into proteins. Although in theory $4^3 = 64$ different triplets are possible, some are redundantly used for specifying the same amino acid. For example, GGU, GGC, GGA, and GGG (for RNA) all lead to glycine. Some triplets are used as start (AUG, for RNA) and stop (UAA, UGA, UAG, for RNA) codons. This code is nearly universal among all living organisms on Earth, although it is quite possible that extraterrestrial life differs in its use of different nucleic acid bases, number of bases in the genetic code (e.g., two instead of three), and the kinds of amino acids to build proteins.

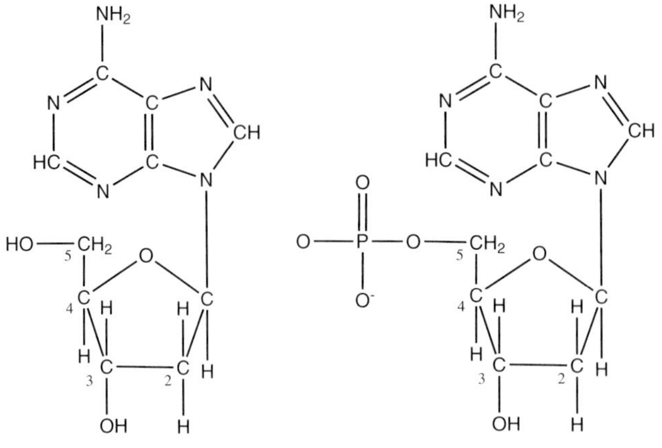

(a)   nucleoside   (b)   nucleotide

Figure 2.8 (a) a nucleoside is formed by the association of the sugar with one of the nucleobases. In this example, the deoxyribose is connected to an adenine base to form the nucleoside deoxyadenosine; (b) a nucleotide is formed by attaching a phosphate group to the number 5 carbon of the sugar in the nucleoside. The phosphate serves as a link to an adjacent nucleotide through its number 3 carbon atom (not shown).

Although the detection of interstellar DNA and RNA (with molecular masses ranging between of 10^{11} and 10^5 Dalton) is almost impossible, it is quite feasible to search for the nucleobases that make up DNA and RNA.

2.4
Summary

Carbon is an extremely versatile element capable of forming a rich array of molecules and solids. Even by its own, it can create compounds of different geometries in the form of graphite, diamond, fullerenes, and carbynes. Together with H, different forms of hydrocarbons, including aliphatics (alkanes, alkenes, alkynes) and aromatics can be formed. Using functional groups (e.g., the methyl group), different classes of organic molecules can be created. Organic molecules in living systems such as carbohydrates, lipids, proteins, and nucleic acids are just larger organic molecules with biological functions. These biological molecules are built upon structural units such as amino acids (for proteins) and purines and pyrimidines (for nucleic acids). Carbon can also be part of amorphous compounds with no definite structure. Carbonaceous amorphous compounds can occur naturally (e.g., coal) or be made artificially (e.g., soot). The search and detection of all these diverse forms of organic matter in different parts of the observable universe represents an interesting and challenging aspect of modern astronomy. In the following chapter, we will discuss the variety of techniques used for these searches and some of the detections obtained.

3
Interstellar Molecules

Molecules in interstellar space can be detected through their electronic, vibrational, or rotational transitions either in emission or in absorption against a background continuum radiation source. Since the electronic transitions lie primarily in the ultraviolet/visible region, vibrational transitions in the infrared regions, and the rotational transitions in the microwave/millimeter-wave regions, detection of these lines requires very different detectors and observing techniques. For molecules well studied in the terrestrial environment, the observed astronomical lines can be identified by comparison with catalogued laboratory measured line lists. For unstable or short-lived molecules, the frequency pattern of a series of astronomical lines can be used to deduce the carrier molecule. In fact, a number of molecular radicals were identified in space before they were synthesized and studied in the laboratory.

In Section 3.1, we discuss the different radiation mechanisms that molecules in space can manifest themselves, allowing their detection by remote observations. In the subsequent Sections 3.2–3.17, the observations of molecules in different organic families are discussed. The ability of telescopes capable of scanning through wide spectral bands make the detection of molecules without prior knowledge of their transition frequencies possible. The possible limits to the detection of large molecules are discussed in Section 3.19.

3.1
Electronic, Vibrational, and Rotational Structures of Molecules

The electronic configuration of a multielectron atoms under Russell–Sanders (L–S) coupling is described by the spectroscopic notation $n^{2S+1}L_J$, where n, S, L, and J are the principal, spin, orbital, and total angular momentum quantum numbers, respectively. Since a molecule has more than one atomic nuclei, it can have vibrational and rotational states in addition to electronic states, and can undergo transitions between these states. Because of the very different energies of the electronic, vibrational, and rotational states, their wave functions are assumed to be separable (the Born–Oppenheimer approximation) and their structures can be discussed separately.

Organic Matter in the Universe, First Edition. Sun Kwok.
© 2012 WILEY-VCH Verlag GmbH & Co. KGaA. Published 2012 by WILEY-VCH Verlag GmbH & Co. KGaA.

3.1.1
Electronic Transitions

The electronic states of molecules are designated by $^{2S+1}\Lambda_{\Omega}$ where S is the total electronic spin, and Λ is the projection of the total electronic orbital angular momenta along the internuclear axis. The upper case Greek letters $\Sigma, \Pi, \Delta, \ldots$ are used to represent $|\Lambda| = 0, 1, 2, \ldots$, analogous to the $S, P, D, \ldots$ notation used in atomic structure. Ω, the projection of the total electronic angular momentum onto the internuclear axis, is given by the sum of Λ and Σ, where Σ (not to be confused with the state designation for $\Lambda = 0$) is the projection of the electronic spin angular momenta along the internuclear axis. The value of Σ ranges from $-S$ to $+S$ in increments of one. For example, a $^2\Pi$ state has $^2\Pi_{3/2}$ and $^2\Pi_{1/2}$ (corresponding to $|\Lambda| = 1$, $\Sigma = \pm 1/2$). Unlike atomic notation, the values of Ω, Λ, and Σ can be added algebraically rather than vectorially because they all refer to the same projection. If the electron wave function of a Σ state changes sign when reflected about any plane passing through both nuclei, it is denoted as Σ^-; if unchanged, then Σ^+. For a molecule with two identical nuclei (e.g., H_2, C_2), the right-subscript g (even) refers to the fact that the electronic part of the molecular wave function of this state remains unchanged on reflection through the center of the molecule, and u (odd) refers to a sign change.

The electronic states of diatomic molecules are also labeled with letters: X is used for the ground state, while $A, B, C, \ldots$ are used for excited states of the same multiplicity $(2S + 1)$ as the ground state. States with a multiplicity different from that of the ground state are labeled with lower case letters (a, b, c, etc.). For example, the ground state of C_2 is $X^1\Sigma_g^+$ and the lowest excited state is $a^3\Pi_u$. Since the ground state has a multiplicity of 1 and the first excited state has a multiplicity of 3, the excited state is labeled as a.

The first indication that organic molecules exist in stars and interstellar space was the detection of electronic absorption bands. The mysterious λ 405 nm band in cometary tails was discovered in the nineteenth century [52] and was identified by A.E. Douglas [53] as the $A^1\Pi_u - X^1\Sigma_g^+$ band of C_3 in 1951. In the spectra of old, carbon-rich stars, the electronic transitions $A^1\Pi_u - X^1\Sigma_g^+$ Phillips System of C_2 and the $A^2\Pi - X^2\Sigma^+$ Red System of CN can be seen, suggesting that these carbon-based molecules can exist in the atmospheres of cool stars. The first interstellar molecules were detected through their electronic transitions seen in absorption against the continuum spectra of bright, hot stars. These include the molecules CN ($B^2\Sigma^+ - X^2\Sigma^+$ at 387.68 nm), CH^+ ($A^1\Pi - X^1\Sigma^+$ at 423.25 nm), and CH ($A^2\Delta - X^2\Pi$ at 430.03 nm).

3.1.2
Vibrational Transitions

For a diatomic molecule vibrating as a simple harmonic oscillator of spring constant k, the vibrational energy level is given by

$$E_v = \left(v + \frac{1}{2} \right) h v_0 , \tag{3.1}$$

where v is the vibrational quantum number and

$$v_0 = \frac{1}{2\pi} \sqrt{\frac{k}{\mu}} \tag{3.2}$$

is the natural oscillator frequency and μ is the reduced mass. From (3.2), we can see that the vibrational frequency decreases with increasing atomic mass. For example, v_0 for a C–H bond is approximately 2.5 times larger than v_0 for the C–C bond. Since the spring constant is higher for a stronger bond, the vibrational frequency is correspondingly higher (e.g., in a triple bond versus a single bond).

For polyatomic molecules, other vibrational modes are possible. In general, a molecule with N atoms have $3N - 6$ number of vibrational modes. These modes can be separated into stretching (when the lengths of the bonds between two atoms change) and bending (when the bond angle between three atoms changes) modes. These modes are labeled as $v_1, v_2, v_3, \ldots$ and so on. For a linear molecule, the number of vibrational modes possible is $3N - 5$, $N - 1$ of which are stretching modes between adjacent atoms. The remaining $2N - 4$ are double-degenerate bending modes, giving $N - 2$ different bending frequencies.

Examples of these vibrational modes can be found in the triatomic linear molecules CO_2. When the C–O bonds of CO_2 shorten and lengthen together in unison, it is called a symmetric stretching mode. Another stretching mode where one bond shortens (O moving towards C) while another bond lengthens (the other O atom moving away from C, is called asymmetric stretching). A third mode for CO_2 is the bending mode when both O atoms move in the same direction off the O–C–O axis. There are two ways that O–C–O can bend, for example, in the plane of the paper and perpendicular to it. These two modes of vibration, however, have the same frequency of vibration and are therefore degenerate. These $3 \times 3 - 5 = 4$ motions constitute the normal modes for CO_2 and generate three vibrational frequencies. These three stretching and bending modes are also found in the linear molecule HCN, where the C–N stretch at 4.8 µm is designated as v_1, the bending mode at 14 µm as v_2, and the C–H stretch at 3 µm as v_3.

We note that most functional groups have vibrational bands at frequencies that do not change significantly from one compound to the other. For example, the C=O band of ketone is in the range of 5.7–5.9 µm, the O–H band of alcohol is in the range of 2.7–2.9 µm, and the C=C band of alkene is in the range of 5.9–6.1 µm. This allows us to determine the presence of functional groups in a complex organic compound by infrared spectroscopy.

The development of near-infrared spectroscopic capability in the late 1970s allowed for the detection of stretching modes of simple molecules, mostly through absorption against infrared bright background sources [54]. Bending modes of molecules generally occur at longer wavelengths, which need to be observed outside of the Earth's atmosphere by satellite-based telescopes.

3.1.3
Rotational Transitions

The rotation of a molecule can be described in terms of motions about the principal axes. If all three moments of inertia (I_A, I_B, I_C, in increasing order) are different, the molecule is referred to as an asymmetric top. If two of these moments of inertia are equal, the molecule is referred to as a symmetric top. For a diatomic molecule such as CO where $I_A = 0$, $I_B = I_C$, the energy of the rotational state J is

$$E_J = h B J(J + 1) \,, \tag{3.3}$$

where the rotational constant B is given by

$$B = \frac{h}{8\pi^2 I_B} \,. \tag{3.4}$$

For electric-dipole transition, only transitions between successive rotational stars are allowed ($\Delta J = \pm 1$) and the transition frequency from an upper state $J + 1$ to a lower state J is given by

$$\nu_{J+1,J} = 2B(J + 1) \,. \tag{3.5}$$

Since real molecules are not rigid rotators, their rotational energy states are approximated by fitting functions using coefficients D, H, … etc:

$$E_J = h \left[B J(J + 1) - D J^2(J + 1)^2 + H J^3(J + 1)^3 - \cdots \right]. \tag{3.6}$$

For nonlinear molecules, additional quantum numbers are needed to specify the rotational states. For a symmetric-top rotator (molecules with two equal moments of inertia), the quantum number K corresponding to the angular momentum projection on the axis of symmetry is needed in addition to the total angular momentum quantum number J. The energy levels of the rotational states (labeled as J_K) are:

$$E_{JK} = B J(J + 1) + (A - B) K^2 \tag{3.7}$$

for a prolate (cigar-like) symmetric top ($I_A < I_B = I_C$), and

$$E_{JK} = B J(J + 1) + (C - B) K^2 \tag{3.8}$$

for an oblate (pancake-like) symmetric top ($I_A = I_B < I_C$). The quantum number K takes on integer values ranging from $-J, -J+1, \ldots, J-1, J$ and A, B, and C are the rotational constants along the three rotation axes:

$$A = \frac{h}{8\pi^2 I_A}$$

$$B = \frac{h}{8\pi^2 I_B}$$

$$C = \frac{h}{8\pi^2 I_C} \tag{3.9}$$

with $A \geq B \geq C$. The electric-dipole selection rules are $\Delta K = 0$, $\Delta J = 0, \pm 1$.

For asymmetric rotators where all of their principal moments of inertia have different values, the quantum numbers K_{-1} and K_1 (corresponding to the limiting cases of prolate and oblate symmetric rotators) are used, and the energy levels are designated as J_{K_{-1}, K_1}. The selection rules are $\Delta J = 0, \pm 1$ and $K_{-1}, K_1 = ++ \leftrightarrow$ $--$ and $-+ \leftrightarrow +-$, where $+$ and $-$ as the evenness or oddness of the K_{-1} and K_1 quantum numbers.

The selection rule for vibrational transitions in a diatomic molecule within the harmonic approximation is $\Delta v = \pm 1$. The rotational transitions within each vibrational transition are organized into branches according to the change in the rotational quantum number (initial minus final in an emission process) ΔJ. The selection rule for one-photon, electric-dipole transitions are $\Delta J = -1$ and $+1$ (called P and R branches, respectively). The frequencies of the transitions in the fundamental mode ($v = 1 \leftrightarrow 0$) are:

$$\nu = \nu_0 + 2(J+1)B \quad J = 0, 1, 2, \ldots \quad \text{for the } R \text{ branch}$$
$$\nu = \nu_0 - 2JB \quad J = 1, 2, 3, \ldots \quad \text{for the } P \text{ branch}, \tag{3.10}$$

where $h\nu_0$ is the energy between the ground rotational states of the vibrational levels. These vibrational-rotational transitions are labeled as $P(J = 1, 2, 3, \ldots)$ and $R(J = 0, 1, 2, \ldots)$ where J is the rotational quantum number of the lower vibrational state.

Molecules with no electric dipole moment (e.g., diatomic molecules with two identical nuclei such as H_2, O_2, and N_2) have no allowed electric-dipole rotational transitions and these molecules have to be studied by other techniques, for example, through their magnetic dipole transitions.

The development of high-frequency radio receivers in the late 1960s led to the detection of rotational transitions of interstellar molecules. Using a Schottky diode receiver mounted on the 11-m NRAO telescope at Kitt Peak, R.W. Wilson, K.B. Jefferts, and A.A. Penzias of Bell Labs reported in the one-line abstract of their paper in 1970: "We have found intense 2.6-mm line radiation from nine galactic sources which we attribute to carbon monoxide" [55]. This represents the beginning of the extensive searches for molecules in space, and the discovery of over 160 (as of 2010) different species of interstellar molecules.

3.1.4
Effects of Electron and Nuclear Spins

The total angular momentum J of a linear molecule is given by the vector sum $J = N + L + S + l$, where N, L, S, and l are the rotational, electronic orbital, electron spin, and vibrational angular momenta, respectively. Vibrational angular momentum is only present in polyatomic molecules. An example of the vibrational angular momentum is the bending modes of linear polyatomic molecules (e.g., HCN, Section 3.7.2). Since most common molecules have no electronic orbital angular momentum and no unpaired electron spin, J can be considered to be equal to $N + l$ for most astrophysical molecules. In that case, the possible quantum numbers for J are $|l|, |l| + 1, \ldots$, and so on. The two common exceptions are O_2 ($X^3\Sigma_g^-$) and NO ($X^2\Pi$). O_2 has unpaired electrons ($S = 1$) and NO has nonzero electronic orbital angular momentum ($\Lambda = 1$) and nonzero spin ($S = 1/2$).

For molecules with nonzero electron spin angular momentum, the total angular momentum has values of $|J - 1|$ to $|J + S|$. For example, the oxygen molecule (O_2) in the electronic ground state of $^3\Sigma_g^-$ ($L = 0, S = 1$) has its rotational levels split into $J = N - 1, 1, N + 1$ levels. These are called fine-structure splitting.

The rotational line of a molecule can be split into hyperfine components as the result of nuclear spins. For example, ^{14}N has a nuclear spin of one, and the rotational states of HCN are split into three hyperfine components ($F = J + I$) $F = J - 1$, J, and $J + 1$. For molecules consisting of more than one atom with a nonzero nuclear spin, the total nuclear spin quantum number (I) is the sum of all the individual nuclear spins. For example, the water (H_2O) molecule has two H atoms, each with nuclear spin of 1/2, leading to values of $I = 0$ or 1. The nuclear spin state with the larger statistical weight is called ortho and smaller weight is call para. In the case of water, the para state corresponds to $I = 0$ and the ortho state to $I = 1$.

3.2
Hydrocarbons

Methane (CH_4) is a spherical top molecule (like C_{60}) and have 4 identical nuclei. CH_4 has $3(5) - 6 = 9$ vibrational modes, 4 of which are stretching modes and 5 are bending modes [56]. Only the ν_3 antisymmetric C–H stretch and the ν_4 bending modes are infrared active. Interstellar CH_4 was first detected in its ν_3 at 3.3 μm and ν_4 at 7.6 μm [57].

In the atmospheres of giant planets, methane is highly abundant and is believed to be the carrier of most of the carbon in these atmospheres. The higher members of the alkane family (e.g., ethane (C_2H_6), propane (C_3H_8)) are seen in planetary atmospheres (Section 7.4), but not yet detected in the interstellar medium.

The first member of the alkenes family ethylene ($H_2C{=}CH_2$) was detected in absorption against the dust infrared continuum of the carbon star IRC+10216 [58]. Ethylene is a planar asymmetric-top and has no permanent dipole moment. The observed lines are the vibrational-rotational transitions $\nu = 0$–1 $0_{00} \rightarrow 1_{10}$ and $5_{05} \rightarrow 5_{15}$ at 10.5 μm.

Acetylene (H–C≡C–H) is the first member of the alkynes family of hydrocarbons. A symmetric linear molecule with four atoms has seven vibrational modes, of which three are stretching modes (symmetric C–H stretch ν_1 at 2.96 μm, C≡C stretch ν_2 at 5.07 μm, antisymmetric C–H stretch ν_3 at 3.03 μm) and four bending modes (doubly degenerate *trans* bend $_\downarrow$H–C≡C–H$^\uparrow$ ν_4 at 16.3 μm and *cis* bend $_\downarrow$H–C≡C–H$_\downarrow$ ν_5 at 13.7 μm).

Interstellar acetylene was discovered through the ν_5 line in absorption against infrared sources in molecular clouds with the NASA IRTF 3-m telescope [59]. The 13.7 μm band is since widely observed in AGB stars and is a key signature of the advanced evolutionary status of an AGB star (Section 6.3).

In addition to acetylene, which is commonly seen in carbon-rich AGB stars, poly-acetylenic chains are detected in protoplanetary nebulae. The fundamental bending modes ν_8 of diacetylene (C_4H_2) and ν_{11} of triacetylene (C_6H_2) at 15.9 and 16.1 μm, respectively, have been detected by ISO in AFGL 618. Also seen in this spectrum are the bending modes of cyanopolyynes, including ν_2 of HCN, ν_5 of HC_3N, and ν_7 of HC_5N.

3.3
Alcohols

Alcohol is defined as a compound that has hydroxyl (OH) groups bonded to carbon atoms. Alcohols can be considered to be related to water in the sense that one of the H atoms in water is replaced by a group (R–O–H). They also have structures similar to water as the R–O–H bond angle has a similar values as water. The class of alcohols does not include phenols where an OH group is attached to an aromatic ring, or enols where the OH group is attached to vinylic carbon. The vibrational spectra of alcohols are characterized by the O–H stretching mode at 2.8–3.0 μm.

3.3.1
Methanol

Interstellar methanol (or methyl alcohol, CH_3OH) was first detected through its 1_1 K-doublet rotation lines at frequency of 834 MHz [60]. Methanol has nearly the same geometry as water, with one H being replaced by a methyl group ($-CH_3$). Since the methyl group has three equivalent H atoms, the combinations of the nuclear spins of these 3 H atoms create symmetry states called A, E. The A state corresponds to all spins being parallel for a total of $I = 3/2$, while the E state has $I = 1/2$. The internal rotation of the methyl group against the HO–C framework is referred to as torsion.

Although methanol is an asymmetric rotor, it is nearly a prolate symmetric top as its internal rotation is so fast that it effectively averages out the asymmetry. As a result, methanol energy level expressions can be approximated by that of a symmetric rotator, and its rotational energy levels can be labeled by a single quantum number K. The mm and submm spectra of interstellar clouds are filled with hundreds of lines from methanol.

3.3.2
Vinyl Alcohol

Interstellar vinyl alcohol ($H_2C=CHOH$) was first found in emission toward Sgr B2 through its rotational transitions [61]. It is a planar molecule with two conformers, depending on which side the O–H group lies relative to the C–H group. Both conformers have been detected in the interstellar medium.

Ethanol (CH_3-CH_2-OH) was first detected in Sgr B2 through its $6_{06}-5_{15}$, $4_{14}-3_{03}$ and $5_{15}-4_{04}$ rotational transitions at 85.265, 90.117 and 104.809 GHz, respectively [62]. In their discovery paper, the authors state that "the alcoholic content of this cloud (Sgr B2), if purged of all impurities and condensed, would yield approximately 10^{28} fifths[1] at 200 proof.[2] This exceeds the total amount of all of man's fermentation efforts since the beginning of recorded history" [62]. This quote attracted a lot of attention in the press, and probably did more to bring public awareness of interstellar chemistry than any other of piece of work.

Ethylene glycol ($HOCH_2CH_2OH$) was detected towards Sgr B2 in several rotational transitions [63]. It is a triple-rotor molecule with three axes, two of CO and one of CC. Twisting (torsion) around these three axes produces ten different conformers. Ethylene glycol as well as other sugar alcohol with up to six C atoms, are found in the Murchison and Murray meteorites [64]. The discovery paper is titled "Interstellar Antifreeze: ethylene glycol", the significance of which may be missed in tropical countries but served its purpose for readers of the Astrophysical Journal in America.

3.4
Carboxylic Acids

Acids are traditionally defined as substances that ionize to give protons when dissolved in water. The most common organic acids are the carboxylic acids which contains the −COOH carboxyl group. Acetic acid (CH_3COOH, the main ingredient of vinegar) is a member of the carboxylic acid family ($C_nH_{2n}O_2$). Substitution of one H atom by the amino group −NH_2 leads to the simplest amino acids glycine (NH_2CH_2COOH). Higher members of the family include propionic acid (CH_3CH_2COOH), butanoic acid ($CH_3CH_2CH_2COOH$), and so on. From a structural point of view, we may consider carboxylic acids to be related to ketones and alcohols for carboxylic acids consist of both C=O and O–H groups.

---

1) A fifth is one fifth of a gallon (slightly more than 750 ml), the typical size of bottle of hard liquor sold in the USA.
2) Proof is defined as $200 \times [CH_3CH_2OH]/([CH_3CH_2OH] + [H_2O])$.

Formic acid (HCOOH) is the simplest organic acid and this slightly asymmetric rotator was observed in space in 1971 through its $1_{11}-1_{10}$ rotational transition at 1638.8 MHz with the NRAO 11-m telescope [65]. This identification was confirmed by the detection of the $2_{11}-2_{12}$ transition at the Effelsberg 100-m telescope [66].

Although the isomer of acetic acid methyl formate (HCOOCH$_3$) is abundant in interstellar clouds, acetic acid was late in being discovered because the large energy difference between the A and E states makes laboratory frequency assignments of doublets difficult [67]. The molecule was discovered by observations with the Berkeley–Illinois–Maryland Association Array (BIMA) and the Owens Valley Millimeter Array, representing the first detection of an interstellar molecule using interferometers [68].

3.5
Aldehydes and Ketones

Aldehydes are molecules containing an O=CH− functional group (called the aldehyde group). Ketones are molecules with a carbonyl group (C=O) attached to two other carbon atoms. Some simple aldehydes include formaldehyde (HCHO), acetaldehyde (CH$_3$CHO), propionaldehyde (CH$_3$CH$_2$CHO), and so on. The carbonyl group makes aldehydes and ketones polar molecules with large dipole moments leading to strong rotational transitions. In the infrared, aldehyde and ketones have strong C=O stretching vibrational modes in the spectral region 5.7–5.8 μm.

3.5.1
Formaldehyde

The λ 6 cm (4830 MHz) $1_{11}-1_{10}$ transition of formaldehyde (H$_2$CO) was one of the first molecular rotational transitions detected in the radio, and was also the first organic polyatomic molecule found in space [23]. Formaldehyde is a near-symmetric rotor with two moments of inertia nearly equal, though much larger than the third. The corresponding rotational constants are $A = 281\,970.672$ MHz, $B = 38\,836.0455$ MHz, $C = 34\,002.2034$ MHz. The small asymmetry about the C–O axis causes a splitting of the rotational levels into K doublets. Since the two H atoms can have parallel or antiparallel nuclear spins, formaldehyde can be in ortho or para states that can be treated as two distinct molecules.

This λ 6 cm line is always seen in absorption, even in dark clouds where no source of continuum radiation exists. This suggests that the molecule is absorbing the 3 K cosmic background radiation itself. Since the excitation temperature of the line is less than 3 K, the line can be considered as an "antimaser", where the upper level of the transition is selectively depopulated (or the lower state selectively over-populated) by a pumping mechanism. One of the first pumping mechanisms considered was collisional pumping. When neutral particles strike the H$_2$CO molecule along the molecule plane, the lower level of the doublet at each rotational level will be favorably excited. Rapid radiative decays with $\Delta J = 1$ will lead to the over population of the 1_{11} lower state and therefore a lowering of the excitation temperature [69, 70].

3.5.2
Cyanoformaldehyde

Cyanoformaldehyde (CNCHO, also named formyl cyanide) is related to formaldehyde by substituting an H atom a CN radical. CNCHO is a planar asymmetric-top molecule made up of the aldehyde and cyano groups. Its rotational transitions $7_{07}-6_{16}$ (8.6 GHz), $8_{08}-7_{17}$ (19.4 GHz), $9_{09}-8_{18}$ (30.3 GHz), and $10_{0,10}-9_{19}$ (41.3 GHz) in emission, and one transition $5_{15}-6_{06}$ (2.1 GHz) in absorption, have been detected in Sgr B2 with the 100-m Green Bank Telescope [71].

3.5.3
Acetaldehyde

Acetaldehyde (CH_3CHO) is an asymmetric molecule with an internal rotor in the form of the methyl group. The nuclear spins of the three H atoms lead to the *A* and *E* symmetry states (Section 3.3.1) and there are no allowed electric-dipole transitions between the *A* and *E* states. Its initial discovery in space and follow-up observations were based on the *K*-doublet transitions $1_{10}-1_{11}$ (1.065 GHz), $2_{11}-2_{12}$ (3.195 GHz), $3_{12}-3_{13}$ (6.390 GHz), and $4_{13}-4_{14}$ (10.649 GHz), and so on [72]. The first rotational transitions detected were the $1_{01}-0_{00}$ transitions of both the *A* and *E* symmetry states [73]. Interestingly, the intensities of the *A* and *E* state lines are about equal in both TMC-1 and L134N.

3.5.4
Propynal, Propenal and Propanal

Propynal (HC≡CCHO), an acetylene derivative (Section 3.10), was detected through its rotational transitions $2_{02}-1_{01}$ (18.65 GHz) and $4_{04}-3_{03}$ (37.29 GHz) rotational transitions in TMC-1 [74]. The $2_{02}-1_{01}$ transition was also later found in Sgr B2 [75].

Interstellar aldehydes molecules propenal (CH_2CHCHO) and propanal (CH_3CH_2CHO) can be formed by hydrogen addition reaction with propynal. Possible identification of these molecules in Sgr B2 were reported in 2004 [75]. The measured transitions are $2_{11}-1_{10}$ (18.221 GHz) and $3_{13}-2_{12}$ (26.079 GHz) for propenal, and $1_{11}-0_{00}$ (21.269 GHz), $2_{02}-1_{01}$ (20.875 GHz), $2_{12}-1_{11}$ (19.690 GHz), $2_{11}-1_{10}$ (22.279 GHz), $3_{03}-2_{12}$ (21.452 GHz), $4_{13}-4_{04}$ (19.182 GHz) for propanal.

3.5.5
Ketene

The class of organic molecule ketene has the structure of RR′C=C=O. It simplest member $H_2C=C=O$ where RR′ are two H atoms is also referred to as ketene.

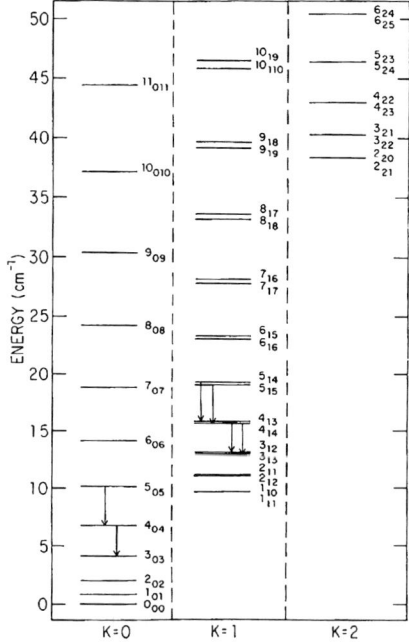

Figure 3.1 The lowest 42 rotational levels of ketene. The K-doubling of the lower J levels in the $K = 1$ and 2 ladders are too small to be shown. The arrows indicate the six detected transitions in the discovery. Figure adapted from [76].

Interstellar ketene was first detected in Sgr B2 with the NRAO 11-m telescope [76]. Ketene is a planar molecule which is very nearly prolate symmetric rotator. The lowest 42 rotational levels are shown in Figure 3.1. Its abundance relative to H_2 is estimated to be $\sim 10^{-10}$–10^{-9}, with an observed ortho/para ratio ~ 3–6 [77].

3.5.6
Acetone

Acetone (CH_3COCH_3) was first detected in Sgr B2 with the IRAM 30-m Telescope [78]. A total of 54 transitions were detected in the high-mass star formation region Orion-KL [79].

Acetone has two identical internal rotors (two methyl groups attached to the CO group) and is therefore structurally similar to methanol. Consequently, the rotational spectrum of acetone is complicated due to interactions between the two methyl tops with each other and with the rigid-body rotation of the molecule. Each rotational transition is split into four components corresponding to the products of A- and E-symmetry states of the CH_3 rotors. Methanol and acetone are among many asymmetric rotator organic molecules detected in the ISM.

3.6
Ethers and Esters

Ether is a class of organic compound that contains an O atom connected to two alkyl groups. The structure of ethers (R–O–R′) is similar to that of water (H–O–H), with the O atom in a sp³ hybridization and forming a tetrahedral bond. Esters have the form of RC=OOR′. An example is ethyl acetate ($CH_3COOCH_2CH_3$).

Dimethyl ether (CH_3OCH_3) is the simplest molecule with two internal rotors. Two methyl (–CH_3) groups rotate around the C–O axis and therefore very similar to acetone. This motion creates four torsional substates labeled as AA, AE, EA, and EE. These substates have slightly different energy levels, leading to each rotational transition being split into four components as in the case of acetone (Section 3.5.6). Dimethyl ether was first detected through its 6_{06}–5_{15} rotational transition at the frequency of 90.9 GHz with the NRAO 11-m telescope [80]. It is widely present in regions of massive star formation, in particular in quiescent sources with gas densities of 10^6–10^8 cm^{-3} and temperature >100 K (known as "hot cores").

Ethyl methyl ether ($CH_3OC_2H_5$) is another asymmetric-top molecule with two methyl group internal rotors. It was detected by the IRAM 30-m and the SEST 15-m telescopes [81]. As this is a large (12 atom) and complex molecule, several transitions are needed to confirm its detection. Since the spectra are contaminated by lines from other organic molecules (most prominent of which is methanol), it is necessary to search for transitions that are free from blending from other lines in order to secure a detection.

Methyl formate (CH_3–O–CHO) was first detected through its 1_{10}–1_{11} A-state transition at 1.610 GHz in Sgr B2 [82]. It has two internal rotors, one along the OC–OC bond and another along the O–CH_3 bond. The former rotor suffers from a large potential barrier which in effect creates two stable conformers, referring to the position of the double-bonded O relative to the methyl group (Figure 3.2).

Methyl formate has a very rich spectrum and together with dimethyl ether and methanol, account for most lines that appear in spectral line surveys (Sec-

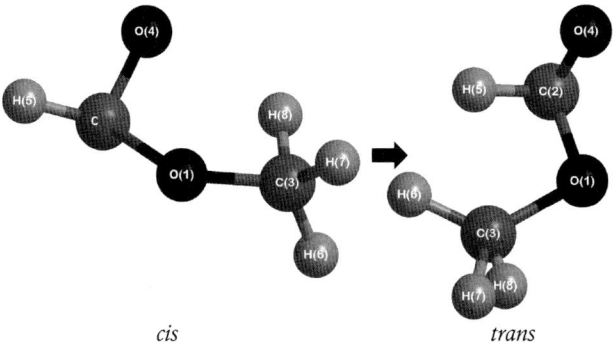

cis   *trans*

Figure 3.2 The two conformers of methyl formate. It takes 13.8 kcal/mol to change from the *cis* state to the *trans* state [83].

tion 3.18) in the millimeter and submm region, earning the nickname of "interstellar weed".

3.7
Amines, Nitriles, and Nitrogen-Containing Molecules

Amines are nitrogen-containing molecules of the forms $-NH_2$, $-NHR$, or $-NR_2$. Some examples are methylamine (CH_3NH_2), and ethylamine ($CH_3CH_2NH_2$). Nitrogen is a key element in all living organisms. It is needed for the synthesis of amino acids, nucleotides, and other biomolecules. The search and detection of N-containing molecules therefore form the foundation of astrobiochemistry. Amines are derivatives of ammonia, as alcohols are derivatives of water. Heterocyclic amines refer to compounds where the N atoms are part of a ring, as in pyridine, pyrrole, quinoline, imidazole (Section 3.15.3).

Nitriles have the form $-C \equiv N$. An example is acetonitrile ($CH_3C \equiv N$). Esters, amides, and nitriles all yield carboxylic acids on hydrolysis and are therefore considered as carboxylic acid derivatives.

3.7.1
Ammonia

The ammonia (NH_3) molecule is a polyatomic symmetric-top molecule that is widely observed in the ISM. Each of the H nuclei is a fermion ($I = 1/2$) and the wave function symmetry under the exchange of two of the three identical H atoms leads to the separation of the molecule into ortho and para forms. Since the rotation of the molecule by $120°$ around the symmetry axis is equivalent to the exchange of two pairs of H nuclei, and it turns out that the molecule is in ortho form if K is a multiple of three (Figure 3.3).

Because of the tetrahedral structure of NH_3, the N atom flips from one side of the plane defined by the 3 H atoms to the other side, like the flipping of an umbrella under a strong wind. These two configurations have a small difference in energy and the flipping motion results in an inversion transition. The transition between two inversion states of the $J_K = 1_1$ level of para-NH_3 at 23.6945 GHz is commonly observed in the ISM.

Each of the inversion doublets is split into hyperfine components due to the interaction between the electric quadrupole moment of the N nucleus and the electric field of the electrons. Weaker magnetic hyperfine interactions associated with the H nuclei further split these hyperfine states and introduce 18 magnetic hyperfine transitions. Both types of hyperfine transitions have been detected in molecular clouds.

Due to the existence of a symmetry axis in NH_3, the molecule has no dipole moment perpendicular to the axis. From the selection rules for symmetric tops ($\Delta K = 0$, $\Delta J = 0, \pm 1$), rotational transitions between different Ks are forbidden and rotational transitions occur down the K ladder between successive Js (Fig-

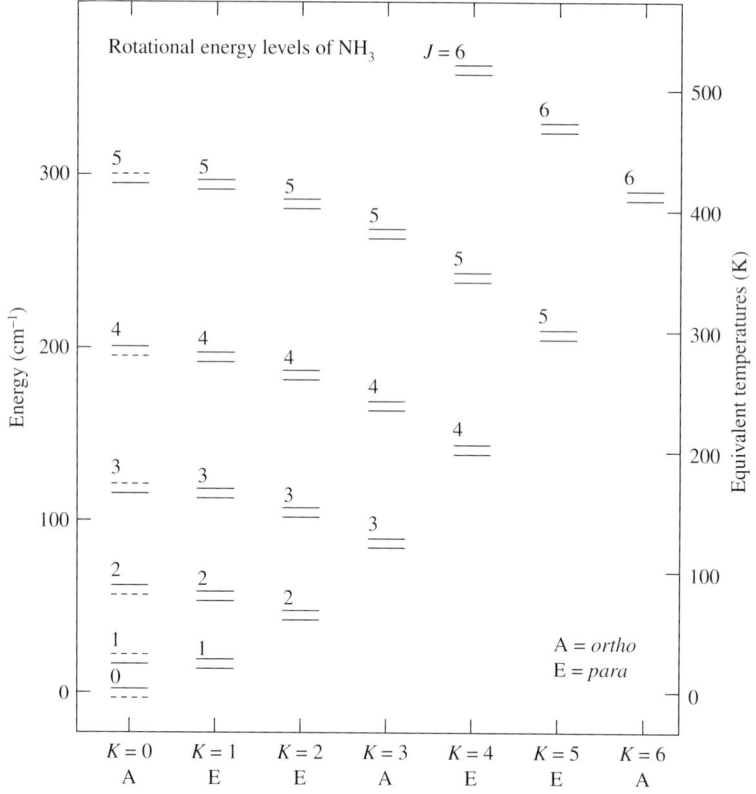

Figure 3.3 The energy level diagram of NH₃ arranged in columns of constant K. Each of the rotational levels is split into two inversion states and are labeled with their J value. The excluded inversion states in $K = 0$ are indicated by dashed lines.

ure 3.3). Most of these transitions occur in the far infrared and are difficult to observe from ground-based telescopes. The ground-state 1_0–0_0 transition of NH₃ at 572.498 15 GHz was detected in 1983 with the KAO [84].

3.7.2
Hydrogen Cyanide

Hydrogen cyanide (HCN) and its isotopic counterpart (HC¹³N) was first detected in 1971 through its ground-state rotational transition ($J = 1$–0) [85]. The molecule is abundantly present in many astronomical environments, from dark clouds to star-forming regions and circumstellar envelopes. HCN represents a member of the hydrogenation series of the cyanide radical, beginning with CN, HC≡N, H₂C≡NH, H₃C−NH₂ (methylamine), and so on.

Because of the nuclear spin of the ¹⁴N atom, the rotational states of HCN are split into three hyperfine components. For example, the $J = 1 \rightarrow 0$ transition of HCN

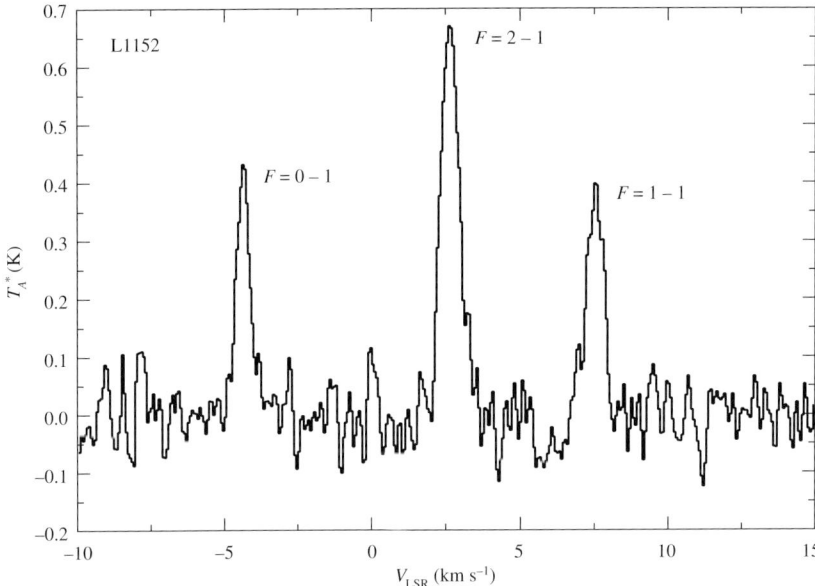

Figure 3.4 The spectrum of the HCN $J = 1 - 0$ line in L1152 showing the three hyperfine components. The actual observed line strengths of the three components do not correspond to the theoretical values of $1:5:3$ (data courtesy of Yong-Sun Park).

has three hyperfine transitions ($\Delta F = 0, \pm 1$, $F = J + I = 2 \rightarrow 1$, $F = 1 \rightarrow 1$, and $F = 0 \rightarrow 1$). The relative intensities of these hyperfine transitions correspond to their respective statistical weights of the upper states, and have values of 5, 3, and 1 respectively. An example of the hyperfine spectrum of HCN is shown in Figure 3.4.

HCN is a linear molecule with three vibrational modes: the ν_1 CN stretch at 4.8 μm, the ν_2 doubling degenerate bending modes at 14 μm, and the ν_3 CH stretch at 3 μm. The double degeneracy of the ν_2 bending mode is lifted when the molecule is bending and rotating simultaneously. This is known as *l*-type doubling. Since the splittings are small, the *l*-type transitions occur at low frequencies and can be observed by cm-wave telescopes. For example, the frequencies of the *l*-type transitions of the first excited bending mode of HCN is approximately given by $\nu = qJ(J + 1)$, where $q \approx 224$ MHz [86].

3.7.3
Methylenimine

The molecule methylenimine is of interest because it is believed to be the precursor of glycine. It is a prolate asymmetric rotor with dipole moments of 1.325 and 1.53 debyes along the two molecular axes. Methylenimine (CH_2NH) was first detected in Sgr B2 through its $1_{10}-1_{11}$ transition at 5.3 GHz [87]. The molecule was also found in molecular clouds W 51, Orion KL and G34.3+0.15 [88] and circumstellar envelopes [89].

3.7.4
Methylamine

Methylamine (CH_3NH_2) is part of the hydrogenation series of cyanide (Section 3.7.2) and substitution of one H atom of methylamine by a carbonyl group COOH leads to glycine. Interstellar methylamine was detected in its $5_{15}-5_{06}$ (73.0 GHz) and $4_{14}-4_{04}$ (86.1 GHz) transitions in 1974 [90].

3.7.5
Cyanamide

Cyanamide (NH_2CN) was detected in its $4_{13}-3_{12}$ (80.5 GHz) and $5_{14}-4_{13}$ (100.6 GHz) transitions with the NRAO 11-m telescope [91]. Cyanamide is a very nearly prolate symmetric rotor. The amine (NH_2) group attached to the end of the N−C≡N axis makes it a nonplanar molecule.

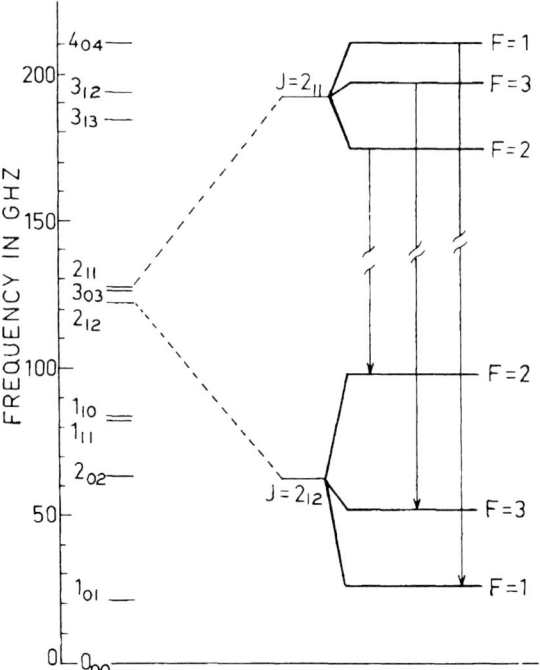

Figure 3.5 Lowest rotational energy levels of formamide. Hyperfine splittings of the 2_{12} and 2_{11} levels are expanded on the right. The three $\Delta F = 0$ transitions are marked. Figure adapted from [92].

3.7.6
Formamide

Formamide (NH_2CHO) is an amide derived from formic acid. Interstellar formamide was detected in its $2_{11}-2_{12}$ rotational transition at 4.620 GHz [92]. The ^{14}N nucleus causes hyperfine splitting (Figure 3.5). Confirmation was obtained by the detection of its $1_{11}-1_{10}$ transition at 1.5399 GHz, including the detection of the $F = 1-1, 1-2, 2-1, 1-0, 2-2$, and $0-1$ hyperfine transitions [93]. A more recent spectrum of formamide in Sgr B2 is shown in Figure 3.6.

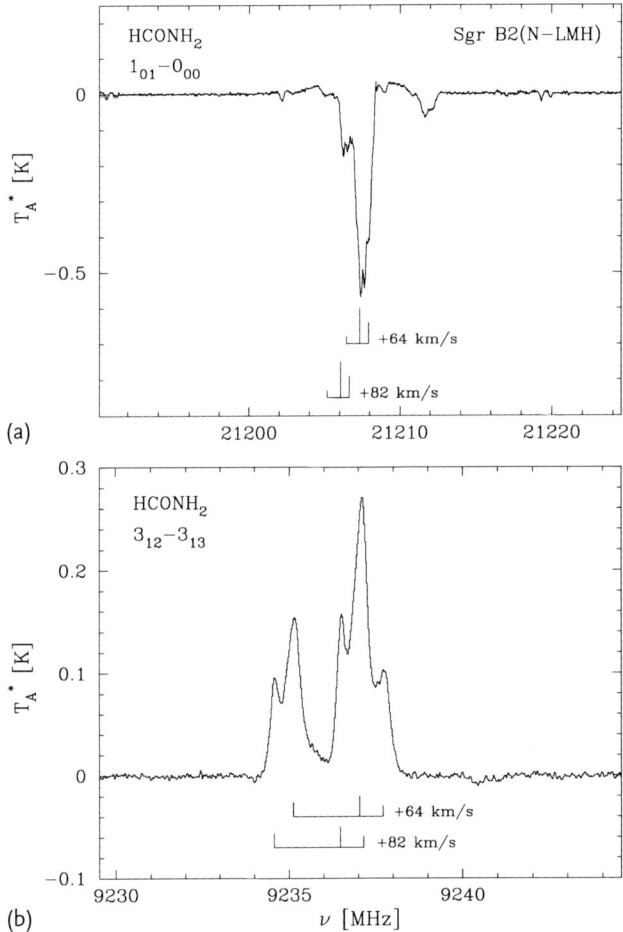

Figure 3.6 The spectrum of the (a) $1_{01}-0_{00}$ and (b) $3_{12}-3_{13}$ transitions of formamide toward Sgr B2. The marks below the spectra show the expected relative intensities of the hyperfine lines. Figure adapted from [94].

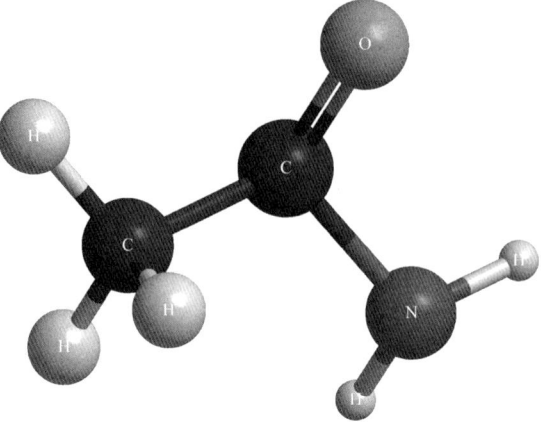

Figure 3.7 A schematic diagram of the molecular structure of acetamide.

3.7.7
Acetamide

Other than formamide, acetamide (CH_3CONH_2) is the only known interstellar molecule containing a peptide bond (Section 2.3.3) as of 2010 [94]. Acetamide was detected with the NRAO 100-m Green Bank Telescope [94]. As in some of the molecules discussed previously, the spectrum is complicated by the internal rotation of the methyl group (Figure 3.7).

3.7.8
Ketenimine

Ketenimine ($H_2C=C=N-H$) is an isomer of methyl cyanide. It was discovered through its rotational transitions $7_{16}-8_{08}$ (41.5 GHz), $8_{19}-9_{09}$ (23.2 GHz), and $9_{18}-10_{0,10}$ (4.9 GHz) with the 100-m Green Bank Telescope [95]. Laboratory microwave spectroscopy has determined its rotational constants to be $A = 201\,443.685(75)$ MHz, $B = 9663.138(2)$ MHz, and $C = 9470.127(2)$ MHz, allowing an astronomical search for its rotational transitions.

3.7.9
Amino Acetonitrile

Amino acetonitrile (NH_2CH_2CN) is considered a possible precursor of glycine in the ISM. Being an 8-atom, nonlinear heavy molecule, it needs many lines to be identified. Eighty-eight transitions in the form of fifty one observed features were detected in the IRAM 30-m Telescope observation of Sgr B2 (N) [96].

3.8
Radicals

Molecules with at least one unpaired electron (called radicals) are highly reactive and unstable and difficult to study in the laboratory setting due to their short lifetimes. However, the low density of interstellar space allows for the condition for the survival of free radicals which can be abundantly present. This is evident as many radicals have been detected in space, including some of the first molecules (e.g., OH and CN) detected. Radicals therefore play an important role in interstellar chemistry.

3.8.1
CH

The CH radical was the first molecule detected in space [16, 17]. It was first found through its electronic transitions at 430.0, 389.0, and 314.6 nm. The detection of its rotational transitions came much later. CH has one unpaired electron and spin-orbit interaction of the one unpaired electron of CH splits the rotational levels N into $^2\Pi_{1/2}$ and $^2\Pi_{3/2}$ ladders ($J = N + \Lambda + S$, $\Lambda = 1$, $S = 1/2$, $N = 1, 2,...$). Each of these J states is split into a Λ doublet (denoted by their parity $+$ or $-$) and further split into hyperfine components $F = J + I$ ($I = 1/2$). In the ground state $^2\Pi_{1/2}$, the allowed hyperfine transitions are $F = 1-0, 1-1, 0-1$ at 3349, 3335, and 3264 MHz, respectively. The frequencies of the hyperfine transitions in the $^2\Pi_{3/2}$ are $F = 2^- \rightarrow 2^+, 1^- \rightarrow 1^+, 1^- \rightarrow 2^+, 2^- \rightarrow 1^-$ at 701.63, 722.1, 724,72, and 704.27 MHz, respectively. The 3 GHz transitions were first detected in 1974 [97], and the 0.7 GHz lines in 1985 [98]. Because of the light weight of CH, its rotational transitions are in the THz range. The $N = 2 \rightarrow 1$ rotational transitions $^2\Pi_{3/2} \rightarrow$ $^2\Pi_{1/2}$ at 149.390 μm ($- \rightarrow +$) and 149.091 μm ($+ \rightarrow -$) were first detected in absorption against Sgr B2 by the KAO in 1987 [99].

3.8.2
CH$^+$

A sharp interstellar absorption line at 423.3 nm discovered in 1937 was identified by A.E. Douglas and G. Herzberg in 1941 as the $R(0)$ line of CH$^+$ [18]. In the circumstellar environment, a series of unidentified emission lines at 422.58, 422.71, 422.95, 423.27, 423.77, and 423.95 nm in the Red Rectangle [100] were identified as the $R(3)$, $R(2)$, $R(1)$, $R(0)$, $Q(1)$ and $Q(2)$ lines of CH$^+$ [101]. CH$^+$ is found to be common in the spectra of post-AGB stars [102].

Because of the low molecular weight, the rotational transitions of CH$^+$ occur in the submm and far infrared region. In the ISO LWS spectrum of NGC 7027, an emission line at 179.62 μm first attributed to H$_2$O [103]. It was later realized that the frequency of this line is in $2:3:4$ ratios with two other unidentified lines at 119.90 and 90.03 μm, and these three lines are identified as the $J = 2-1, 3-2,$

and 4−3 rotational transitions of CH^+ [104]. The $J = 1$–0 359 μm transition was detected by the Herschel Space Observatory [105].

3.8.3
The Methylene Radical

Methylene (CH_2) is an asymmetric top molecule with two unpaired electrons. It has an electronic ground state of $^3B_1$ and is the simplest neutral polyatomic molecule with a triplet electron ground state. The molecule has a H–C–H equilibrium bond angle, is 133.9308(21) degrees and is floppy due to its shallow potential energy curve. It has a permanent dipole moment of 0.57 debye. In spite of its simplicity, it took Gerhard Herzberg almost 20 years to determine the electronic structure of the methylene (CH_2) radical [106].

Due to $S = 1$, each of the rotation levels N_{K_{-1}, K_1} is split into fine-structure states of $J = N − 1, N, N + 1$. Since each of the two H atoms have a nuclear spin of

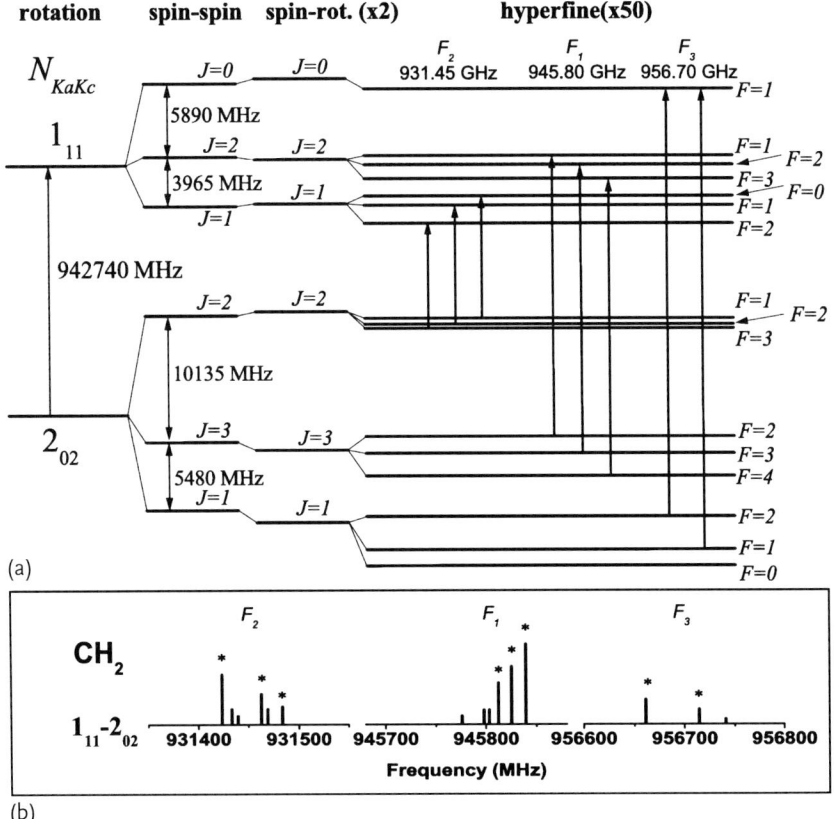

(a)

(b)

Figure 3.8 (a) A schematic diagram illustrating the splitting of the 1_{11}–2_{02} rotational transition of ortho-CH_2 by fine-structure and hyperfine interactions. (b) shows the frequency distributions of the hyperfine transitions. Figure adapted from [107].

$I = 1/2$, CH_2 can be in either ortho ($I = 1$) or para ($I = 0$) forms. For the ortho form, each of the fine structure states is further split into hyperfine states of $F = J - 1, J, J + 1$, whereas the para form has no hyperfine splitting. Selection rules for rotational transitions are $\Delta J = 0, \pm 1$ and $\Delta F = 0, \pm 1$. All ortho states have $K_{-1} + K_1$ being even, and para states have $K_{-1} + K_1$ being odd. As an example, the ground state of para-CH_2 1_{01} is split into fine structure states of $J = 0, 2, 1$ (in increasing energy) and the first excited state of para-CH_2 1_{10} is split into $J = 1, 2, 0$. From the selection rule, there are six allowed fine-structure lines in the rotational transition $1_{10} - 1_{01}$. For ortho-CH_2, the rotational state 2_{02} is split into fine-structure states of $J = 1, 3, 2$ and each is further split by hyperfine splitting as shown in Figure 3.8.

The lowest rotational levels of CH_2 are shown in Figure 3.9. The first excited state of ortho-CH_2 is 2_{02}, which is ~ 67 K above the ground state 0_{00}. For para-CH_2, the lowest state is 1_{01}, which is ~ 23 K above 0_{00}. Laboratory measurements of the transition frequencies began with the detection of the $4_{04} - 3_{13}$ of ortho-CH_2 at 70 GHz [108] and followed by the $2_{12} - 3_{03}$ and $5_{05} - 4_{14}$ of para-CH_2 at 440–445 and 592–594 GHz, respectively [109]. Because of the floppiness of the molecule, the frequencies of the lower transitions cannot be reliably extrapolated from these measurements. The first measurements in the THz range were the $1_{11} - 2_{02}$ transition of ortho-CH_2 at 943 GHz [107], the $2_{11} - 2_{02}$ transition of para-CH_2 at 1.954 GHz and $1_{10} - 1_{01}$ ground-state transition of para-CH_2 at 1.915 GHz [110].

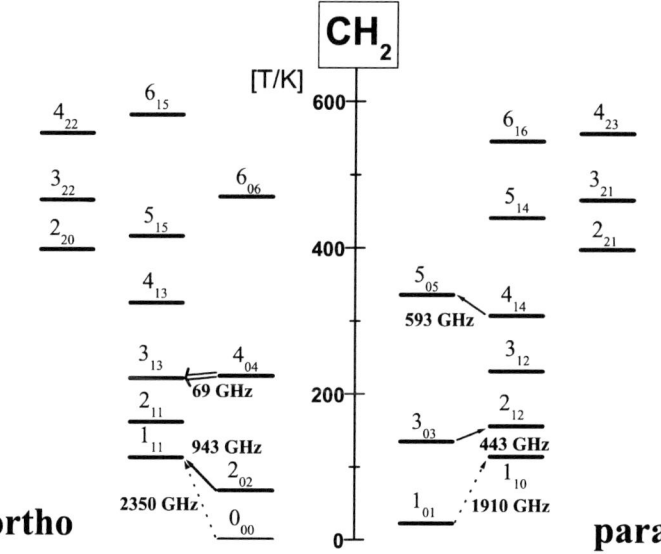

Figure 3.9 Low-lying rotational levels of CH_2. The quantum number N_{K_{-1}, K_1} of each rotational state is labeled, but the fine structure states or the hyperfine states are not shown. Also given are the frequencies of some of the laboratory-measured transitions. The energy scale along the vertical axis is measured in kelvin. Figure adapted from [107].

Most of the transitions in Figure 3.9 are in the far infrared and are not accessible from the ground. The lowest rotational transition observable from the ground is $4_{04}-3_{13}$, with three rotational transitions between the fine-structure components ($J = 5-4$ at 68.37 GHz, $J = 4-3$ at 70.68 GHz, and $J = 3-2$ at 69.01 GHz). These transitions, as well as the $F = 6-5, 5-4, 4-3$ hyperfine components of the $J = 5-4$ transition, were detected in Orion KL in 1995 [111]. The low-lying far infrared transitions $1_{11}-0_{00}$ (127.646 µm) and $3_{13}-2_{02}$ (93.662 µm) of ortho-CH_2 and the $2_{12}-1_{01}$ (107.720 µm) of para-CH_2 were detected in Sgr B2 in absorption by ISO [112].

The electronic bands of $CH_2\,^3A_2 - X^3B_1$ (at λ 141.58 nm) and $^3\Pi_u - {}^3\Sigma_u^-$ (at λ 141.01 nm) were detected in absorption against the UV spectra of stars HD 154368 and ζ Oph by the Hubble Space Telescope Goddard High Resolution Spectrometer (GHRS) [113].

3.8.4
Methyl Radical

Due to its symmetric planar structure, CH_3 does not have an electric dipole moment and therefore has no permitted rotational transitions. The ν_1 vibrational mode corresponds to simultaneous stretching of all three protons moving away from the center of mass. Since the electric dipole moment is unchanged in ν_1, it is infrared inactive. The asymmetric stretching mode ν_3 (3.16 µm) corresponds to the stretching of one (or two) C–H bond while the other two (or one) is contracting. The strongest vibrational mode is the ν_2 out-of-plane bending mode.

The nuclear spins of the three H atoms can couple together to form an ortho ($I = 3/2$) or a para ($I = 1/2$) state of the molecule, with statistical weights of $(2I + 1) = 4$ and 2 respectively. Since the methyl radical is a fermi particle, the $K = 3, 6, 9, \ldots$ states are associated with the ortho state (quartets), with the other K states associated with the para ($I = 1/2$) state (doublets). The Q branch of ν_2 at 16.5 µm was detected by ISO in Sgr A^* [114].

When the H is replaced by deuterium, the resultant radical CH_2D^+ is asymmetric and therefore has a rotational spectrum. As deuterium is bound to CH_2D^+ more tightly than H to CH_3^+, the abundance of the former can exceed that of the latter in cold clouds. The rotational transitions $1_{01}-0_{00}$ (280 GHz), $2_{11}-2_{12}$ (200 GHz) and $1_{10}-1_{11}$ (67 GHz) have been detected in NGC 6334 [115].

Hydrocarbon radicals are believed to play an important role in the photochemistry of the atmospheres of giant planets. Ethane (C_2H_6), for example, is formed by the reaction of two methyl (CH_3) radicals. Radicals such as CH, CH_2, and CH_3 in planetary atmospheres are the result of photodissociation of methane (CH_4), the most abundant hydrocarbon in these atmospheres. While methane, ethane, and acetylene have been observed in planetary atmospheres from ground-based and Voyager observations, the hydrocarbon radicals were first detected by ISO [116].

3.9
Carbon Chains

Carbon chains that have been detected in space include cyanopolyynes (HC_nN), isocyanopolyynes ($HC_{2n}NC$), methylpolyynes ($H_3C_{2n+1}H$), methylcyanopolyynes ($H_3C_{2n}N$), cumulene carbenes (H_2C_n), ring-chain carbenes ($C_{2n+1}H_2$, $H_{2n}N$), and free radicals C_nN. These asymmetrical carbon chains are highly polar molecules and their large dipole moments allow their rotational transitions to be detected by millimeter-wave spectroscopy. In fact, the simpler carbon chains (e.g., CCH, CCCH, ..., C_6H) were discovered in the ISM before they were synthesized in the laboratory.

3.9.1
Carbynes

Polyynyl radical chains (C_nH) have one nonbonding electron and cumulene carbenes (e.g., H_2C_3 and H_2C_4) have two nonbonding electrons. The unpaired electron of C_nH leads to a $^2\Pi$ electronic ground state. For $C_{2n}H$, C_2H and C_4H have $^2\Sigma_g^+$ ground states, but switch to $^2\Pi$ for the longer chains. Coupling of the spin angular momentum of the electron to the orbital angular momentum and the molecular rotational angular momentum splits the $^2\Pi$ state into two fine-structure components $^2\Pi_{1/2}$ and $^2\Pi_{3/2}$, each in turn is split into two components by Λ doubling. For $C_{2n}H$, $^2\Pi_{3/2}$ has lower energy, whereas $^2\Pi_{1/2}$ is lower in energy for $C_{2n+1}H$.

C_3H has two isomers, the linear (l-C_3H) and cyclic (c-C_3H). The $^2\Pi$ $J = 7/2-5/2, 9/2-7/2, 13/2-11/2$ and $15/2-13/2$ λ-doublet transitions of l-C_3H were detected in IRC+10216 and the $J = 3/2-1/2$ hyperfine-structure $F = 2-1, 1-0$, $2-1, 1-0$ components were detected in TMC-1 [117] (Figure 3.10). The astronomical detection of c-C_3H is discussed in the section on rings (Section 3.11). From the frequencies of four successive doublets seen in IRC+10216, it was determined that these transitions correspond to the $N = 9-8, 10-9, 11-10, 12-11$ rotational transitions of C_4H radical, from which a rotational constant B_0 of 4758.48 ± 0.10 MHz was derived [118]. The highest members detected are C_7H [119] and C_8H [120].

The cumulene carbenes are nearly symmetric rotators and have rotational structures similar to that of formaldehyde (H_2CO) (Section 3.5.1). For each $\Delta J = 1$ rotational transition, there are three lines corresponding to $\Delta K = 0, \pm1$. Among the cumulene carbenes detected are H_2CC (vinylidene), H_2CCC (propadienylidene), and H_2CCCC (butatrienylidene). Figure 3.11 shows the energy diagram for H_2CCCC and H_2CCC. Since the two molecules have identical H atoms, the para levels have even values of K and the ortho levels odd values of K (shown as separate columns in Figure 3.11. The ortho $K = 1$ levels are about 15 K above the para ($K = 0$) levels for both molecules.

The class of long-chain molecules cyanopolyynes (or cyanoacetylenes, HC_nN, $n = 3, 5, 7, 9, ...$) have triple bond structures as in carbynes but also have the cyano group to terminate the acetylenic chains. This stabilizes the molecule and cyanopolyynes are commonly detected in the circumstellar envelopes of evolved

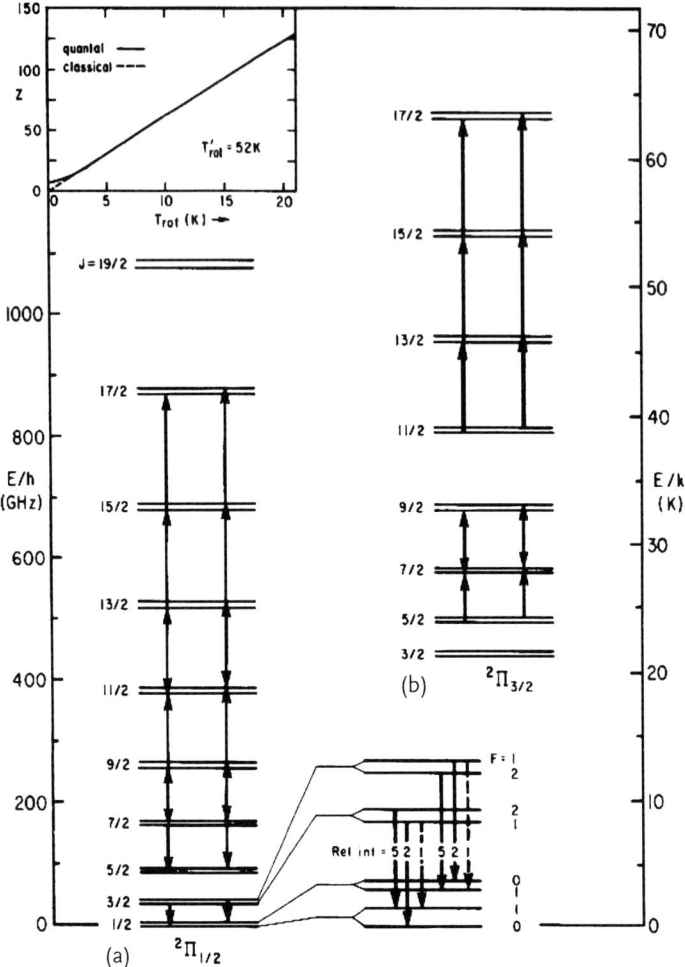

Figure 3.10 The lower rotational levels of C$_3$H. Column (a) is the $^2\Pi_{1/2}$ and the column (b) is the $^2\Pi_{3/2}$ ladder. The doublets in each rotational level are due to Λ doubling. For the $^2\Pi_{1/2}$ $J = 3/2-1/2$ transition, the hyperfine splittings are also shown. Two energy scales (frequency on the left and temperature on the right) are shown on the vertical axes. Figure adapted from [117].

stars. The largest cyanopolyynes detected in space are HC$_{11}$N, found in the cold molecular cloud of TMC-1 (Figure 3.12) [122]. With an atomic weight of 147, it holds the record as the heaviest molecule in space.

Ethinylisocyanide (HCCNC), an isomer of cyanoacetylene (HCCCN), is also expected to have a linear structure and a fairly large electric dipole moment. It is less stable, with an energy of 0.82 eV higher than that of HCCCN. It was first detected through its $J = 4-3$, $J = 5-4$, and $J = 9-8$ rotational transitions in the spectral line survey of TMC-1 with the Nobeyama 45-m telescope [123]. Another

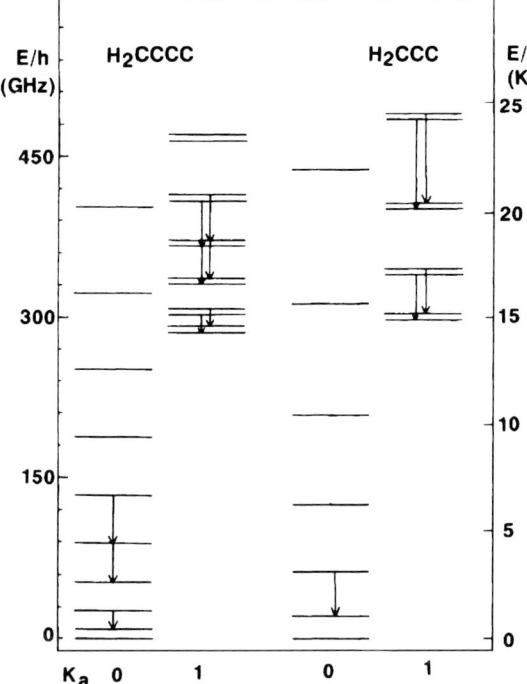

Figure 3.11 Energy diagrams of H_2CCCC and H_2CCC with the observed rotational transitions marked. The columns correspond to para ($K = 0$) and ortho ($K = 1$) states of the two molecules. The lowest para transitions are $2_{02}-1_{01}$ (17.863 910 GHz) for H_2CCCC and $2_{02}-1_{01}$ (41.584 693 GHz) for H_2CCC. The lowest ortho transitions are $2_{12}-1_{11}$ (17.788 576 GHz) $2_{11}-1_{10}$ (17.863 910 GHz) for H_2CCCC and $2_{12}-1_{11}$ (41.198 345 GHz), $2_{11}-1_{10}$ (41.967 686 GHz) for H_2CCC. The energy spacings of the $\Delta K = \pm 1$ transitions are exaggerated for clarity. Figure adapted from [121].

isomer, HNCCC, with an energy of 2.2 eV higher than HCCCN, is detected in its $J = 3-2$, $J = 4-3$, $J = 5-4$ rotational transitions in the same line survey [124]. The HCCNC lines were first reported as unidentified lines in the survey, but later identified as the result of laboratory spectroscopy work giving a better determination of the rotational constant of the molecule.

3.9.2
Carbon Chain Ions

Models of ion molecule reactions in space [125] has led to interest in the detection of molecular ions in the ISM and in circumstellar envelopes of stars. Theoretical calculations have suggested the abundance of negatively charged carbon chains (e.g., C_nH^-) can be very high [126, 127]. A series of unidentified lines discovered in the spectral-line survey of IRC+10216 [128] was identified as due to C_6H^- [129]. The detection of other members of the series C_4H^- [130] and C_8H^- [131] soon

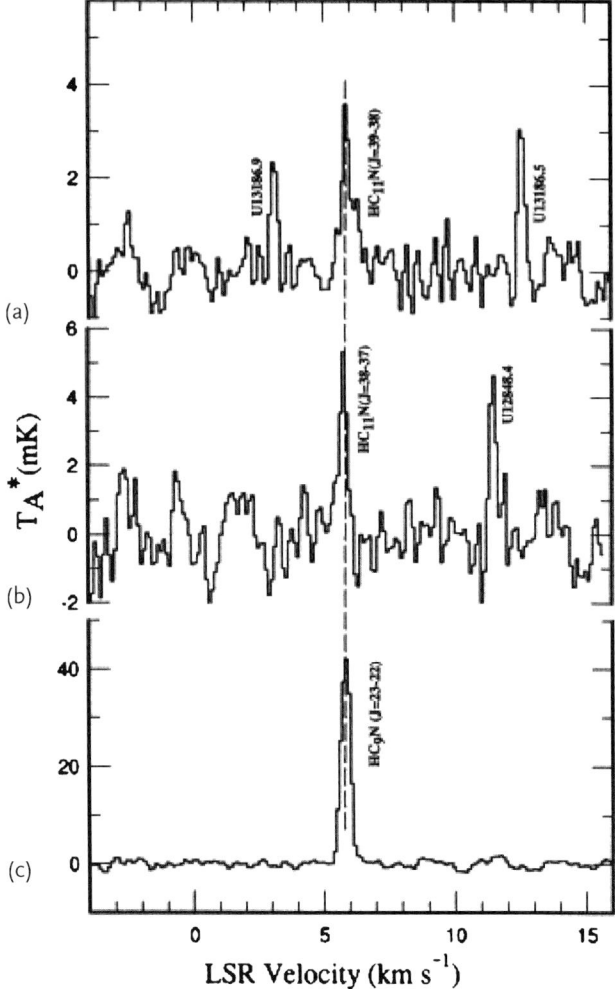

Figure 3.12 The spectra of two detected rotational transitions ($J = 39-38, 38-37$) of the heaviest known interstellar molecule $HC_{11}N$ (a), (b). Also shown for comparison is the spectrum of HC_9N (c) in the same source TMC-1. Figure adapted from [122].

followed. It is quite likely that other ions for example, CN^-, C_3N^-, C_3H^-, and so on may also be detectable.

3.9.3
Pure Carbon Chains

Carbon chains such as C_n have no permanent electric dipole moment and cannot be detected by radio telescopes through their rotational transitions. However, they have asymmetric stretching modes around ~ 5 μm and bending modes

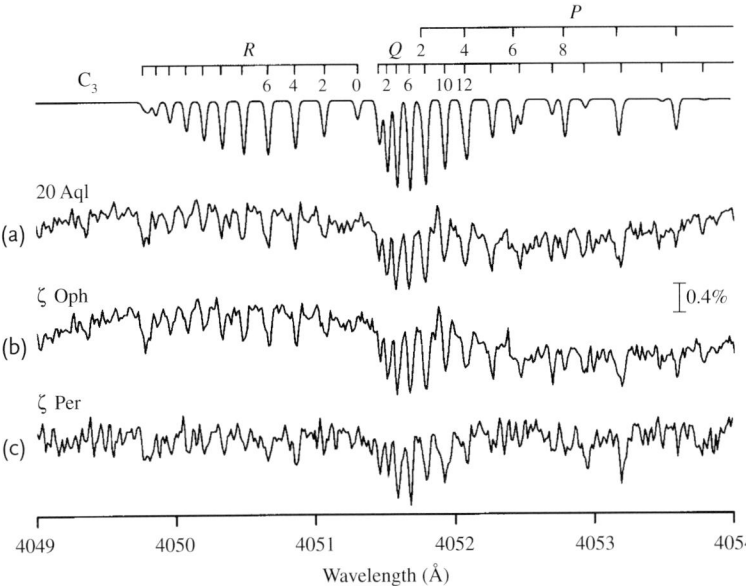

Figure 3.13 Electronic transitions of interstellar molecules can be detected in absorption against the continuum of background stars. This figure shows the rotational lines in the P, Q, and R bands of the $A^1\Pi_u - X^1\Sigma_g^+$ transition of C_3 in the spectra of (a) 20 Aql, (b) ζ Oph, and (c) ζ Per. Figure adapted from [137].

around $\sim$100 μm which can be detected by infrared spectroscopy. For example, the antisymmetric stretching modes of C_3 (ν_3 at 4.90 μm or 2040 cm^{-1}) and C_5 (ν_3 at 4.62 μm or 2164 cm^{-1}) have been detected in absorption in warm environments such as the circumstellar envelopes of carbon stars [132, 133]. Laboratory measurements suggest a C_6 ν_4 stretching mode at 5.1 μm (1960 cm^{-1}).

Triatomic carbon (C_3) has a vibrational bending mode ν_2 $(0, 1^1, 0) \leftarrow (0, 0^0, 0)$ near 158 μm [134] and this transition has been detected by KAO [134] and ISO [135] in absorption against the dust continuum of Sgr B2. For the higher members of the pure carbon chains, the bending mode frequencies are poorly known. The ν_7 bending mode of C_5 at 104.8 cm^{-1} ($\sim$95 μm) and the ν_9 bending mode of C_6 at 108 cm^{-1} ($\sim$93 μm) have been suggested to be responsible for a feature around 98 μm in NGC 7027 [136].

Carbon chains in the diffuse ISM have also been searched through their electronic transitions in absorption against the spectrum of bright stars. So far, only C_3 ($A^1\Pi_u - X^1\Sigma^+$ at 405.2 nm) has been detected (Figure 3.13) [137] with C_4 ($^3\Sigma_u^- - ^3\Sigma_g^-$ at 378.9 nm) and C_5 ($^1\Pi_u - ^1\Sigma_g^+$ at 510.9 nm), yielding only upper limits [138, 139].

3.10
Acetylene Derivatives

The ends of the acetylene chain can be replaced by a methyl and/or a CN group, leading to the series of methylpolyynes ($CH_3C_{2n}H$), cyanoacetylenes ($HC_{2n+1}N$), and methylcyanopolyyne ($CH_3(C\equiv C)_nCN$). Methylacetylene ($CH_3C\equiv CH$, propyne) is a symmetric top and was first discovered through its $J = 5_0-4_0$ rotational transition at 85.5 GHz in Sgr B2 [85]. The 2_0-1_0 and the 2_1-1_1 lines at 34.183 42 and 34.182 76 GHz were later observed in TMC-1 [140]. Two higher members of the series methyldiacetylene CH_3C_4H ($H_3C-C\equiv C-C\equiv C-H$) and methytriacetylene CH_3C_6H ($H_3C-C\equiv C-C\equiv C-C\equiv C-H$) have now been seen [141]. Figure 3.14 shows the $K = 0$ and $K = 1$ components of the $J = 6-5$ rotational transition of CH_3C_4H, the higher K members being too weak to be detected.

The first member of the methylcyanopolyynes methyl cyanide (CH_3CN) was first detected in Sgr B2 through its $J = 6-5$ rotational transition at 110 GHz [142]. The rotational transition of this prolate symmetric top molecule is split into J different lines corresponding to different values of $K(\leq J)$ as the result of centrifugal distortion (Figure 3.15). The spectrum is therefore complicated with the exception of the $J = 1-0$ transition as only the $K = 0-0$ transition is allowed [143].

Each of these lines are further split by hyperfine coupling due to the nuclear spin of ^{14}N ($I = 1$). For $J = 0$, $F = J + I = 0 + 1 = 1$ and $J = 1$, $F = 1 + 1 = 0, 1$, or 2. From the selection rule $\Delta F = 0, \pm 1$, three hyperfine transitions $F = 1 \rightarrow 1$ (18.3967 GHz), $2 \rightarrow 1$ (18.3980 GHz), and $0 \rightarrow 1$ (19.3999 GHz) are possible. Under local thermodynamical equilibrium (LTE), the relative strengths of these three hyperfine components are proportional to the statistical weights of the

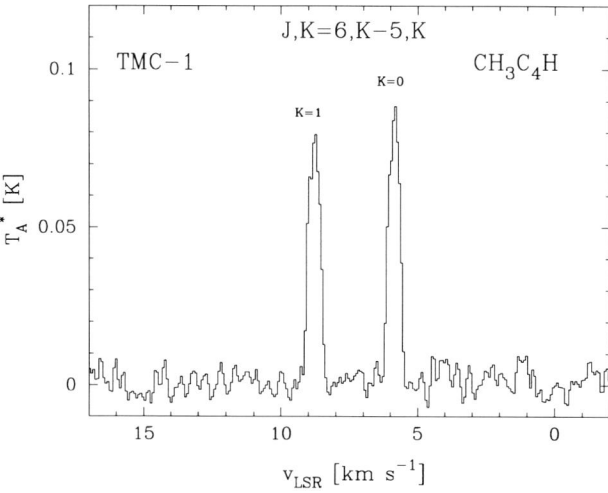

Figure 3.14 Spectrum of methyldiacetylene toward TMC-1 showing the 6_1-5_1 and the 6_0-5_0 rotational transitions. Figure adapted from [141].

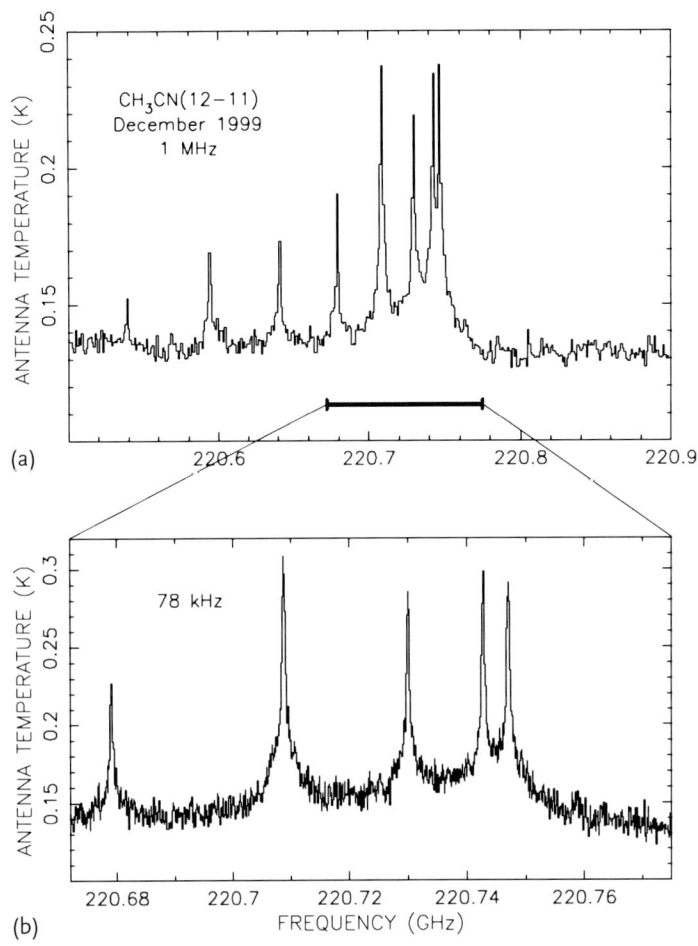

Figure 3.15 (a) The spectrum of the $J = 12 - 11$ rotational transition of CH_3CN in Titan. The seven emission features in (a) at 220.747, 220.743, 220.730, 220.709, 220.679, 220.641, and 220.594 GHz correspond to $K = 0, 1, \ldots, 6$. An expanded view of the $K = 0, \ldots, 4$ lines are shown (b). Figure adapted from [144].

upper states, or $3:5:1$. These hyperfine transitions of CH_3CN were first detected in TMC-1 in 1982 [143].

Higher members of the series methylcyanoacetylene CH_3C_3N [145] and methyl-cyanodiacetylene CH_3C_5N [146] have been detected in TMC-1 (Figure 3.16). Cyanoallene (CH_2CCHCN) is an isomer of methylcyanoacetylene and its rotational transitions $4_{14}-3_{13}, 4_{04}-3_{03}, 5_{15}-4_{14}$, and $5_{05}-4_{04}$ have been detected in TMC-1 [147].

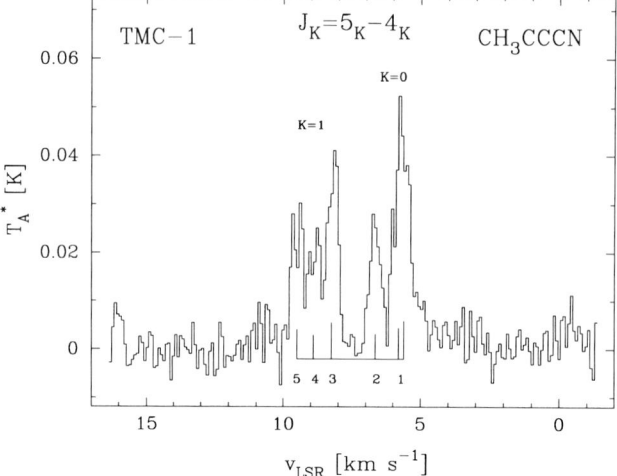

Figure 3.16 The $J_K = 5_0 - 4_0$ and $5_1 - 4_1$ rotational transitions of CH_3CCCN in TMC-1. The tick marks below the spectrum indicate the expected positions of the hyperfine transitions. Figure adapted from [147].

3.11
Rings

Many biomolecules contain ring structures. The simplest ring molecule has 3 C atoms arranged in a triangular form (e.g., cyclopropane C_3H_6), but most ring compounds have 5 or 6 C atoms (e.g., benzene C_6H_6). Sometimes one or more of the C atoms are replaced by O or N (e.g., purine and pyrimidine, Figure 2.6). In order to distinguish ring molecules from isomers with nonring structures, they are given a cyclic designation such as c-C_3H.

3.11.1
Propynl

The propynl radical c-C_3H (also known as cyclopropynylidyne) is the simplest ring observed in space. It consists of 3 C atoms in a triangle with the H atom attached to one corner and is an isomer of the linear CCCH. The rotational transitions of this asymmetric rotor is split into fine and hyperfine structures (Figure 3.17). c-C_3H was first detected through its $N = 2_{12} - 1_{11}$, $J = 5/2 - 3/2$, $F = 3 - 2$ (91.494 35 GHz), $F = 2 - 1$ (91.497 59 GHz) and $J = 3/2 - 1/2$, $F = 2 - 1$ (91.699 49 GHz), $F = 1 - 0$ (91.692 82 GHz) transitions in TMC-1 [148]. Its $N = 1_{10} - 1_{11}$, $J = 3/2 - 3/2$, $F = 2 - 2$ and $F = 1 - 1$ hyperfine transitions was detected in a large number of molecular clouds [149].

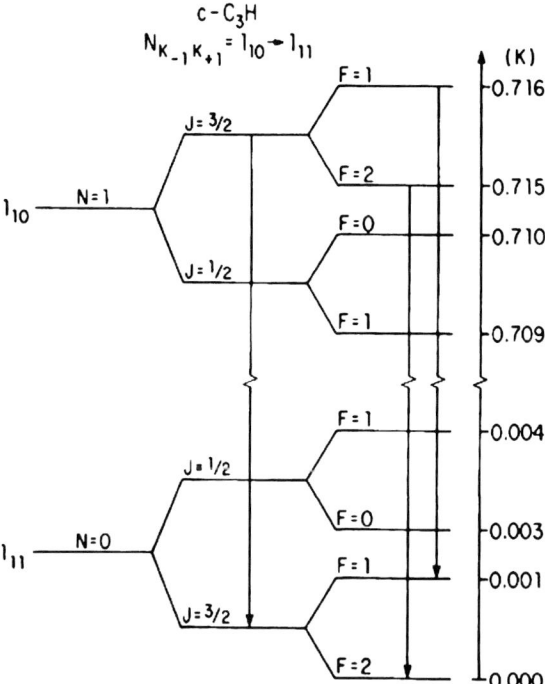

Figure 3.17 Energy diagrams of c-C_3H showing the fine and hyperfine structures of the 1_{10}–1_{11} rotational transition. Figure adapted from [149].

3.11.2
Cyclopropenylidene

The most abundant ring molecule in interstellar and circumstellar environments is cyclopropenylidene (C_3H_2). It is a planar three-member ring molecule (Figure 3.18) with two hydrogen atoms attached to the ring. Generally speaking, small rings suffer from internal strain and are not expected to be stable. C_3H_2 is able to offset some of this strain energy through the delocation of the electron density. Since the two H atoms attached to the two C atoms connected by a double bond are equivalent, they create the para and ortho forms of the molecule. The quantum numbers (J_{K_{-1},K_1}) of the rotational states, the separation of the rotational states into ortho and para groups, and the selection rules of the transitions, are all very similar to those of water. The two unpaired electrons on the third C makes this an exceptionally polar molecule ($\mu = 3.43$ debye) [150].

As a five-atom nonlinear molecule, C_3H_2 has $3(5) - 6 = 9$ vibrational modes. C_3H_2 was first found in the laboratory through its four infrared bands [152], which frequencies were confirmed by quantum calculations [153]. The discovery of interstellar C_3H_2 is the result of laboratory identification of a number of previously unidentified astronomical lines, in particular the strong and ubiquitous lines at 85.338 and 18.343 GHz [154]. The 18.343 GHz line, identified as the ground-state

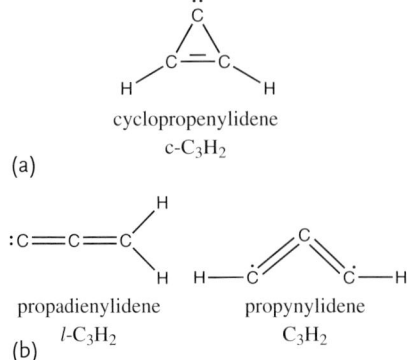

cyclopropenylidene
c-C_3H_2

(a)

propadienylidene
l-C_3H_2

propynylidene
C_3H_2

(b)

Figure 3.18 The schematic structure of cyclopropenylidene (a) and its two isomers (b). Figure adapted from [151].

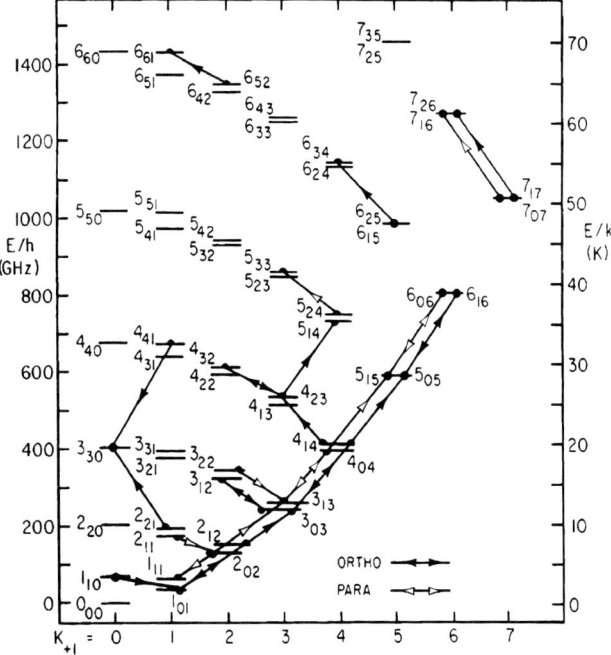

Figure 3.19 The energy level diagram of C_3H_2 showing some of the detected astronomical transitions (downward arrows) and the laboratory transitions (upward arrows) used by [154] to identify C_3H_2 in space. Figure adapted from [154].

1_{10}–1_{01} transition of ortho-C_3H_2 (Figure 3.19), is one of the strongest interstellar molecular lines at short cm-wavelengths. The counterpart of this line in water, a much lighter molecule, is at 557 GHz. This line is seen in emission in cold dust clouds and circumstellar envelopes, in absorption towards the galactic center and H II regions [155].

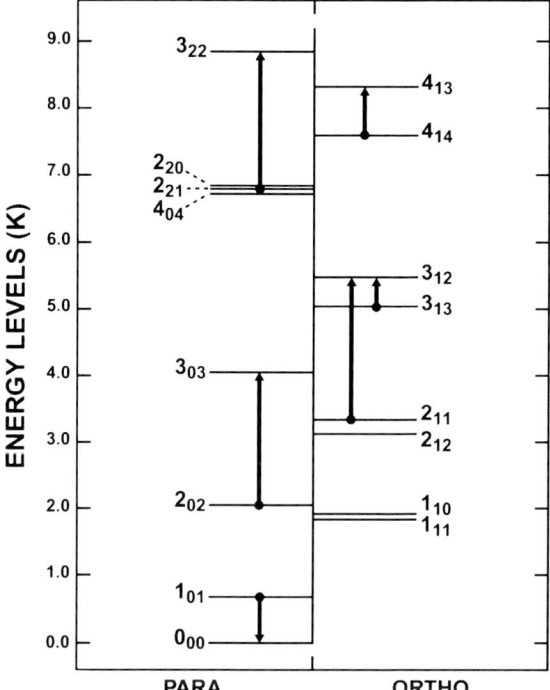

Figure 3.20 The energy level diagram of c-H_2C_3O showing some of the detected transitions. Figure adapted from [156].

3.11.3
Cyclopropenone

Cyclopropenone (c-H_2C_3O) is formed by an O atom attaching to the divalent C in c-C_3H_2 through a double bond. As in c-C_3H_2, the spins of the H atoms give rise to para and ortho forms of c-H_2C_3O and the conversion from one to the other via radiative or collisional processes is strictly forbidden. Some of the detected rotational transitions in both ortho and para states are shown in Figure 3.20.

3.11.4
Ethylene Oxide and Propylene Oxide

The discovery of interstellar ethylene oxide (c-C_2H_4O) is based on the identification of ten rotational transitions [157]. The c-C_2H_4O molecule is a three-membered C–O–C ring with two H atoms attached to each of the C corner in a planar structure (Figure 3.21). It is an asymmetric oblate rotator. Being a seven-atom nonlinear molecule, ethylene oxide has $3(7) - 6 = 15$ nondegenerate vibrational modes, of which twelve are infrared active.

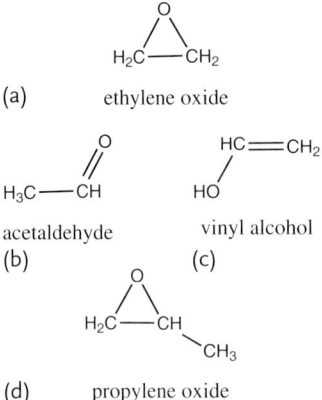

(a)　　　　ethylene oxide

acetaldehyde　　　vinyl alcohol
(b)　　　　　　(c)

(d)　　　propylene oxide

Figure 3.21 Chemical structure of the ethylene oxide (a) and its isomers acetaldehyde (b) and vinyl alcohol (c). The structure of the homologue of ethylene oxide propylene oxide is also shown (d).

A homologue of ethylene oxide is propylene oxide (c-CH_3H_6O), where one of the H atom in ethylene oxide is replaced by a methyl group (Figure 3.21). Although its isomer propanal (CH_3CH_2CHO, Section 3.21) has been detected, the search for the ring molecule propylene oxide has not yet been successful [158].

3.12
Phosphorus Containing Molecules

Phosphorus is the seventeenth most abundant element in the Universe (eleventh in the Earth's crust). The solar abundance of phosphorus is 2.7×10^{-7} relative to hydrogen and nearly identical to that of chlorine. Phosphorous is in the same chemical family as nitrogen, but with an abundance 300 times smaller. In biochemistry, phosphorus plays a central role in metabolic processes and is a key element in the origin of life (Table 3.2). It is surprising that phosphates play such a major role in biochemistry in spite of the fact that the abundance of phosphorus is so low on Earth and in the Universe. This constraint is manifested in the observation that the availability of P is the chief limiting factor of growth in agriculture. It is possible that the biosphere has used up all the prebiotic phosphorus as the present content of phosphate on Earth is in calcium phosphate rock, which is derived from fossils.

In diffuse interstellar clouds, phosphorus exists almost entirely in the form of P^+ and is undepleted in warm neutral clouds [159, 160]. In slightly cooler diffuse clouds, of phosphorus the abundance is 0.3 of the solar value [160]. Interestingly, in the giant planets, phosphorous is found in the form of phosphine (PH_3) and the volume mixing ratio is about 10^{-6} on Jupiter [161]. As the main constituent of the Jovian atmosphere is molecular hydrogen, essentially all of the phosphorus is present in the form of PH_3.

In the interstellar and circumstellar environment, the only known P-bearing molecules are PN [162, 163], CP [164], HCP [165], and PO [166, 167]. Since P is an important element in the biochemistry of living things, for example, being involved in the storage and transfer of information (nuclei acids), energy transfer (adeninosine triphosphates, ATP; and guanosine triphosphates, GTP), membrane structure (phospholipids), and signal transduction (cyclic nucleotides and inositol-polyphosphates), the detection and abundance determination of P-bearing molecules is of great interest. Possible candidates for searches include PH, PH_2, PH_3, PS, PO, PO_2, PO_3, HC_nP, HP_nN, and so on.

3.12.1
PH

Based simply on abundances, PH is a likely candidate for detection by rotational emission from interstellar clouds or circumstellar envelopes. The lowest energy levels of the PH radical in the $X^3\Sigma^-$ state are shown in Figure 3.22. Since both P and H have nuclear spins of 1/2, the appropriate angular momenta are defined as $F_1 = J + I(P)$ and $F = F_1 + I(H)$, in which $J = N + S$, N and S are

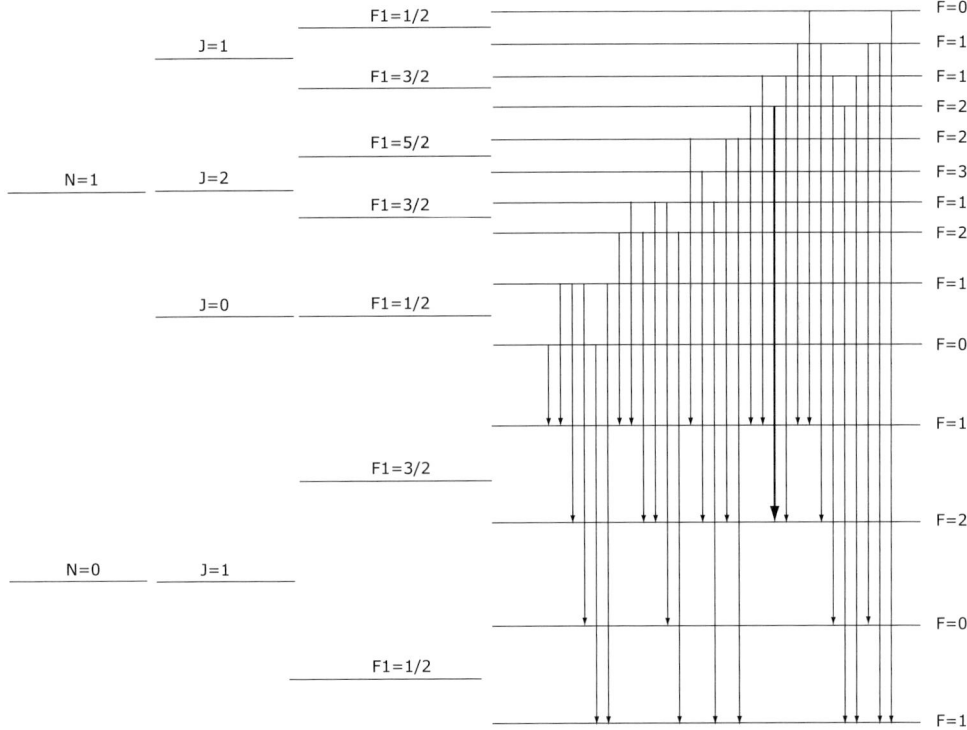

Figure 3.22 The energy level diagram of PH showing the lowest states and possible transitions. The thick-line arrow marks the $J = 1-1$, $F = 2-2$, $F_1 = 3/2-3/2$ transition at 553.362 85 GHz that has been searched for by the Odin satellite.

the rotational and total spin angular momenta, respectively. Since $S = 1$, each rotational level (labeled by the quantum number N) with $N > 0$ is split into three fine-structure components (labeled by J) with $J = N, N \pm 1$. Each of the three fine structure levels is doubled by the hyperfine structure due to the P nuclear spin of $1/2$, labeled by $F_1 = J \pm 1/2$. Each of the F_1 levels is doubled again by proton nuclear hyperfine structure with $F = F_1 \pm 1/2$ (Figure 3.22).

Because of its relatively large rotational constant ($B = 252.200\,81$ GHz), the rotational transitions of PH occur in the submillimeter part of the electromagnetic spectrum, making them difficult to observe astronomically. Laboratory measurements of the $N = 1-0$ transitions of PH in the 423–554 GHz region, and the $N = 2-1$ lines near 1 THz have been recorded. For astronomical detection, most of the transitions of PH are blocked by strong telluric absorption and the most favorable line at 553.362 85 GHz can only be seen from orbit or high-flying aircraft.

3.13
Polycyclic Aromatic Hydrocarbons

Polycyclic aromatic hydrocarbon (PAH) molecules are benzene rings of sp^2-hybridized C atoms linked to each other in a plane, with H atoms or other groups saturating the outer bonds of peripheral C atoms. PAH molecules can be divided into two general classes: compact PAHs where some C atoms belong to three rings (e.g., pyrene $C_{16}H_{10}$, coronene $C_{24}H_{12}$, voalene $C_{32}H_{14}$) and noncompact PAHs where no C atom belongs to more than two rings. The noncompact class includes linear chains (e.g., naphthalene $C_{10}H_8$, tetracene $C_{18}H_{12}$, pentacene $C_{22}H_{14}$) and nonlinear or branched forms (e.g., phenanthrene $C_{14}H_{10}$, chrysene $C_{18}H_{12}$) [168] (Figure 3.23). Some examples of PAH molecules are listed in Table 3.1 and shown in Figure 3.24.

As can be seen in Table 3.1, PAH molecules are very stable. Laboratory measured 3–75 μm absorption spectra of the vibration modes of many different PAHs have been catalogued [169]. A list of common PAHs and their vibrational frequencies can be found in: http://astrochemistry.ca.astro.it/database (accessed 5 July 2011).

Although the unidentified infrared emission (UIE) features in the mid-infrared have been widely attributed to PAHs (Section 9.3), they actually correspond to C–C and C–H vibrational modes that are characteristics of the chemical bonds and can be produced by many different aromatic molecules and compounds. Rotational spectra of cold, neutral PAHs have recently been obtained in the laboratory, therefore opening the possibility of their detection by mm/submm telescopes [172]. However, PAHs generally have very low dipole moments and the lines are expected to be weak. An example of the rotational spectrum of the PAH fluorene ($C_{13}H_{10}$) is shown in Figure 3.25. The bending modes in the far infrared corresponding to skeleton motions are dependent on the size and shape of the molecules, and their detection would allow for the identification of specific PAH molecules.

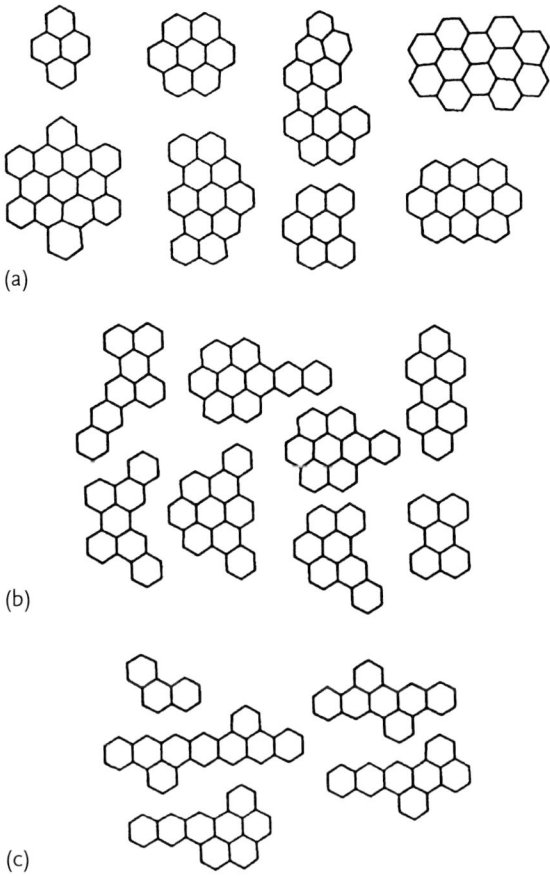

(a)

(b)

(c)

Figure 3.23 Examples of compact (a), intermediate (b), and noncompact (c) PAHs. Figure adapted from [169].

3.14
Molecules Containing Trace Elements

Although six elements (H, C, N, O, S and P) make up 98% of all living tissue, a number of other trace elements are also important for life. These include sodium (Na), chlorine (Cl), potassium (K), fluorine (F), calcium (Ca), iodine (I), magnesium (Mg) and iron (Fe). Some other elements are also found in the human body, including boron (B), aluminum (Al), silicon (Si), chromium (Cr), manganese (Mn), copper (Cu), zinc (Zn), selenium (Se), strontium (Sr), molybdenum (Mo), silver (Ag), tin (Sn), lead (Pb), nickel (Ni), bromine (Br), vandadium (V), and so on (Table 3.2). These trace elements are essential to the function of specific proteins. For example, iron plays an important role in the oxygen-transportation of hemoglobin molecules, zinc holds together the insulin molecules in the form of a hexamer, and manganese in enzymes. In spite of their relatively low cosmic abun-

Table 3.1 Properties of some PAHs. Adapted from [170].

Molecule	Formula	C/H ratio[a]	Number of vibrational modes	C–H dissociation energy
Benzene	C_6H_6	1.00	30	41 600
Naphthalene	$C_{10}H_8$	1.25	48	53 000
Anthracene	$C_{14}H_{10}$	1.40	66	65 600
Pyrene	$C_{16}H_{10}$	1.60	72	69 500
Chrysene	$C_{18}H_{12}$	1.50	84	77 200
Perylene	$C_{20}H_{12}$	1.67	90	83 300
Benzoperylene	$C_{22}H_{12}$	1.83	96	(88 000)
Coronene	$C_{24}H_{12}$	2.00	102	(92 000)
Ovalene	$C_{32}H_{14}$	2.29	132	($\geq$110 000)
Hexabenzo-coronene	$C_{42}H_{18}$	2.33	174	(>110 000)

[a] Minimum vibrational energy content required for the C–H bond rupture rate to equal the radiative relaxation rate.

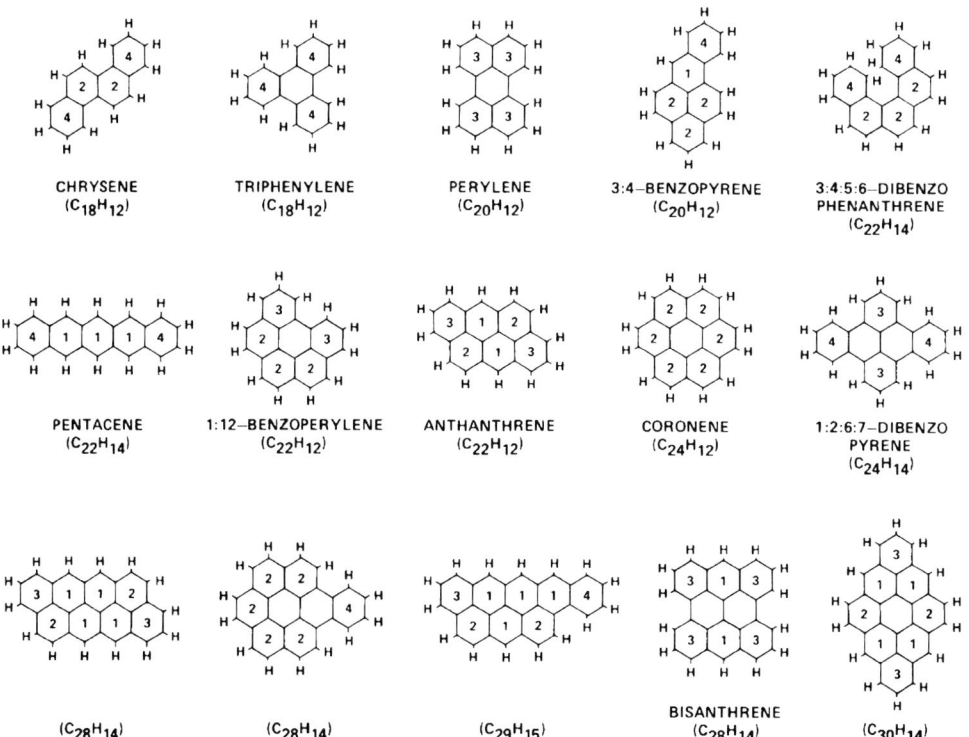

Figure 3.24 Examples of some PAH molecules. The number inside each ring indicates the number of exposed H edges of the ring. Figure adapted from [171].

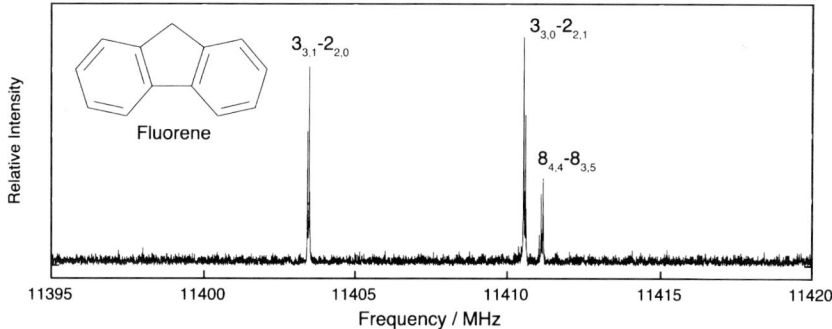

Figure 3.25 Laboratory spectrum of fluorene showing the $3_{31}-2_{20}$, $3_{30}-2_{21}$ pair of rotational transitions. Also shown is the $8_{44}-8_{35}$ Q-branch ($\Delta J = 0$) transition. Figure adapted from [172].

Table 3.2 Chemical elements in the human body.

Element	Percent by mass
Oxygen	65
Carbon	18
Hydrogen	10
Nitrogen	3
Calcium	1.5
Phosphorus	1.2
Potassium	0.2
Sulfur	0.2
Chlorine	0.2
Sodium	0.2
Magnesium	0.05
Iron, Cobalt, Copper, Iodine	<0.05 each
Selenium, Fluorine	<0.01 each

dance, molecules containing trace elements have been detected by astronomical spectroscopic means. Examples of these molecules include NaCl, KCl, AlCl, AlF, NaCN, MgNC, and MgCN. In molecular clouds, refractory molecules are believed to be heavily depleted in the gas phase, and the only commonly found species are SiO and SiS. However, in circumstellar envelopes, a much richer variety of refractory molecules are found.

3.14.1
Metal Hydrides

Since H is the most abundant element in space, metal hydrides are naturally expected to be widely present in the ISM. Although the electronic transitions of CrH and FeH have been detected in the atmosphere of late M stars, the rotational transitions of diatomic metal hydrides, for example, NaH, CaH, MgH, have been searched without success [173]. The ground-state rotational transitions of these light, fast rotators occur in the submm part of the spectrum. For example, the $J = 1-0$ transition frequencies for ^{23}NaH and ^{39}KH are 289.864 and 202.282 GHz, respectively. Since both the Na and K nuclei have nuclear spin of $I = 3/2$, the $J = 1-0$ transition of these molecules are split into three hyperfine components, $F = 5/2-3/2$, $F = 3/2-3/2$, and $F = 1/2-3/2$. These hyperfine splittings broaden the line by approximately 2.5 MHz.

3.14.2
Halides and Cyanides

A number of halides and cyanides have been detected in circumstellar envelopes. Possible candidates for astronomical searches of halides include AlCl, AlF, NaCl, KCl, and for cyanides or isocyanides, MgNC, NaCN, MgCN, AlNC [174]. AlNC is a linear molecule with a $^1\Sigma^+$ ground electronic state. The derived fractional abundance of this molecule from its rotational transitions observed in IRC+10216 is $f(AlNC/H_2) \sim 3 \times 10^{-10}$, much lower than the abundances of halides ($f(AlF/H_2) \sim 1.5 \times 10^{-7}$, $f(AlCl/H_2) \sim 2.2 \times 10^{-7}$). The total abundance of aluminum in molecules is therefore only a small fraction of the element, assuming that the cosmic abundance of Al (3×10^{-6}) is applicable to the circumstellar envelopes of AGB stars.

Since some transition metal elements (e.g., Fe, Ni, Cr, Mn, etc.) also have relatively high cosmic abundances, it is likely that molecules containing these elements also exist in the ISM. For example, the nickel monochloride (NiCl) radical has a ground electronic state of $^2\Pi_{3/2}$ similar to that of OH, and has rotational transitions in the submillimeter region. The $J = 35.5\,1/2-34.5\,1/2$ transition of ^{58}Ni^{35}Cl has been observed to have a frequency of 385.719 696 GHz in the laboratory. The rotational spectra of other transition metal carbides (e.g., FeC, CoC, NiC, etc.) have been recorded in the laboratory [175]. Since the NiCl molecule is highly ionic, most of the unpaired electrons are near the Ni nucleus with negligible electron spin density near the Cl nucleus. Consequently, the hyperfine structure is small and difficult to observe.

3.14.3
Calcium Carbide

Among the common metals, calcium is the only element with no molecular species detected as of 2010. Since the cosmic abundance of Ca is similar to those of Al and

Na, the absence of Ca-bearing molecules in the ISM suggests that Ca may be heavily depleted in the gas phase due to condensation onto grains. One likely molecular carrier of Ca is calcium monocarbide (CaC). The ground electronic state of this radical is $^3\Sigma^-$, which has two unpaired electrons giving a total electron spin of $S = 1$. Each rotational transition therefore has three fine-structure components, $J = N - 1$, N, and $N + 1$. Laboratory measurements suggest a rotational constant of 10.338 GHz, giving rise to detectable transitions of $N = 4-3$ and $N = 5-4$ at 82.7 and 103.4 GHz, respectively [176].

3.15
Biomolecules

As the detection of biomolecules (Section 2.3) by remote observations is difficult at present, most astronomical efforts have been focused on the search for the building blocks of proteins, carbohydrates, nucleic acids as well as membrane lipids.

3.15.1
Amino Acids

Glycine is the simplest α-amino acid and has several conformers and their molecular structures are shown in Figure 3.26. It is a prolate asymmetric rotor and most astronomical searches have been focused on rotational transitions from conformer I which has the lowest energy, and conformer II which has a much larger dipole moment [177]. Conformer I glycine has an a-type dipole moment of $\mu_a = 0.911(6)$ debye and a b-type one of $\mu_b = 0.697(10)$ debye.

The first claimed detection of glycine was made by Kuan *et al.* [179] who detected 27 lines of glycine conformer I in 19 different spectral bands toward Orion KL, Sgr B2, and W 51. However, only three transitions are commonly detected in all three objects. Also at least one of the detected lines $15_{1,15}-14_{1,14}$ at 900.4313 GHz could be confused with transition of acetic acid (HCOOCH$_3$, Section 3.4). It has been suggested that other lines may also be confused with vibrationally excited vinyl cyanide (CH$_2$CHCN) and ethyl cyanide (CH$_3$CH$_2$CN) [180].

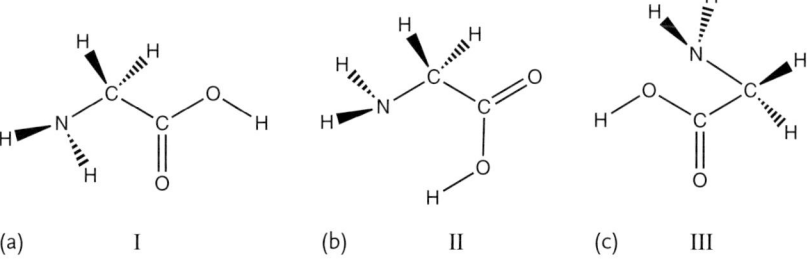

(a)   I   (b)   II   (c)   III

Figure 3.26 Molecular structures of the three conformers of glycine, that is, I (a), II (b), and III (c) [178].

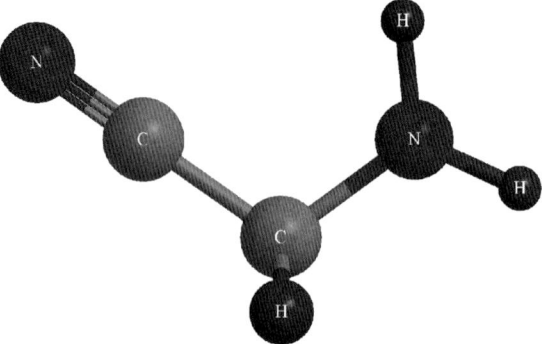

Figure 3.27 Chemical structure of the molecule amino acetonitrile.

Amino acetonitrile (NH_2CH_2CN), a molecule containing both the nitrile and amino groups (Figure 3.27), is considered as a possible precursor of glycine. It was detected in Sgr B2 with the IRAM 30-m telescope [96]. A total of 51 transitions of the molecule were detected and nine of these features are confirmed by interferometric observations at IRAM Plateau de Bure Interferometer (PdBI) and the Australian Telescope Compact Array (ATCA). The emission region is found to be confined to a compact object of $2''$ in size.

3.15.2
Sugars

Glycolaldehyde (CH_2OHCHO) was first detected in Sgr B2 through six rotational transitions in the millimeter-wave region with the Kitt Peak 12-m telescope [181]. This initial identification was later confirmed by the observations of lower frequency transitions ($1_{10}-1_{01}, 2_{11}-2_{02}, 3_{12}-3_{03}, 4_{13}-4_{04}$ at 13.48, 15.18, 17.98, and 22.14 GHz respectively) with the Green Bank 100-m Telescope [182]. Higher transitions $20_{2,18}-19_{3,17}, 14_{0,14}-13_{1,13}$, and $10_{19}-9_{28}$ at 220.467 387, 143.640 94, and 103.667 91 GHz are detected in the massive star forming region G31.41+0.31 with the IRAM Plateau de Bure Interferometer [183].

Although the discovery paper carries the title "Interstellar glycolaldehyde: the First Sugar" [181], simple sugars, or monosaccharides, are aldehyde or ketone derivatives of straight chain polyhydroxy (containing more than one OH group) alcohols with at least three C atoms. Strictly speaking glycolaldehyde, with two C atoms, does not qualify as a sugar. Monosaccharides are separated into poly-dydroxylated aldehydes ($H-[CHOH]_k-CHO$, called aldoses) and polyhydroxy-lated ketones ($H-[CHOH]_l-CO-[CHOH]_m-H$, called ketoses). An example of aldoses is glyceraldehyde ($CH_2OHCHOHCHO$) and an example of ketoses is 1,3-dihydroxyacetone ($CO(CH_2OH)_2$, DHA), both are sugars containing three C atoms. Glyceraldehyde can therefore be considered as the "first sugar". Glyceraldehyde is also the simplest chiral carbohydrate as it has two stereoisomeric forms. Glyceraldehyde has been searched for, but not detected [75]. DHA has stable isomers such as dimethyl carbonate [$(CH_3O)_2CO$] and methyl glycolate (CH_3OCOCH_2OH). DHA is a near-prolate asymmetric top with the only dipole moment lying along the b-axis ($\mu_b = 1.765$ debye). Although the detection of DHA in Sgr B2 has been reported [184], it has not been confirmed [185].

The detection of glycoladldehyde is significant because it can react with propenal to form ribose, a central constituent of RNA.

3.15.3
Nucleic Acids

The detection of a wide variety of organic molecules, in particular those with ring structures, has raised the possibility of the existence of other biochemically important ring molecules. Planar rings containing other heavy elements (N, O, S) in addition to C play a fundamental role in biochemistry. Examples of such rings include furan (C_4H_4O), pyrrole (C_4H_5N), and imidazole ($C_3H_4N_2$) (Figure 3.28). These are five-membered rings with one O atom (furan), one N atom (pyrrole) and two N atoms (imidazole) as members of the ring. Furan forms the basis of simple sugars ribose and deoxyribose, the backbone molecules of RNA and DNA (Section 3.15.3). Pyrrole serves as a precursor for the side train of the amino acid proline and is a constituent of the heme groups in hemoglobin and chlorophyll. Imidazole is formed as a side chain in the amino acid histidine and the biomolecule histamine. Searches for these heterocyclic compounds through their rotational transitions have not yet been successful and upper limits have been established for furan, pyrrole [186, 187].

The N-containing aromatic compounds pyrimidine (c-$C_4H_4N_2$) and purine (c-$C_5H_4N_4$) are both planar molecules where the N atoms have replaced the C atoms in the rings. They are the parents of the nucleobases that constitute the structural units of DNA and RNA. Pyrimidine is the base for cytosine (DNA and RNA), thymine (DNA), and uracil (RNA), whereas purine is the basis for adenine and guanine (in both DNA and RNA) (Figure 2.6).

Astronomical observations of the rotational transitions of these ring molecules in the interstellar medium include searches for pyrimidine [188, 189], imidazole [187], pyrrole and furan [186]. Due to the complexity of the molecules, the molecular population is spread over a number of rotational levels. This is illustrated in Fig-

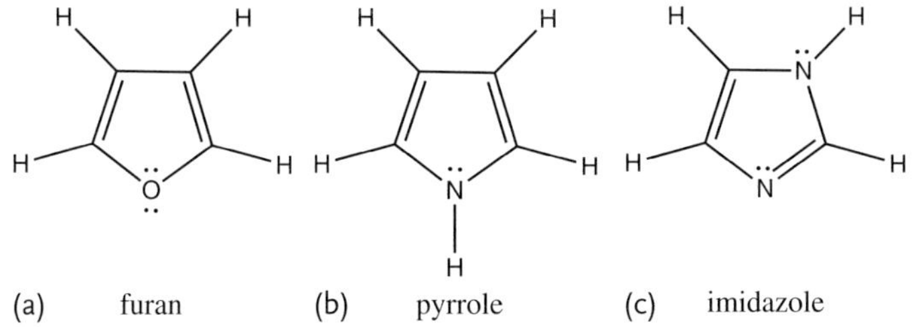

(a) furan (b) pyrrole (c) imidazole

Figure 3.28 Chemical structures of the molecules furan (a), pyrrole (b), and imidazole (c).

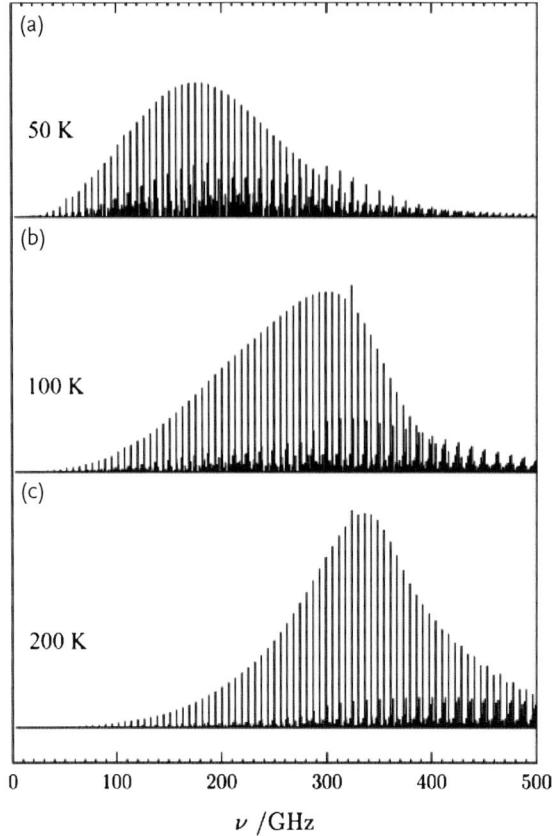

Figure 3.29 Simulated relative line intensity distribution of pyrimidine spectra between 0–500 GHz for excitation temperatures of 50 (a), 100 (b), and 200 K (c). Figure adapted from [189].

ure 3.29 where the relative line intensities of pyrimidine over the frequency range of 0–500 Hz are plotted as a function of excitation temperature.

3.16
Diamonds

Diamond is a semiconductor with band gaps at 7 and 5.45 eV. Since only photons of energies greater than these values can raise electrons across these gaps, diamond therefore absorbs strongly in the UV but poorly in the visible. Because of its symmetric structure, vibrational transitions are largely forbidden and crystalline diamonds have an almost featureless spectrum. Although long suspected to be present in space, diamonds were only unambiguously identified in a stellar spectrum when Guillois *et al.* [190] were able to match a pair of unidentified infrared emission features with the CH stretching modes of H-covered nanodiamonds [191]. Figure 3.30 shows a comparison of the infrared spectra of two Herbig Ae/Be stars with the laboratory absorbance spectra of diamond nanocrystals prepared using microwave plasma chemical vapor deposition. The two major features are due to C–H stretches of hydrides attached to the crystalline surface.

Although diamonds were found in meteorites as early as 1888 [193], diamonds of presolar origin were only identified in meteorites by their anomalous isotopic ratios in 1987 [194]. Presolar diamonds are found to be much more abundant than presolar SiC or graphite, accounting for ~6% of the total carbon in the Murchison meteorite. The typical size of diamonds is between 1–10 nm and are therefore referred to as nanodiamonds. Electron diffraction studies show that these nanodiamonds are face-centered cubic crystals and are generally similar to terrestrial macroscopic diamonds.

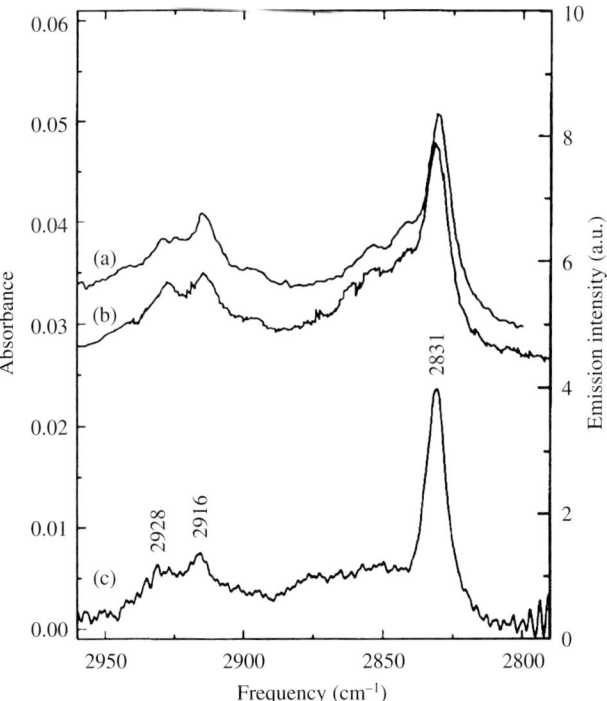

Figure 3.30 The infrared spectra of Elias 1 (curve a) and HD 97048 (curve b) compared to the laboratory absorption spectrum of 0.1 μm hydrogenated diamond nanocrystals (curve c) taken at 950 K. The 3.43 and 3.53 μm features are identified as CH stretches on C(100) and C(111) surfaces. Figure adapted from [192].

3.17
Fullerenes

The C_{60} molecule has icosahedral symmetry and has 46 vibrational frequencies. Due to the center of symmetry, the first overtone transitions ($v = 2 \leftrightarrow 0$) are forbidden. The anharmonicity of the molecule allows transitions to change more than one quantum number and many combinational modes are possible. For example, from the ground state, there are 380 binary combinational modes and the vibrational spectrum of C_{60} is therefore very complex [195, 196]. When macroscopic quantities of C_{60} was synthesized in the form of a solid [197], it became possible to measure the infrared spectrum of the substance. For the purpose of astronomical identification of C_{60}, it would be particularly useful to measure the gas-phase spectrum of the molecule. The laboratory measurement of the strongest vibrational modes of C_{60} is given in Table 3.3. These bands shift with temperature, for example, the position of the v_{26} band shifts from 8.40 µm at 0 K to 8.58 µm at 1000 K. The width of the feature is the result of a combination of rotational structure and the contribution from hot vibrational lines, scaling roughly with the square root of the gas temperature. For example, a FWHM of about 0.04 µm is expected for these bands. Similar measurements for C_{60}^+ have also been made, with strong bands showing at 7.1 and 7.5 µm [198].

Astronomical searches for the spectroscopic signatures of C_{60} began soon after the laboratory synthesis of the molecule. Searches for the 386 nm feature have been made in the protoplanetary nebula AFGL 2688 [200], and along the lines of sight to reddened stars [201]. The 8.6 µm infrared feature of C_{60} has been searched in the carbon-rich star R Cor Bor and the AGB star IRC+10216 without success [202]. It is suspected that the dominant ionization state of C_{60} in the diffuse ISM may be C_{60}^+, and the diffuse interstellar bands (Section 8.2) at 958 and 963 nm have been suggested to be due to C_{60}^+ [203].

Confirmed detection of C_{60} was made with the Spitzer Space Telescope in the planetary nebulae Tc 1 [199], M 1–20, M 1–12, K3–54, SMP SMC 16 [204], the protoplanetary nebula IRAS 01005+7910 [205], and the reflection nebula NGC 7023 [206]. Tc 1 is a young planetary nebula with a low-temperature ($\sim30\,000$ K) central star, and C_{60} was detected in the central carbon-rich, hydrogen-poor region. It is interesting to note that while all the infrared active bands of C_{60} and C_{70} are

Table 3.3 Frequencies of the fundamental bands of C_{60}. From [195, 196, 199].

Mode	Gas (0 K)		Gas ($\sim$1065 K)		Solid ($\sim$300 K)	
	v (cm^{-1})	λ (µm)	v (cm^{-1})	λ (µm)	v (cm^{-1})	λ (µm)
v_{25}	1436	6.97	1406.9	7.11	1429	7.00
v_{26}	1191	8.40	1169.1	8.55	1183	8.45
v_{27}	574	17.41	570.3	17.53	577	17.33
v_{28}	531	18.82	527.1	18.97	528	18.94

present in the spectrum (Figure 3.31), there is no AIB present except the broad 8 and 12 μm emission plateaus (Section 6.2) and the 30 μm emission feature (Section 8.5). In contrast to Tc 1, the AIB and plateau emission features are prominently present in IRAS 01005+7910. The detection of C_{60} in a protoplanetary nebula is particularly significant as it suggests that the molecule is synthesized soon after the end of AGB evolution. The detection of C_{60} therefore stimulates interest in the possible synthesis of other carbon allotropes such as carbon nanotubes, carbon onions, and nanodiamonds in the circumstellar environment [207].

From the strengths of the features, it is estimated that $\sim 6 \times 10^{-8}\ M_{\odot}$ of C_{60} and $\sim 5 \times 10^{-8}\ M_{\odot}$ of C_{70} are present in Tc 1, accounting for $\sim 1.5\%$ of the available carbon [199]. We can see from Figure 3.31 that the widths of the C_{60} are very broad ($\sim 0.3\ \mu m$), which corresponds to a wavenumber width of $\sim 10\ cm^{-1}$ for the 18 μm band. This compares well with the laboratory measured widths of $\sim 13\ cm^{-1}$ of gas-phase C_{60}. Although it is likely that the observed infrared bands arise from C_{60} molecules in the solid phase excited by UV light, a gas-phase origin of the molecule cannot be ruled out.

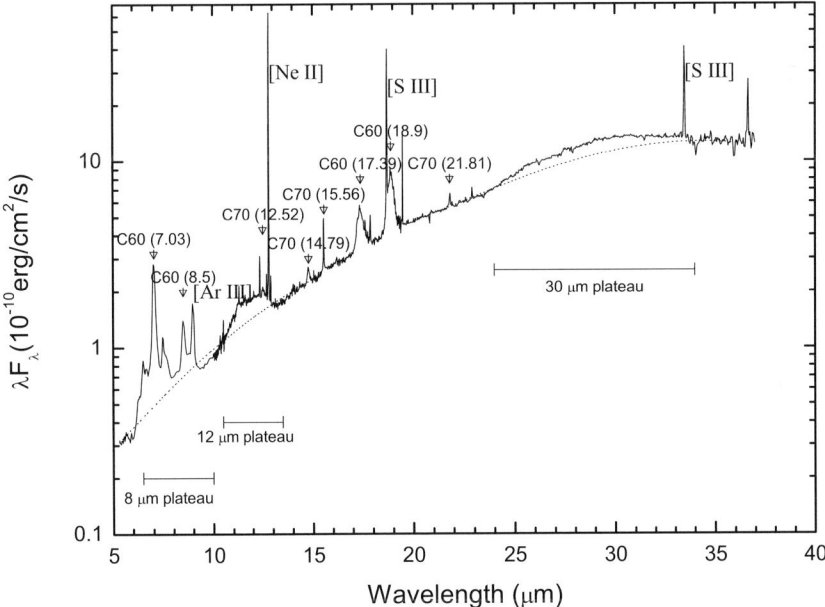

Figure 3.31 Spitzer IRS spectrum of the young planetary nebula Tc 1 showing the detection of C_{60} and C_{70}. The detected bands are marked with arrows. The narrow lines are atomic lines. A dotted line is plotted to indicate the approxi- mate level of the continuum. Three broad features are detected, the 8 and 12 μm plateaus and the 30 μm emission feature. IRS data courtesy of Jan Cami and Jeronimo Bernard-Salas.

3.18
Spectroscopic Scans

A number of astronomical sources known to be rich in molecular content have been subjected to spectroscopic scans covering a wide wavelength range by repeated spectroscopic observations with the same telescope and spectrometer. Examples of spectral scans include a 72–91 GHz surveys of Orion KL and IRC+10216 with the 20-m telescope of the Onsala Space Observatory, with a detection of 170 lines from 24 molecules detected in Orion KL and 45 lines from 12 molecules from IRC+10216 [208]. A 215–247 GHz survey of Orion KL was conducted with the 10.4-m telescope at Owens Valley Radio Observatory, with 544 lines detected of which 517 lines are identified with 25 molecules [209]. A survey in the 330–360 GHz was conducted with the newly resurfaced NRAO Kitt Peak 12-m Telescope, resulting in the detection of 180 lines [210]. The spectral survey was extended to the submm region with the 15-m James Clerk Maxwell Telescope in the 455–507 GHz range and a total of 254 lines are detected in Orion KL [211]. An even higher frequency survey between 607–725 GHz carried out at the Caltech

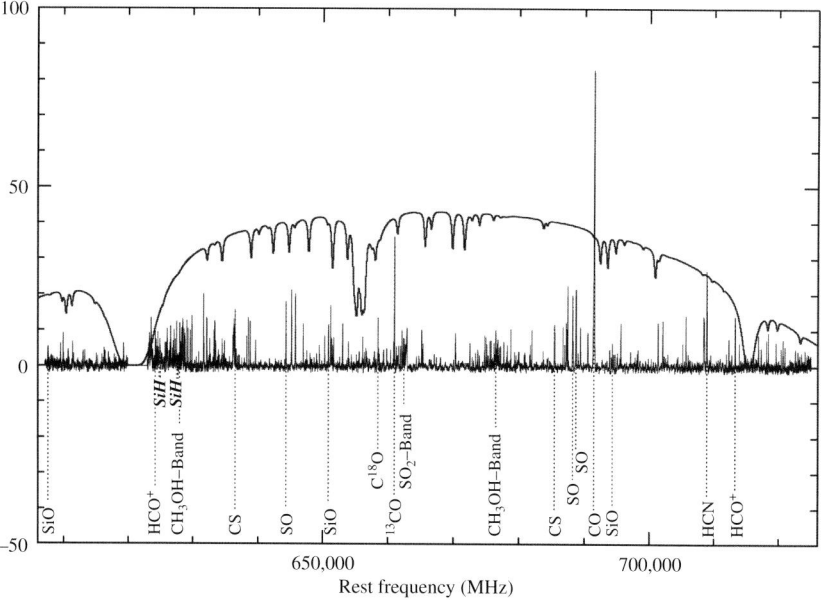

Figure 3.32 The spectrum of Orion-KL from 607 to 725 GHz taken at the Caltech Submillimeter Telescope. Overlaid is a plot of the zenith atmospheric transmission for λ 1 mm of precipitable water vapor at Mauna Kea (altitude 4200 m). This region covers the 450 μm window, bracketed by broad atmospheric absorption by water at 557 and 752 GHz. The gaps in this spectrum at 620.7 and 715.4 GHz are due to atmospheric H_2O and O_2, respectively. The ordinate is the antenna temperature (in K) for the spectrum and in % for the atmospheric transmission. The molecules responsible for some of the strongest lines are marked by dotted lines below. Figure adapted from [212].

Submillimeter Telescope, where 1064 spectral features consisting of 2032 partially blended lines are detected (Figure 3.32). Out of this number, there are 155 unidentified features [212]. Among the identified lines, the molecules CH_3OH and SO_2 contribute the largest number of lines. The highest frequency (780–900 GHz) line survey of Orion KL was carried out at CSO, where a total of 541 features from 26 molecular species were detected [213].

Because of absorption by atmospheric H_2O and O_2, part of the submm spectral window is not accessible from the ground. The first spectral scan in the submillimeter region was carried out by the Odin satellite, which is equipped with four tunable receivers in the submm region. A spectral survey of Orion KL was carried out covering the spectral range 486–492 GHz and 541–577 GHz [214]. Over 1000 lines have been identified from 40 difference molecular species. Organic molecules detected include methanol (CH_3OH, $^{13}CH_3OH$), methyl cyanide (CH_3CN), dimethyl ether (CH_3OCH_3), formaldehyde (H_2CO, $H_2^{13}CO$, HDCO), thio-formaldehyde (H_2CS), thioformyl cation (HCS^+), and hydrogen isocyanide (HNC) (Figure 3.33).

The galactic center region Sgr B2 was surveyed in the 70–150 GHz region with the Bell Labs 7-m telescope [216]. This was followed by a survey in the same frequency range at the NRAO with a detection of over 700 lines [217]. Subsequent surveys include a survey in the 8.8–50 GHz range conducted at the Nobeyama Radio Telescope [218] and a 4–6 GHz survey with the 305 m Arecibo Observatory. A survey at the λ 3-mm band with the NRAO 12-m telescope and the BIMA array [219]

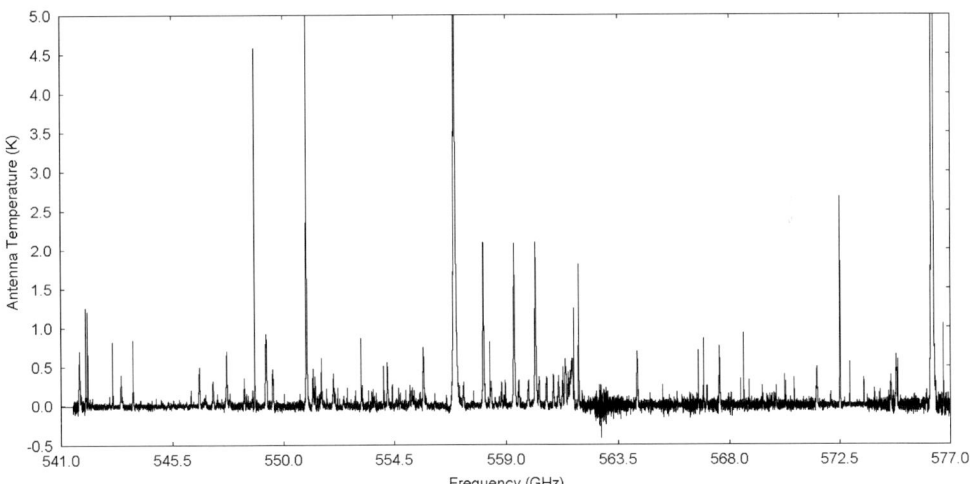

Figure 3.33 Spectrum of Orion KL from 541–577 GHz observed with the Odin satellite. Several organic molecules are present, including methanol, methyl cyanide, dimethyl ether, formaldehyde, thio-formaldehyde, thioformyl cation and hydrogen isocyanide. The strongest lines in this spectrum are CO $J = 5$–4 (576.27 GHz), $^{13}CO\ J = 5$–4 (550.93 GHz), $C^{18}O\ J = 5$–4 (548.83 GHz), and $H_2O\ 1_{10}$–1_{01} (556.94 GHz). Figure adapted from [215].

resulted in the detection of 218 lines, including 97 identified molecular transitions from 18 molecules.

A spectral line survey of W 51 from 17.6–22.0 GHz was carried out at the NRAO 11-m telescope [220]. The organic-rich hot core G327.3-0.6 was surveyed in the λ 1–3 mm region with the Swedish-ESO Submillimeter Telescope (SEST), where many nitrile species are found.

The Heterodyne Instrument for Far Infrared (HIFI) on board of Herschel Space Observatory was used to perform spectroscopic survey of five star formation regions L1157-B1, IRAS 16293-2422, OMC2-FIR4, AFGL 2591 and NGC 6334I. With high spectral resolution (1 MHz) and broad spectral coverage (555–636 GHz), over 500 lines are detected in NGC 6334I, 70% of which originate from methanol [221]. New molecules detected include H_2O^+, OH^+, H_2Cl^+, D_2O, and ND.

For stellar sources, the most extensively observed object is the carbon-rich AGB star IRC+10216 (CW Leo). This object was discovered in the IRC 2 μm sky survey and is extremely faint in the visible due to circumstellar extinction. In addition to solid particles, the circumstellar envelope of IRC+10216 also contains many

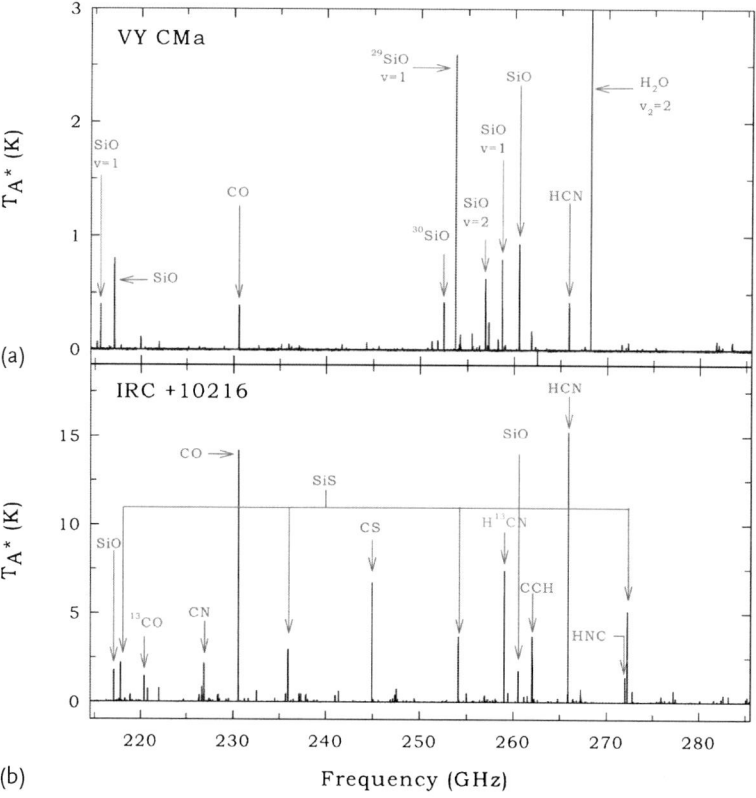

Figure 3.34 A comparison between the 210–290 GHz spectra of the O-rich star VY CMa (a) and the C-rich AGB star IRC+10216 (b). Figure adapted from [89].

molecular species. Spectral scans in the millimeter-wavelengths began with the λ 3–4 mm survey at Onsala [222], followed by the Nobeyama λ 6–10 mm survey [128], λ 0.8 mm survey at JCMT [223] and CSO [224], λ 2 mm survey at IRAM 30-m [225], and the λ 1 and 2-mm survey at Arizona Radio Observatory [89, 226]. In the most recent survey, a total of 717 lines were observed, arising from 32 different molecules, confirming that IRC+10216 is a very rich molecular source [89].

For oxygen-rich stars, the object with the most number of known molecular lines is VY CMa [227]. In the λ 1-mm survey of VY CMa, a total of 130 molecular features corresponding to 18 different species were detected. Detected molecules include oxides and sulfides as well as radicals (SO, CN, NS, PO, AlO), and molecular ions (HCO$^+$) [89]. Figure 3.34 shows a comparison between the spectra of VY CMa and

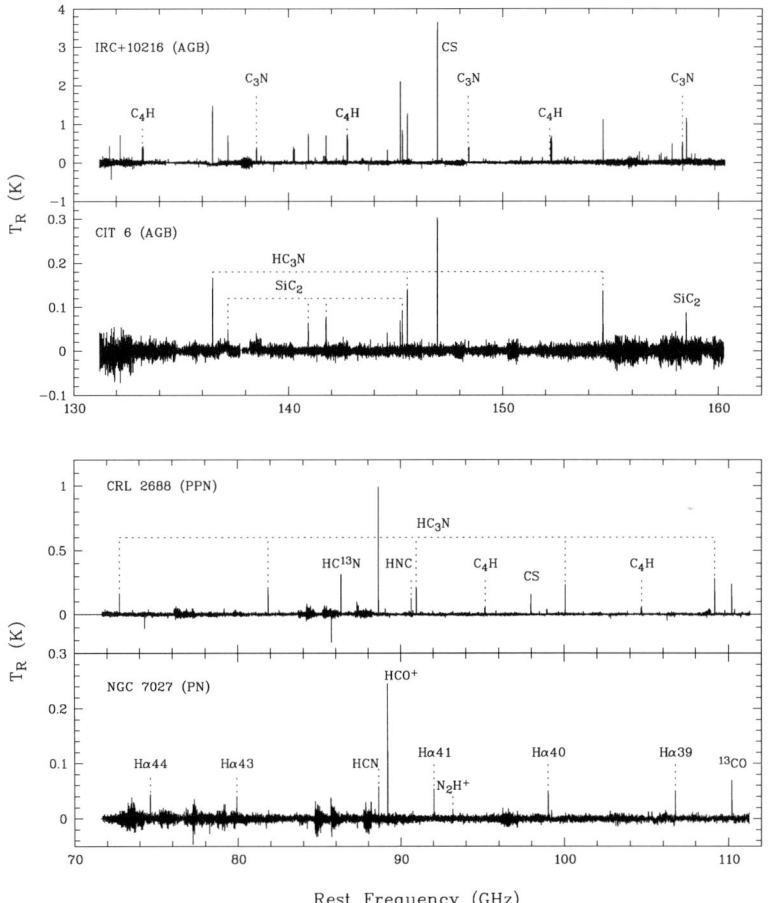

Figure 3.35 Evolution of the molecular content of the circumstellar envelopes of late-type stars, from the AGB (IRC+10216, CIT 6), to proto-planetary nebula (CRL 2688), to plane-tary nebula (NGC 7027). These spectra (70–110 GHz) were taken with the *Arizona Radio Observatory* 12-m telescope. Figure adapted from [231].

IRC+10216 over the same spectral range. The strongest line in the spectrum of VY CMa is the $\nu = 2\ 6_{52}-7_{43}$ line of H_2O whereas the strongest lines in the C-rich IRC+10216 are the $J = 3-2$ of HCN and $J = 2-1$ of CO.

Since molecules are continuously being synthesized in the outflow of evolved stars, we can study the chemical evolution by performing spectral scans of objects in consecutive phases of stellar evolution. Since the evolutionary time scales of these phases are very short (10^4-10^5 yr, $< 10^3$ yr, 10^3-10^4 yr for AGB stars, protoplanetary nebulae, and planetary nebulae, respectively), chemical reaction time scales are very well constrained by these time scales. Spectral scans of the molecular envelopes of the AGB star IRC+10216, CIT 6 and AFGL 3068, the protoplanetary nebula AFGL 2688, and the planetary nebula NGC 7027 have been carried out at the Arizona Radio Observatory [228–230] (Figure 3.35). A total of more than 500 emission features are detected in the survey, showing that the sources in different evolutionary stages have remarkably different chemical composition. As a star evolves from AGB stage to protoplanetary nebula, the abundances of Si-bearing

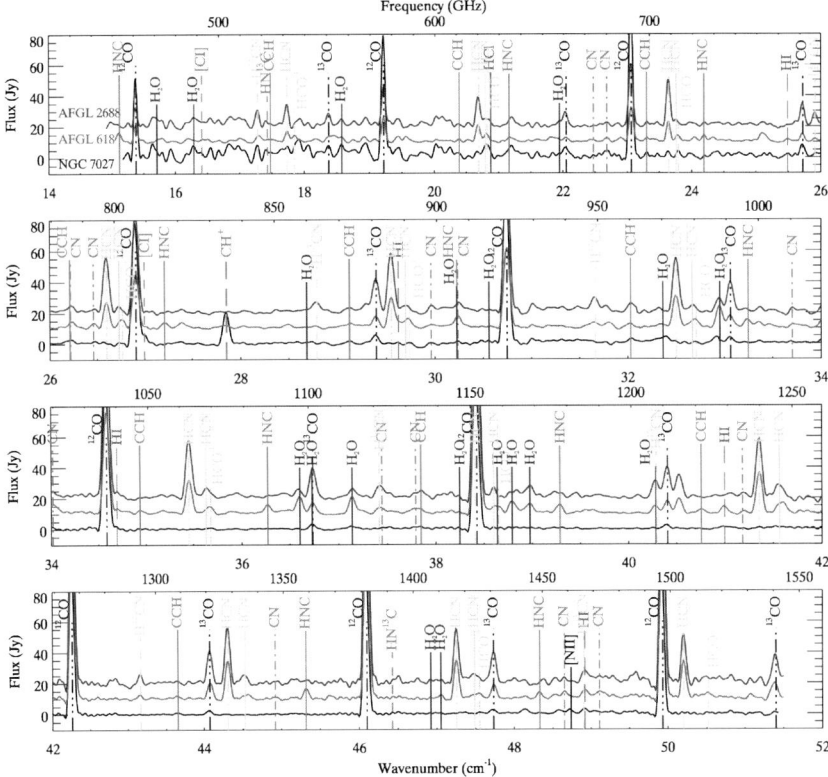

Figure 3.36 Combined Herschel SPIRE FTS short and long waveband spectra of AFGL 2688 (top curve), AFGL 618 (middle curve), and NGC 7027 (bottom curve). The spectra have their respective continua removed and the flux scales offset for display clarity. The wavelength coverage is from 671 to 194 μm. Figure adapted from [105].

molecules (SiO, SiCC, and SiS) decrease, while the abundances of some long-chain molecules, for example, CH_3CN, C_4H, and HC_3N, increase. After further evolution to planetary nebula, the abundances of neutral molecules dramatically decrease, and the emission from molecular ions becomes more intense.

The wide wavelength coverage and high sensitivity of the Herschel Space Observatory allow the molecular spectra of circumstellar envelopes to be observed. Figure 3.36 shows the Herschel SPIRE Fourier transform spectrometer observations of three C-rich protoplanetary nebulae and planetary nebulae. Over 150 lines originating from 18 different molecular species are detected. The detected lines include the high ($J = 4-3$ to $J = 13-12$) rotational transitions of CO, the $J = 1-0$ (359 µm) rotational transition of CH^+ as well as lines from CN, HCN, HNC, CCH, HCO^+, and so on.

3.18.1
Unidentified Lines

When a spectral feature is detected in an astronomical spectrum, its frequency is checked against compilations of molecular line catalogues. The major line catalogues in use are the Cologne Database for Molecular Spectroscopy (http://www.ph1.uni-koeln.de/vorhersagen/, accessed 5 July 2011), the NIST recommended rest frequencies for observed interstellar molecular microwave transitions (http://physics.nist.gov/cgi-bin/micro/table5/start.pl, accessed 5 July 2011) and the JPL Catalogue (http://spec.jpl.nasa.gov/ftp/pub/catalog/catform.html, accessed 5 July 2011). When no counterpart is found, astronomers then try to see whether a group of lines belong to a particular pattern, which in turn can be used to deduce the possible molecular carrier. After such exercises, the remaining lines are called "unidentified lines" and tabulations of astronomical spectral lines of unknown origin can be found in the NIST catalogue by specifying "unidentified" in the molecule field.

3.18.2
All-Sky Spectral Scans

The earliest form of all-sky spectral scans is objective prism surveys where a prism is placed in front of a telescope and very low spectral resolution optical spectrum of every bright star will be recorded when a wide patch of sky is observed by a telescope. The objective prism survey was useful in the discovery of celestial objects with strong atomic emission lines. Since this type of survey is totally unbiased, it has the potential to discover new, unknown emission features or features in unknown or new kinds of celestial sources.

The first large-scale infrared spectroscopic survey was done by the IRAS satellite where its Low Resolution Spectrometer (LRS) automatically recorded a spectrum when an infrared source brighter than a certain limit is detected. The LRS is a slitless spectrometer covering the 7.5 to 23 µm range with a spectral resolution of ∼20–60. As the IRAS satellite surveyed the entire sky, LRS obtained a total of

170 000 individual spectral scans corresponding to $\sim$50 000 sources. The spectra of 11 224 sources (corresponding to a flux limit of 7 Jy at 12 μm) were processed and classified [232]. The spectra were classified as S (for stellar photospheric), F (featureless but flatter than photospheric), E (showing the 9.7 and 18 μm silicate feature in emission), A (silicate features in absorption), C (showing the 11.3 μm SiC feature), P (showing the AIB features), H (showing a red continuum), U (showing unusual features), L (showing atomic line emissions), and I (for noisy or incomplete spectra). The IRAS LRS survey shows that while inorganic silicate and silicate carbide features are very common in AGB stars (with 4000 and 700 sources detected respectively), the AIB features are found in planetary nebulae, reflection nebulae, H$\text{\scriptsize II}$ regions, and galaxies. Most interestingly, the survey discovered a strong, unknown emission feature at 21 μm in four previously unclassified IRAS sources (Section 8.5). The origin of this feature has remained unidentified 20 years after its discovery, but is believed to be due to a carbonaceous compound.

The next mid-infrared spectral survey was performed by the Infrared Telescope in Space (IRTS) satellite. The Mid-Infrared Spectral (MIRS) instrument covered the spectral range of 4.5–11.7 μm with a spectral resolution of $\sim$0.3 μm. Over the one month lifetime of the satellite, thousands of point sources were observed, covering about 7% of the whole sky. Out of these observations, 129 spectra with signal-to-noise ratios greater than 3 were extracted [233].

3.19
Search for Large, Complex Molecules

As the size of the molecules decreases, we expect that the individual rotational lines to be correspondingly weaker. For an optically thin line, the integrated brightness temperature (T_b) of an emission line is given by

$$\int T_b d\nu = T_x \tau_\nu \,, \tag{3.11}$$

where T_x is the excitation temperature of the transition defined by

$$\frac{n_i}{n_j} = \frac{g_i}{g_j} e^{-h\nu_{ij}/kT_{x_{ij}}} \,, \tag{3.12}$$

and g_i is the degree of degeneracy of state i. The optical depth of the line is given by

$$\tau_\nu = \int n_j \left(1 - e^{-h\nu/kT_x} \right) \frac{g_i}{g_j} \left(\frac{c^2}{8\pi\nu^2} \right) A_{ij} \phi_\nu ds \,, \tag{3.13}$$

where ν is the transition frequency, A_{ij} is the spontaneous transition rate and ϕ_ν is the line profile, and ds is the length element along the line of sight. The population (n_j) of a rotational level j can be related to the total molecular population (n_T) by

$$n_j = \frac{n_T}{Z(T)} g_i e^{-E_j/kT_x} \,, \tag{3.14}$$

where

$$Z(T) = \sum_{1}^{\infty} g_i e^{-E_i/kT_x} , \qquad (3.15)$$

is the partition function at temperature T_x, and the population distribution is assumed to be in thermodynamical equilibrium. For a symmetric molecule where the partition function is small, the line is strong. For example, although a heavy molecule such as HC_9N may have 11 atoms and a molecular weight of 123 and therefore is likely to be less abundant, its rotational transitions are strong because it is a linear molecule. For another molecule with a similar molecular weight but a more complex structure, the population of the molecule is distributed over a much large number of rotational states and the population in each rotational level is small, resulting in weaker lines.

Observationally, the detectability of a line improves with decreasing noise in the spectrum, which can be decreased by increasing the aperture of the telescope or by decreasing the noise temperature of the receiver. One can compensate for the weakness of rotational lines by going to higher frequency, as the spontaneous emission coefficient A_{ij} increases with v_{ij}^3. However, in general, the noise of receivers are also higher at higher frequencies, partly negating this advantage.

For simple molecules, their identification by radio spectroscopic means can be extremely robust. Due to the low-temperature conditions in the ISM, the observed lines are very sharp and can match laboratory measurements to the accuracy of a few parts in 10^7 [234]. However, the identification process becomes increasingly difficult for larger molecules. As the complexity of a molecule increases (both in terms of the number of atoms and the geometry), the number of excitable energy levels also increases. This spreads the molecular population thinly, resulting in weaker lines. If a molecule has different conformers each has a different energy, then the population of the molecule will be diluted by the different states of the different conformers.

Although the weakness of the lines can be overcome by improving receiver sensitivity, astronomical spectroscopy runs into the problem of line confusion as every frequency interval is filled with molecular lines, often from smaller species such as methanol. For example, the λ 3 mm region of Sgr B2 has a spectral line density of 6.06 lines per 100 MHz, half of which are unidentified [219]. The identification of a complex molecule therefore requires the measurements of many transitions to be certain. The following criteria are proposed by Herbst and van Dishoeck [235] for the unequivocal identification of a complex molecule: (i) rest frequency of the molecule known to 1 in 10^7; (ii) observed frequency of a nonblended line agree with the rest frequency; and (iii) all expected lines in the spectrum are observed with their predicted intensity ratios. Eventually, we may run into a natural limit for the detection of large molecules through their rotational transitions by single-dish telescopes. This leads to the suggestion that the failure to detect complex organic molecules through their rotational transitions does not necessarily imply the absence or low abundance of the molecule. Interferometric observations separating the emitting regions of different molecules are needed to reduce the confusion.

3.20
Summary

In short, 40 years since the first detection of CO by mm-wave spectroscopy, the number of gas-phase molecules detected in space through their rotational transitions has expanded rapidly. With the development of infrared spectroscopy where molecules can also be detected through their vibrational transitions, practically all families of organic compounds (Section 2.1) have been detected in space. These include hydrocarbons (e.g., methane CH_4, acetylene C_2H_2, ethylene C_2H_4), alcohols (e.g., methanol CH_3OH, ethanol C_2H_5OH, vinyl alcohol $H_2C=CHOH$), acids (e.g., formic acid $HCOOH$, acetic acid CH_3COOH), aldehydes (e.g., formaldehyde H_2CO, acetaldehyde CH_3CHO, propenal $CH_2=CHCHO$, propanal CH_3CH_2CHO), ketones (e.g., ethenone $H_2C=CO$, acetone, CH_3COCH_3), amines (e.g., methylamine CH_3NH_2, cyanamide NH_2CN, formamide NH_2CHO), ethers (e.g., dimethyl ether CH_3OCH_3, ethyl methyl ether $CH_3OC_2H_5$), and so on. Active searches are also underway for prebiotic molecules leading to the formation of proteins, carbohydrates, nucleic acids, and lipids. These include searches for the first sugar glyceraldehyde ($CH_2OHCHOHCHO$), the simplest amino acid glycine (NH_2CH_2COOH) and the parents of the bases that constitute the structural units of DNA and RNA, for example, purine (c-$C_5H_4N_4$) and pyrimidine (c-$C_4H_4N_2$).

These observations have demonstrated the richness of chemical processes in the interstellar medium. However, since our ability of detecting large, complex molecules through rotational transitions is limited by the present state of mm-wave astronomical techniques, the actual complexity of organic molecules could be much higher than is currently known. A sense of the possible complexity and richness of extraterrestrial organic compounds can be seen in meteorites, where millions of diverse chemical structures are likely to be present (Section 7.5). Comparison between Solar System (Chapter 7), stellar (Chapter 6) and interstellar (Chapter 4) studies is one way to improve our understanding of organic chemistry in space.

4
Organic Molecules in the Interstellar Medium

The interstellar medium (ISM) is a diverse environment. The most easily visible parts are emission nebulae consisting of gaseous materials photoionized by hot stars. The emission of recombination lines from H and He and collisionally excited lines from heavy elements make these nebulae bright in the visible wavelengths. These discrete emission nebulae are concentrated in the spiral arms and are often associated with regions of recent star formation. There are also colder clouds that mainly consist of neutral, molecular gas. These clouds may appear to be "dark" in visible light but can be detected through rotational and vibrational lines of molecules in the infrared and millimeter wavelengths and from infrared continuum radiation from solid-state grains in the clouds. The interface between the warm and cold neutral components is defined by the photodissociation of molecules by diffuse galactic starlight, and such photodissociation regions (PDR) can be sites of active chemical processes.

4.1
Dark Clouds

The existence of dark clouds were first indicated by the absence of visible light in regions densely populated by stars and can be quantitatively identified through star counts. When a region of the sky has far lower stellar densities than neighboring areas, it is assumed that absorption by a foreground dust cloud is responsible. On a smaller scale, there are compact, dense clouds known as Bok globules. They are often spherical in shape and have higher densities than dark clouds.

   An example of a dark cloud is the Taurus Molecular Cloud (TMC-1). TMC-1 has the shape of a ridge elongated in the northwest-southeast direction (Figure 4.1). Since the extinction coefficient decreases with increasing wavelength, dark clouds become less opaque in infrared wavelengths. The fact that TMC-1 is still dark in this 8 μm image (Figure 4.1) indicates that this is an extremely dense cloud. Because of the low gas temperature in the cloud ($\sim$10 K), the observed molecular lines have extremely small line widths ($< 0.4\ \mathrm{km\,s^{-1}}$). Such small linewidths allow the transition frequencies to be precisely measured, giving better determination of the molecular constants.

Organic Matter in the Universe, First Edition. Sun Kwok.
© 2012 WILEY-VCH Verlag GmbH & Co. KGaA. Published 2012 by WILEY-VCH Verlag GmbH & Co. KGaA.

Figure 4.1 Spitzer IRAC 8 µm image of TMC-1. The dark region suggests that the cloud is so dense that it is opaque even at the long wavelength of 8 µm.

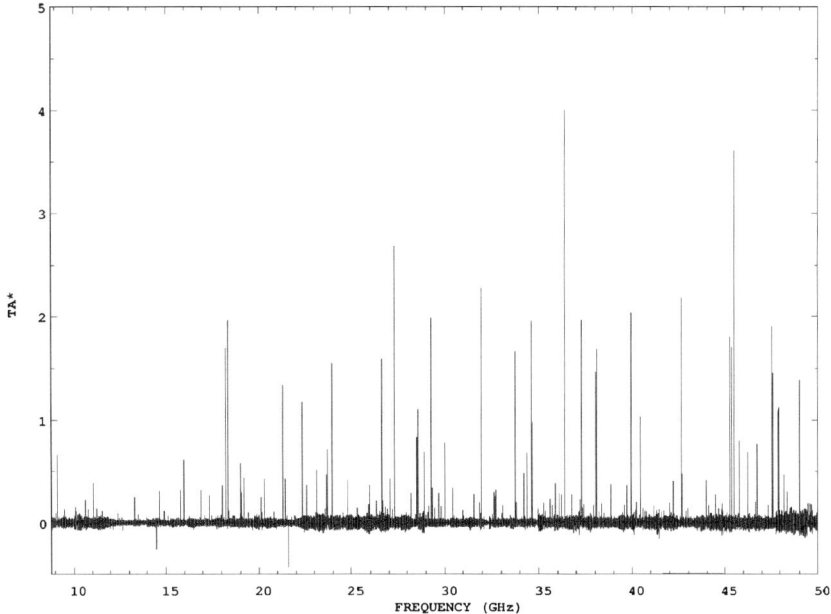

Figure 4.2 The spectrum of TMC-1 in the 8.8–50 GHz region taken by the 45-m Nobeyama Radio Observatory. Figure adapted from [237].

TMC-1 is known to contain several different carbon chains, including cyanopolyynes ($HC_{2n-1}N$), methylcyanopolyynes ($CH_3C_{2n}H$), and methylcyanopolyynes ($CH_3C_{2n+1}N$) (Section 3.10). Mapping exercises have shown that the distribution of cyanopolyynes peaks at position RA(1950) $04^h\ 38^m\ 38^s$, Dec(1950) $25°\ 35'\ 45''$. Because of its rich molecular content, TMC-1 is the target of a number of molecular line surveys [218, 236]. Figure 4.2 shows a combined spectrum of TMC-1 taken at the 45-m Nobeyama Radio Observatory. The spectrum contains 414 lines from 38 molecules, half of which are linear carbon-chain molecules and their derivatives. Only one line remains unidentified.

4.2
High-Mass Star Formation Regions

Molecular clouds that are bright in the infrared and molecular line emissions are called giant molecular clouds. They have higher gas temperatures (50–100 K) than dark clouds, suggesting that there are internal heating sources inside the cloud. These heating sources can be identified by near-infrared imaging and many of them turn out to be newly formed massive stars. Molecular clouds under gravitational collapse lead to the formation of hot cores, which have typical temperatures of $\sim$100–300 K and densities of $\sim 10^7$ cm^{-3}. Given such physical conditions, hot cores are the most chemically rich sites. From dynamical calculations, the lifetimes of hot cores are estimated to be $\sim 10^5$ yr.

4.2.1
Sagittarius B2

Sgr B2 is a massive star-forming region at a distance of 7.1 kpc and within 300 pc of the galactic center. The region is made up of several different sources labeled as Sgr B2 North-Large Molecule Heimat (Sgr B2(N-LMH)) due to the large molecules detected there, Sgr B2 (M), Sgr B2 (OH), Sgr B2 (S), and Sgr B2 (NW). Sgr B2 is also a strong far-infrared continuum source with the emission peaking near 80 µm. Since our line of sight from the Sun to Sgr B2 crosses the main galactic spiral arms, molecules in the cold clouds in the arms can be detected in absorption against the bright continuum emission of Sgr B2.

The infrared spectrum of Sgr B2 is dominated by thermal emission from solid-state particles. The infrared continuum emission peaks spatially near Sgr B2 and spectrally at 80 µm. The cloud is opaque even at far infrared wavelengths, with the 100 µm optical depth estimated as 3.8 $\pm$ 0.4 [136]. A spectral line survey of Sgr B2 over the spectral region 47–196 µm using the ISO LWS has identified NH_3, NH_2, NH, H_2O, OH, H_3O^+, CH, CH_2, C_3, HF and H_2D^+ in addition to atomic emission lines such as OI, CII, OIII, NII and NIII [238]. Because the optical forbidden lines arise from hot, photoionized regions, they appear in emission above the infrared continuum. However, for the molecules in the molecular cloud where the gas kinetic temperature is comparable to the dust temperature (<50 K), the molecular rotational lines often appear in absorption against the dust continuum. This is particularly true for far-infrared lines where the radiative decay rates are high and the molecules are often not thermalized at the kinetic temperature [239].

4.2.2
Orion Nebula

The Orion Nebula (M 42) is an emission nebula photoionized by a group of bright, young, massive stars (the Trapezium stars). Surrounding the optical nebula is a much more extended (~4°) region of neutral gas called the Orion Molecular Cloud. At a distance of 450 pc, the Orion Molecular Cloud is the nearest region of massive star formation. Embedded within the Orion Molecular Cloud are two very bright infrared objects, the Kleinmann–Low (KL) and the Becklin–Neugebauer (BN) regions. When imaged in higher spatial resolution, the Orion BN–KL objects split up into a number of bright infrared sources, the brightest of which are labeled IRc 1-9. Kinematic studies have identified several regions, including a hot core, a ridge of quiescent gas, and a compact (~15″) ridge [240]. Orion KL is known to be one of the richest sources of molecular lines in the Galaxy and has been subjected to more than 20 spectral-line surveys (Section 3.18).

The 11.3 μm AIB feature was first detected in Orion through the 10 μm atmospheric window [241]. The AIB features are prominent in the mid-infrared spectrum of Orion, as can be seen in its ISO spectrum shown in Figure 4.3.

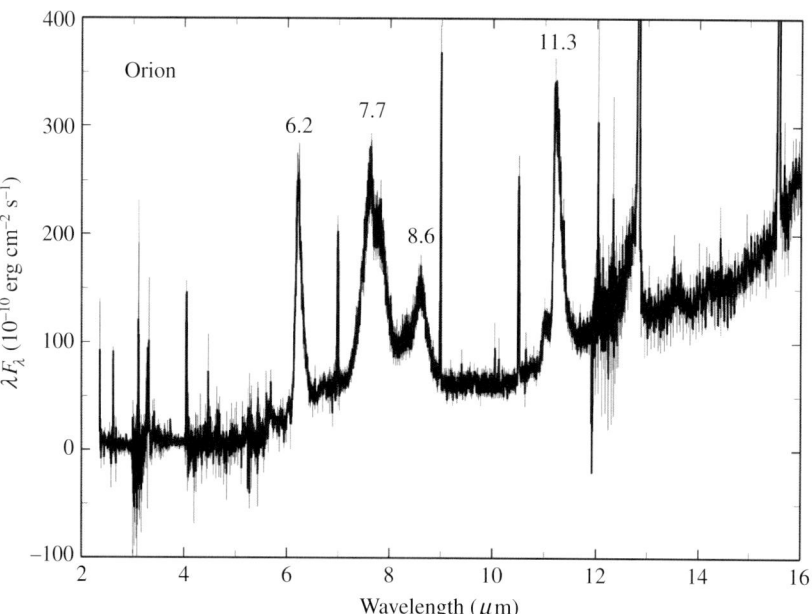

Figure 4.3 ISO SWS spectrum of the Orion bar shows strong AIB emission at 6.2, 7.7, 8.6 and 11.3 μm (marked). The unmarked narrow lines are atomic emission lines.

4.3
Reflection Nebulae

Reflection nebulae are gaseous nebulae whose visible brightness is due to scattered starlight from embedded stars. Photospheric continuum radiation from the embedded stars is scattered by solid particles in the nebulae. Since the scattering coefficient is larger at shorter wavelengths, reflection nebulae often have a blueish appearance. Reflection nebulae are different from emission nebulae such as H ii regions and planetary nebulae in that the central stars are not hot enough to emit in the UV for the photoionization of the nebular gas. Consequently, the optical spectra of reflection nebulae are continuous spectra in contrast to the spectra dominated by emission lines in emission nebulae. A list of commonly observed reflection nebulae is given by van den Bergh [242].

Since the solid particles in reflection nebulae are heated by starlight, they also self radiate in the infrared [243, 244]. The IRAS all-sky survey allows for the first time a systematic study of the self emission properties of reflection nebulae, whose physical properties could previously only be inferred from optical observations [245]. Infrared spectroscopic observations reveal that reflection nebulae also possess strong AIB emission as those observed in H ii regions and planetary nebulae. Comparison of the observed spectra show that the AIB features have remarkably similar profiles in different reflection nebulae, even though they are heated by central stars of very different temperatures. This is illustrated in Figure 4.4 where the normalized spec-

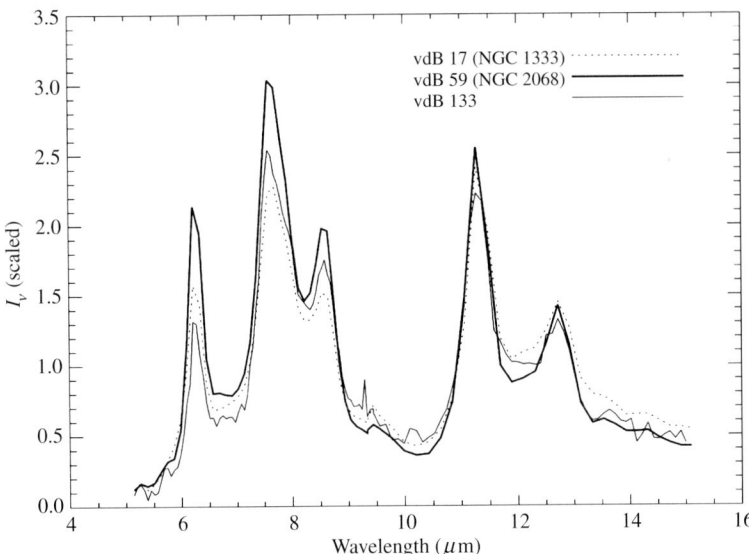

Figure 4.4 ISOCAM CVF spectra of three reflection nebulae showing the 6.2, 7.7, 8.6, 11.3, and 12.7 μm AIB features. Each spectrum has been divided by their respective IRAS 12 μm broadband flux to show that the features have similar peak wavelengths, spectral shapes, and continuum levels independent of the temperatures of the illuminating stars. Figure adapted from [246].

tra of reflection nebulae NGC 1333, NGC 2068, and vdB 133 are superimposed. They show consistent profiles in spite of the fact that these three nebulae are heated by central stars of temperatures of 11 000, 19 000, and 6800 K respectively. This suggests that the AIB features are excited by a wide range of photon energies, in particular by visible photons.

4.4
Diffuse Interstellar Medium

Diffuse interstellar clouds are low-density regions in the ISM whose structure lacks a definite morphology and are semitransparent in the visible ($A_V \sim 1$). When exposed to starlight, the clouds are often in atomic form. In the inner regions where the gas is shielded from starlight, simple molecules can form. The first molecules detected in diffuse clouds are H_2, HD, OH, and CO through their absorption lines in the UV, for example, by the Copernicus satellite. Sometimes the term "translucent clouds" is used to refer to the class of interstellar clouds between diffuse and molecular clouds ($A_V \sim 2-5$). Molecules in translucent clouds, in addition to being observed through their absorption line spectra, can also be studied in emission with millimeter-wave techniques. At high galactic latitudes, diffuse clouds are exposed to less UV radiation because most hot (early-type) stars are concentrated on the galactic plane. As a result, these high-latitude clouds have a different chemistry than diffuse or translucent clouds.

In the diffuse interstellar medium, interstellar matter (gas and solids) can be located far from energy sources (i.e., stars). Lacking energy input and incapable of self radiating, the molecules and solids can still be detected through their ability to absorb background radiation. Spectroscopic observations of distant stars and galaxies can reveal intervening interstellar matter along our line of sight. The most widely observed spectral features are the 10 μm feature due to amorphous silicates and the 220 nm feature due to a carbonaceous solid (Section 8.3).

Absorption features at 3.0 and 3.4 μm were first detected in the spectrum of a Galactic Center source IRS 7, bringing up the possibility that there are other solid-state species in the diffuse ISM [247–249]. The 3.4 μm absorption feature was detected in several more lines of sight [250–253] and is thought to arise from the C–H stretching modes of methyl ($-CH_3$) and methylene ($-CH_2$) groups in aliphatic hydrocarbon materials [254]. This is the first indication that aliphatic organic matter is widely present in the diffuse ISM. An example of the spectrum towards GCS3 is shown in Figure 4.5. To underline the importance of this discovery, it is estimated from the strength of the 3.4 μm feature that 5–30% of the carbon atom is in the carrier of this band [251, 255].

The 3.4 μm feature is made of both symmetric and asymmetric stretching modes. For a methyl ($-CH_3$) group, the symmetric mode corresponds to a vibration in which all 3 H atoms are stretching in phase, whereas the asymmetric modes correspond to one bond contracting while the other two are extending or vice versa (see Figure 4.6). In general, the asymmetric modes are stronger than the symmetric mode.

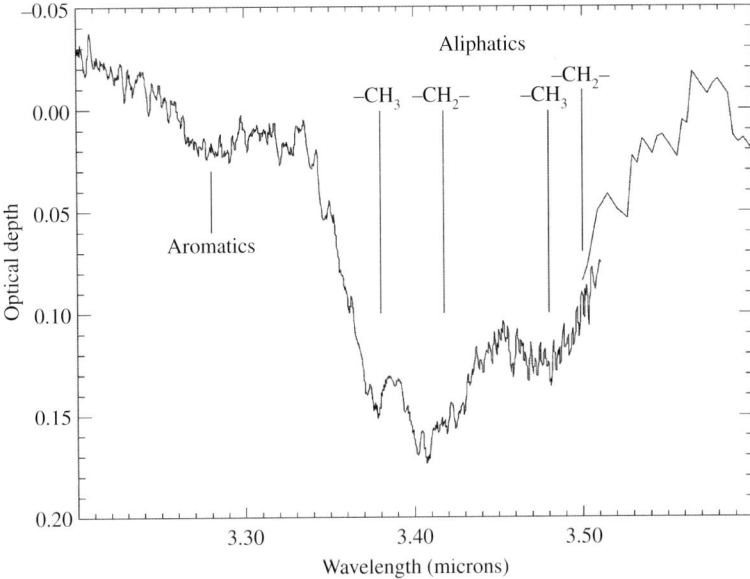

Figure 4.5 ISO SWS spectrum (continuum divided) of the GCS3 (located at ∼14 arcmin north-east of the Galactic Center) showing the aliphatic hydrocarbon features at 3.4 μm and the aromatic C–H stretch at 3.28 μm. Figure adapted from [256].

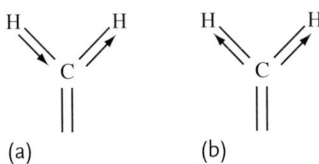

Figure 4.6 Illustrations of the asymmetric (a) and symmetric (b) stretching vibrations of the methyl group.

In addition to the 3.4 μm feature, features at 6.8, and 7.2 μm have also been observed and they are generally interpreted as being due to the asymmetric and symmetric deformation modes of $-CH_3$ and $-CH_2$ groups of aliphatic materials [256]. While the 3.4, 6.8, and 7.2 μm features are widely seen in the diffuse ISM, they have never been detected in dense molecular clouds. Whether this difference is due to production or destruction is not clear [257].

Excess emission in the 12 μm band of IRAS in the diffuse interstellar medium has been suspected to be caused by AIB emission [258]. The first evidence that AIB emission is part of the general diffuse galactic emission was found in balloon experiments [259] and confirmed by the observations with the IRTS satellite [260, 261]. While the galactic plane emits a near-infrared continuum due to integrated starlight, an excess at 3.3 μm can clearly be seen (Figure 4.7). The UIE features at 6.2, 7.7, and 11.3 μm have also been detected in cirrus clouds (Figure 4.8).

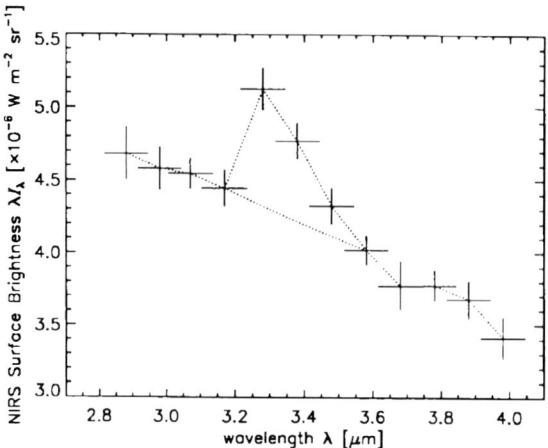

Figure 4.7 IRTS NIRS spectrum of the diffuse emission averaged over the galactic plane ($47° \, 30' < 48°$, $|b| < 15'$). Figure adapted from [260].

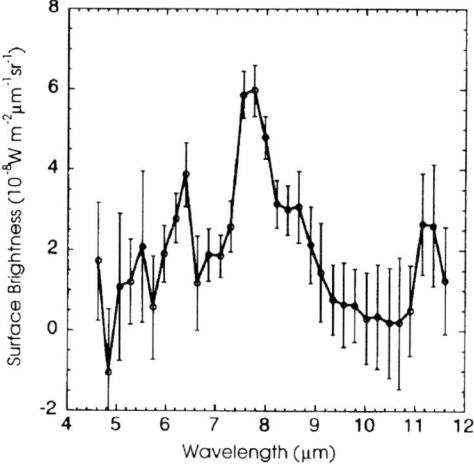

Figure 4.8 IRTS/MIRS spectrum of a cirrus cloud at $\ell \sim 100°$ and $b \sim 13°$. The 6.2, 7.7, and 11.3 μm UIE bands can clearly be seen. Figure adapted from [262].

4.5
Cirrus Clouds

The first evidence for "cirrus" clouds was the observation of faint reflection nebulosity at high galactic latitude. The fact that such diffuse clouds are detected at the IRAS 12 and 25 μm bands suggests that the grain temperature must be high (100–500 K). Due to the lack of heating sources at high galactic latitudes, these grains cannot be in thermodynamic equilibrium. One possible explanation is that these are very small grains heated by single UV photons, leading to high transient temperatures [263]. It is estimated that approximately 20% of the sky is covered by infrared cirrus. Broad-band photometry and spectroscopy of diffuse, high-latitude cirrus clouds reveal an optical emission over scattered starlight and diffuse galactic light. Optical imaging of large high galactic latitude clouds shows strong excesses in the red band which can be attributed to the Extended Red Emission (ERE, Section 8.4). The amount of this red excess represents ∼30% of the optical surface brightness of these clouds. While the visual colors of high galactic latitude clouds are determined by scattering of integrated starlight of the Galaxy, the excess in the red is due to a separate photoluminescence process. It is estimated that the carriers of ERE are responsible for 20–30% of the total dust absorption of visual/UV photons [264]. Although the exact nature of the carrier of ERE is unknown, it is quite possible that it is an organic compound. The strong signal of the ERE over large-scale structures suggests that this compound is widely and abundantly present in the galactic diffuse ISM.

4.6
Summary

Infrared observations have shown that organic materials are not only found in dense interstellar clouds under hard (e.g., HII regions) and low (e.g., reflection nebulae) radiation environments, they are also present in cold, dense clouds. Most interestingly, they are widespread in the diffuse interstellar medium when density is very low and the radiation background is also low. The AIB features are seen on the galactic plane as well as in the galactic halo. If cirrus clouds and ERE are indications of the distribution of organic matter, then they are distributed widely in our own Galaxy, not just near the presence of stars. Future large-scale spectroscopic surveys of the entire sky would provide a more precise picture of the distribution of organic matter in our Galaxy.

5
Organic Compounds in Galaxies

The common accepted view is that galaxies are just a collection of stars where the energy output of the galaxies is dominated by starlight. With the development of infrared and millimeter-wave astronomy, we now realize that molecular gas makes up a significant fraction of galaxies' mass and some types of galaxies have most of the energy output emitted by the solid-state component. The latter point is obvious in the spectral energy distributions of galaxies as shown in Figure 5.1. For spectral energy distributions plotted with the y axis in units of $\lambda F_\lambda = \nu F_\nu$, the absolute

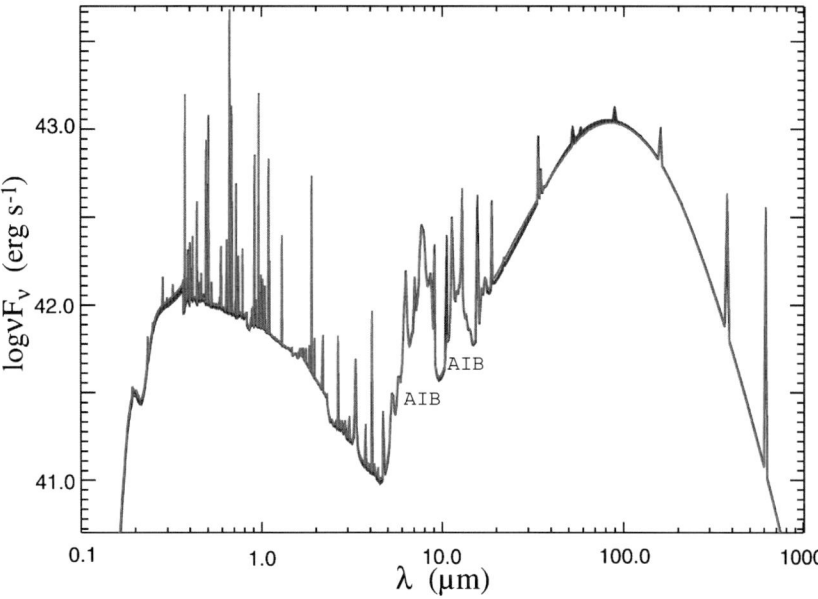

Figure 5.1 Model spectral energy distributions (SED) of starburst galaxies showing the importance of AIB emissions in the energy output of these galaxies. The SED has two major components: an optical component (peaking at ~0.4 μm) due to photoionized gas and an infrared component (peaking at ~100 μm) due to emission by solid-state particles. The narrow features are atomic lines and the broad features between 5 and 20 μm marked as AIB are due to aromatic compounds. Figure adapted from [266].

Organic Matter in the Universe, First Edition. Sun Kwok.
© 2012 WILEY-VCH Verlag GmbH & Co. KGaA. Published 2012 by WILEY-VCH Verlag GmbH & Co. KGaA.

value of the peak of blackbodies is proportional to the total flux output of the black-body.[1] It is clear that in active galaxies, such as, starburst galaxies, the solid-state component is responsible for most of the energy output.

Since external galaxies are usually small in angular size, we can obtain the overall molecular content of the galaxy from a single spectroscopic observation. From the brightness of molecular lines and bands, we can derive the relative abundance of specific molecular species. From the integrated fluxes of the lines and bands, they can be compared to the total energy output in the stellar and dust components of the galaxies. By observing distant galaxies, we are also studying galaxies early in the evolution of the Universe. Rotational transitions of CO have been seen in quasars with redshifts (z) greater than 6, suggesting that stellar nucleosynthesis and molecular formation have taken place before the Universe was 10^9 yr old.

While galaxies certainly contain inorganic solid-state components such as silicates (Figure 5.2), organic solids are also present. Strong AIB features similar to those observed in the Galactic ISM are seen in starburst galaxies (Figure 5.3), showing that the interstellar medium of galaxies is also rich in organic content.

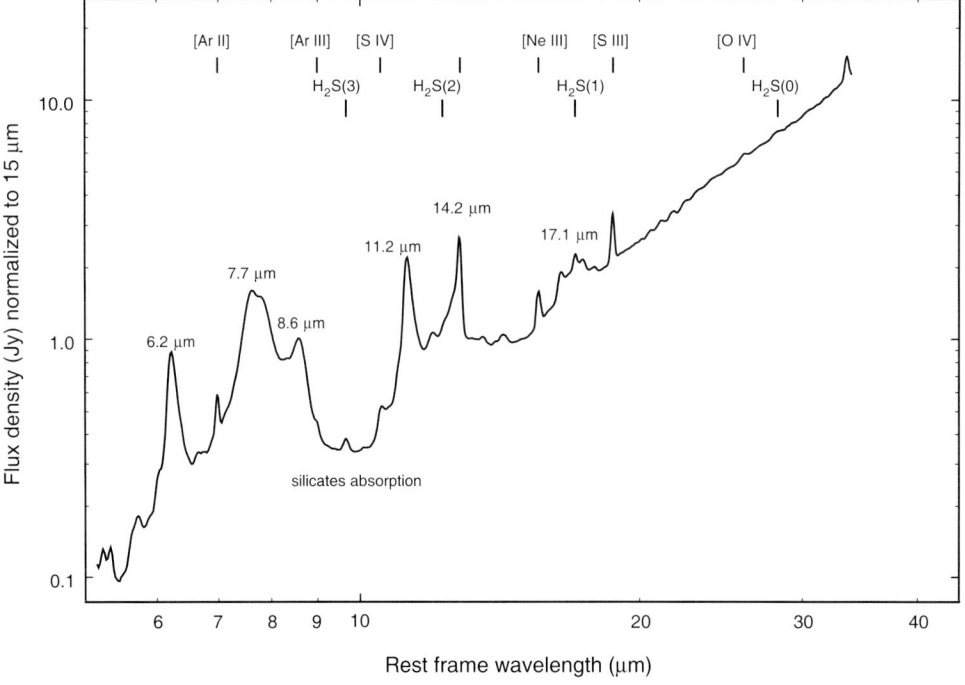

Figure 5.2 Averaged Spitzer IRS spectrum of 13 starburst galaxies. The AIB features as well as some atomic and H_2 molecular lines are marked. The strong broad absorption feature at 10 μm is due to amorphous silicates. Figure adapted from [267].

1) $(\lambda B_\lambda)_{\mathrm{max}} = 0.736 B(T)$, where B is the Planck function at temperature T integrated over all wavelengths [265]. The observed flux F is related to B by $F = \pi \theta^2 B$, where θ is the angular size of the blackbody.

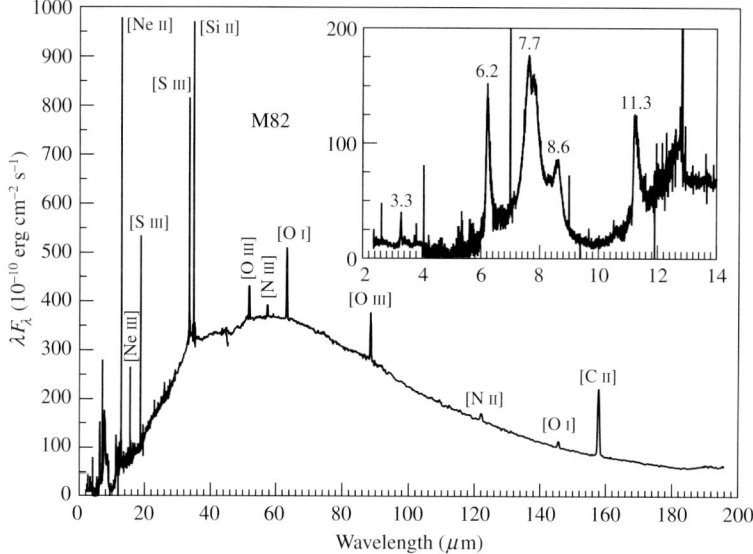

Figure 5.3 The ISO SWS and LWS spectrum of M 82. M 82 is the brightest galaxy in the infrared and is considered as a prototype of starburst galaxies. At a distance of 3.63 Mpc, the total infrared luminosity is $3.8 \times 10^{10} L_{\odot}$. In addition to strong dust continuum (peaking at $\sim$60 μm, corresponding to a temperature of $\sim$50 K) and fine-structure forbidden lines (narrow features marked by square brackets), the spectrum also shows very strong AIB features (insert).

5.1
Aromatic Compounds in Galaxies

The common presence of aromatic compounds in external galaxies was first recognized by ISO observations. Strong AIB features are seen in starburst galaxies such as M 82 (color plate C 1) and NGC 1068 [268, 269]. The wavelengths and the profiles of the features (as seen in the insert of Figure 5.3) are very similar to the AIB features seen in our Galaxy, suggesting that they are emitted from a similar chemical carrier.

The increased sensitivity of the Spitzer IRS instrument over the ISO SWS has resulted in a much larger sample of galaxies detected with AIB emissions [267] (Figure 5.2). The AIB features have been detected by Spitzer in distant luminous infrared galaxies to redshift $\simeq$ 2. The large equivalent widths of the AIB features show that a significant fraction of the total energy output of these galaxies is emitted in these features (Figure 5.1). This clearly shows that complex organic compounds are highly abundant in these galaxies. In our nearest galactic neighbor, the Magellanic Clouds, objects showing the AIB features can be individually spatially resolved. Figure 5.4 shows the Spitzer IRS spectrum of the planetary nebula SMP LMC 58 in the Large Magellanic Clouds. In spite of the metal-poor environment, planetary nebulae in the Magellanic Clouds seem to be able to synthesize aromatic compounds as easily as those in the Milky Way [270].

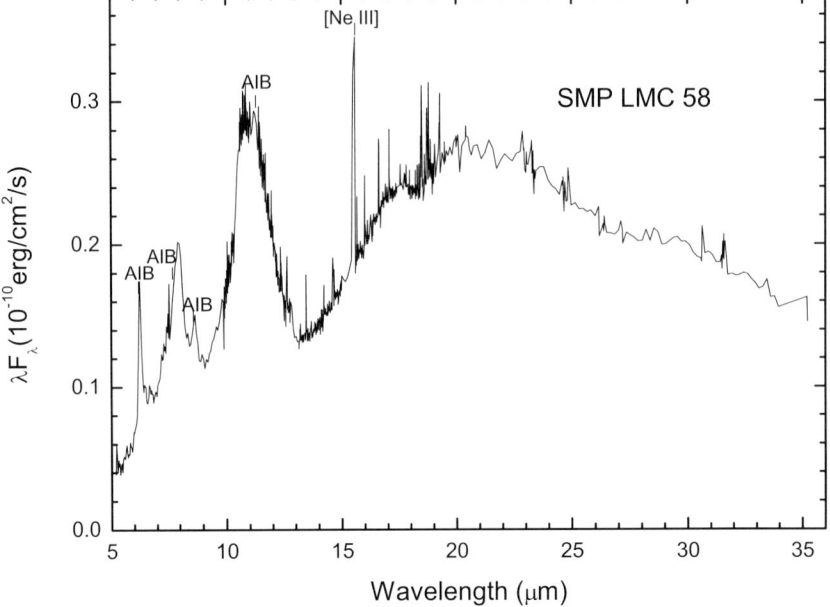

Figure 5.4 Spitzer IRS spectrum of the Large Magellanic Cloud planetary nebula SMP LMC 58, showing that aromatic compounds are also commonly produced by planetary nebulae in the metal-poor environment of the Magellanic Clouds.

5.2
The Aliphatic Component

The 3.4 μm aliphatic feature was first detected in external galaxies NGC 1068, IRAS 08572+3915, NGC 5506, NGC 3094, 7172, and 7479 with the United Kingdom Infrared Telescope (UKIRT) [271, 272]. The profiles of the feature are similar to those observed in the diffuse ISM and are therefore likely to share the same origins (Section 4.4). Figure 5.5 shows the spectra of infrared luminous galaxies IC 694 and NGC 6240 where the 3.3 and 3.4 μm features are clearly detected.

The 6.9 μm feature, due to a mixture of the $-CH_2-$ bend and $-CH_3$ antisymmetric bending modes [274], has also been detected in external galaxies (Figure 5.6). This feature is particularly strong in the distant ($z = 0.0583$) infrared galaxy IRAS 08572+3915. UKIRT and Spitzer Space Telescope Infrared Spectrograph (IRS) observations of this galaxy show a strong 3.4 and 6.9 μm features, suggesting the presence of aliphatic material [275]. Fittings of the profiles of the 3.4 and 6.9 μm features by the stretching and bending modes of CH_2 and CH_3 are shown in Figure 5.7. From these fittings, it is estimated that ~15% of carbon atoms are in the sp^3 with a CH_2 to CH_3 ratio of ~2. The absence of AIB features in this galaxy suggests that aliphatic materials are dominant in the diffuse ISM of this galaxy.

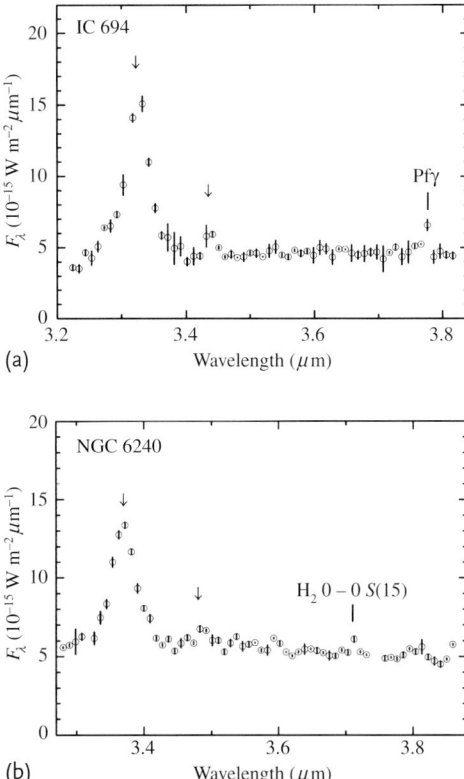

(a)

(b)

Figure 5.5 The UKIRT cooled grating spectrometer spectra of (a) IRLGs IC 694 ($z = 0.010$) and (b) NGC 6240 ($z = 0.024$), showing the 3.4 μm aliphatic emission feature. The positions of the redshifted 3.3 μm aromatic and the 3.4 μm aliphatic features are marked by the arrows. Figure adapted from [273].

Under the AIB features seen in galaxies are two broad emission plateaus with peaks at approximately 8 and 12 μm. These features are easily seen in the spectra shown in Figures 5.1 and 5.3. The shapes of these broad plateau features are similar to those seen in protoplanetary nebulae (Section 6.2).

5.3
Other Organics

The mysterious Extended Red Emission (ERE) seen in circumstellar environment of evolved stars and in the diffuse ISM of the Galaxy is likely to arise from an organic compound (Section 8.4). Figure 5.8 shows that ERE is also prominent in the optical spectrum of galaxies. Assuming that the ERE is the result of photoluminescence, such a strong excess suggests that this organic carrier is widely present in the diffuse ISM of other galaxies and has efficiently converted starlight into red emission.

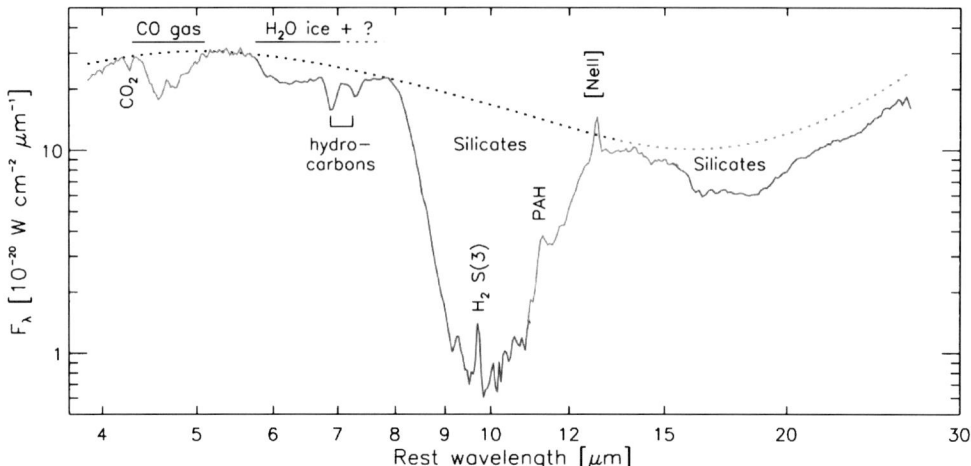

Figure 5.6 Spitzer Space Telescope Infrared Spectrograph (IRS) spectrum of the ultraluminous infrared galaxy IRAS F00183−7111. In addition to the strong silicate absorption features at 10 and 18 μm, aliphatic features at 6.8 and 7.2 μm in absorption can also be seen. Figure adapted from [276].

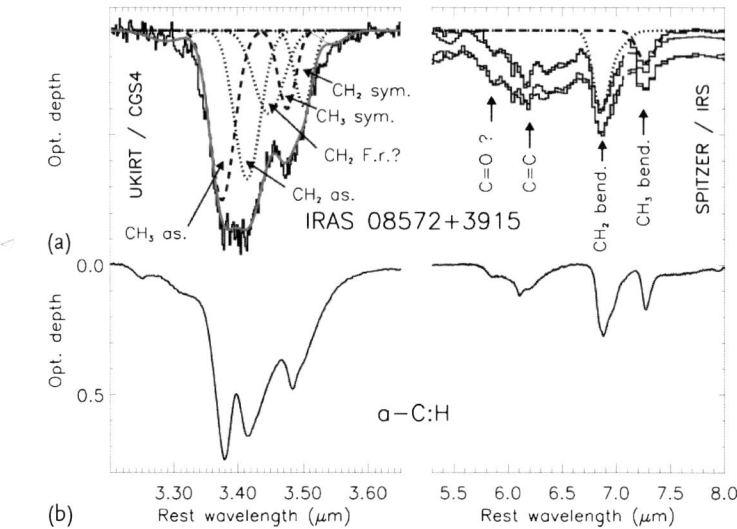

Figure 5.7 (a) Fitting of the 3.4 and 6.9 μm absorption features in the spectrum of IRAS 08572+3915 by the stretching and bending modes of CH_2 and CH_3. (b) shows a comparison spectrum of hydrogenated amorphous carbon (HAC) in these wavelengths. Figure adapted from [275].

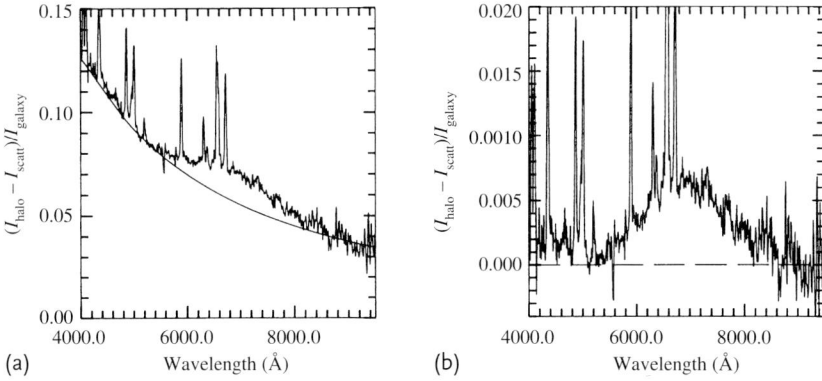

Figure 5.8 (a) an excess of red continuum emission can clearly be seen in the halo of M 82 after dividing the observed spectrum by that of the galaxy. (b) the profile of the ERE emission after subtraction of a scattering continuum. Figure adapted from [277].

5.4
Summary

Infrared and submillimeter-wave photometry and spectroscopic observations of galaxies have shown that molecular species and solid-state compounds were synthesized very early in the history of the Universe. The detection of AIB features in galaxies with $z \sim 2$ suggests that complex aromatic compounds were already present in the Universe as early as 10 billion years ago. Since life as we know it took more than a billion years to develop on Earth, these extragalactic complex organics must have been synthesized abiologically, directly from simple molecules. The wide presence of aliphatics and ERE in the diffuse ISM of galaxies also shows that the process of chemical synthesis must be extremely efficient and common. In contrast to the popular (and indeed reasonable) belief that organics are tied to life, astronomical observations show that the existence of complex organics preceded the emergence of life.

6
Synthesis of Organic Compounds in the Late Stages of Stellar Evolution

6.1
Molecular Synthesis in the Stellar Wind

The element carbon (C), which is the basic atom in the structure of organic compounds, is synthesized in asymptotic giant branch (AGB) stars through helium-burning (triple-α) reaction. The newly formed C atoms are then dredged up from the core to the surface of the star. Due to the low temperature of the photosphere of AGB stars, the C atoms can react with other atoms to form molecules such as carbon monoxide (CO). When the abundance of C overtakes that of oxygen (O), the surplus of C atoms after the formation of CO can be used to form other carbon-based molecules such as C_2, C_3, and CN. The spectral characteristics of these molecules in the photospheric spectra defines the class of AGB stars which we call carbon stars. Although we do not yet have a complete understanding of which AGB stars (e.g., which mass range of progenitors) will evolve into carbon stars, we generally assume that they represent an evolved phase of AGB evolution.

In addition to photospheric spectroscopic signatures, circumstellar spectral properties can also be used to identify carbon stars. AGB stars undergo extensive mass loss through a stellar wind driven by radiation pressure on grains [278] and the mass loss rates of evolved O-rich or C-rich AGB stars can reach as high as $10^{-4}\,M_\odot\,\mathrm{yr}^{-1}$. Chemical reactions in the stellar wind allow for the synthesis of other more complex gas-phase molecules whose rotational transitions can be detected by millimeter-wave or submillimeter-wave spectroscopic observations. To date, more than 60 molecular species have been detected, and they include inorganics (e.g., CO, SiO, SiS, NH_3, AlCl, etc.), organics (C_2H_2, CH_4, H_2CO, CH_3CN, etc.), radicals (CN, C_2H, C_3, HCO^+, etc.), chains (e.g., HCN, HC_3N, HC_5N, etc.), and rings (C_3H_2). Since the dynamical lifetime of the envelope is $\sim 10^4\,\mathrm{yr}$, the chemical reactions that lead to formation of these species must be shorter than this time scale. Interferometric observations also allow mapping of the molecular emission regions and put further constraints on the reaction zone.

Solid-state particles, both amorphous and crystalline, are also found condensing in the stellar winds of carbon stars. The most common solid-state condensates are amorphous silicates and silicon carbide (SiC), whose lattice vibrational modes at 10/18 μm and 11.3 μm are widely detected in O-rich and C-rich AGB stars

Organic Matter in the Universe, First Edition. Sun Kwok.
© 2012 WILEY-VCH Verlag GmbH & Co. KGaA. Published 2012 by WILEY-VCH Verlag GmbH & Co. KGaA.

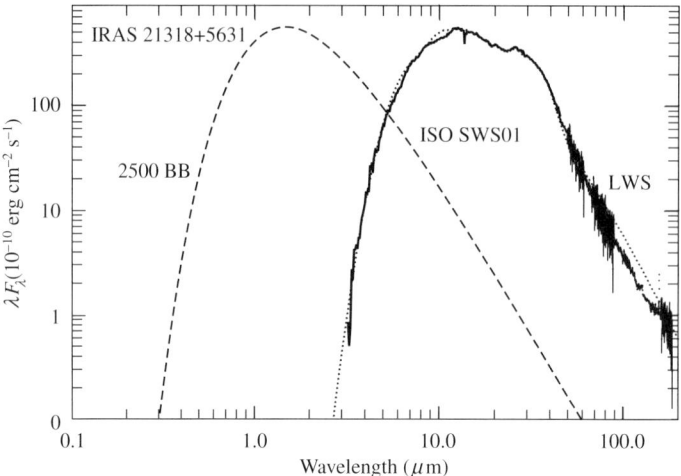

Figure 6.1 IRAS 21318+5631 is an example of an evolved carbon star so obscured by its own ejected circumstellar dust envelope that the central star is totally undetectable in the optical region. Its infrared spectrum (solid lines: ISO SWS01 and ISO LWS) is completely due to dust emission and has a color temperature of 300 K. The dotted line represents the theoretical fit to the spectrum based on a one-dimensional radiation transfer model with a hidden 2500 K central star (dashed line) as the energy source. The absorption feature near the peak of the spectrum is the 13.7 μm band of acetylene, a molecule synthesized in the circumstellar envelope near the end of AGB evolution. Figure adapted from [291].

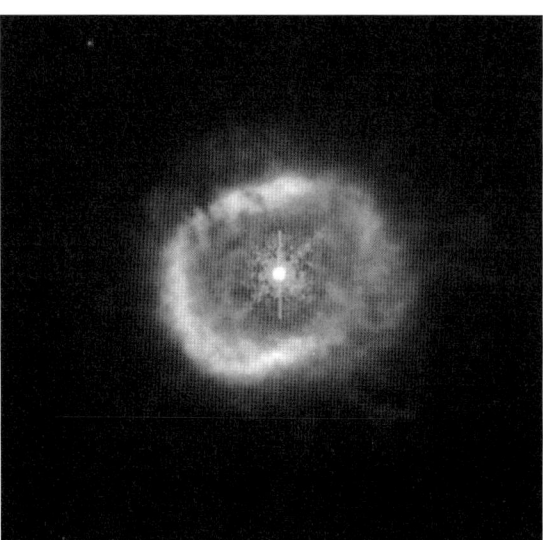

Figure 6.2 Color composite image of HST WFPC 2 observations of BD+30° 3639, another planetary nebula with rich organic content. For a color version of this figure, please see the Color Plates at the beginning of the book.

respectively [232]. Since solid particles have a high opacity to visible light, their condensation in the circumstellar envelopes can cause large obscuration to the light from the central star. The intercepted starlight heats up the solid particles, which in turn cool by self radiating in the infrared. As the ejection rate increases in the late stages of AGB evolution, the envelope can contain so much solid material that the star is completely obscured by dust extinction. Such stars will have no optical counterpart and can only be identified by infrared observations. An example of such infrared stars discovered in the IRAS all-sky survey is IRAS 21318+5631 (Figure 6.1). This carbon star has such a dense dust envelope that the central star suffers from 360 magnitudes of extinction in the visible (A_v) and is completely obscured [279]. Its infrared spectrum has a color temperature of 300 K and is the result of emission from a dust envelope heated by a hidden central star of 2500 K. The identification of these infrared objects as extremely evolved AGB stars and the spectroscopic observations of these objects have led to the realization that these stars are prolific molecular factories.

The high rate of mass loss on the AGB will eventually deplete the hydrogen envelope of the star and gradually expose the hot core. As the envelope thins to below $10^{-3} M_{\odot}$, we can see deeper into the stellar atmosphere. Due to the shrink-

Figure 6.3 Color composite image of HST WFPC 3 observations of the planetary nebula NGC 6302. The bright color lobes are regions of photoionized gas and the molecular and solid materials are in the dark regions near the waist of the bipolar nebulosity. Credit: NASA, ESA, and the Hubble SM4 ERO Team. For a color version of this figure, please see the Color Plates at the beginning of the book.

ing of the photospheric radius, the effective temperature of the star will increase.[1] Through a combination of H-shell burning and mass loss by stellar wind, the envelope will continue to thin, leading to further increases in stellar temperature. When the stellar temperature reaches 20 000 K, the ultraviolet photon output from the star will begin to photoionize the circumstellar envelope, creating a planetary nebula [280]. Several images of planetary nebulae are shown in the color plates (Figures 1.4, 6.2, 6.3). The intervening phase between the end of AGB and the onset of photoionization is called the protoplanetary nebula phase [281, 282].

Unlike stars which have a continuous spectrum, planetary nebulae have spectra dominated by strong emission lines, primarily recombination lines of H and He, and collisionally excited lines of ions of heavy elements (O, S, Ne, Si, etc). Although planetary nebulae are bright visible objects because of their emission-line

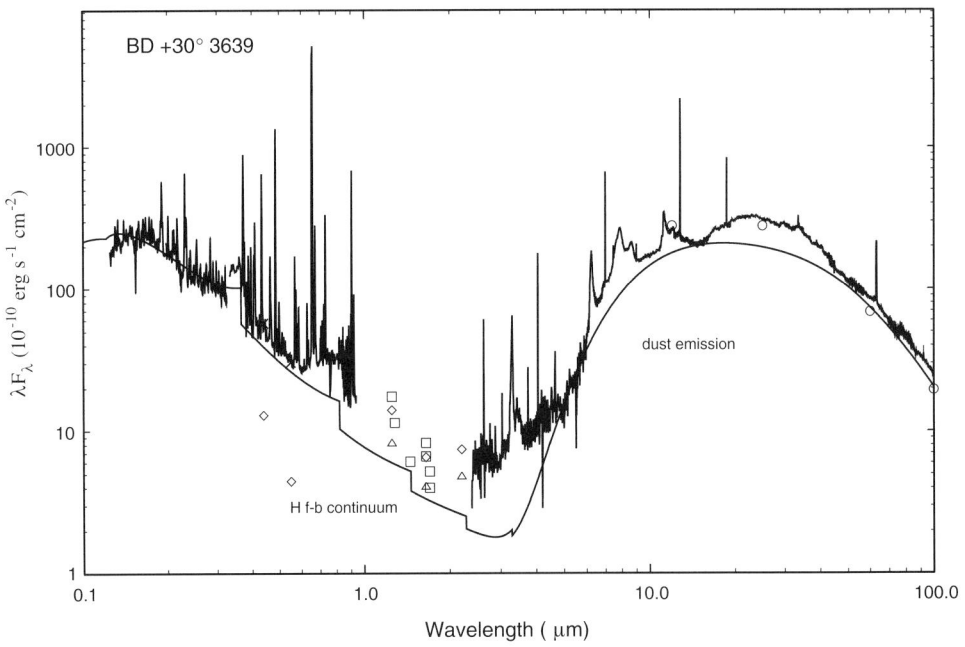

Figure 6.4 The spectral energy distribution of the PN BD+30° 3639 (Figure 6.2) showing the spectral richness of the PN phenomenon. Recombination lines of H and He and collisionally excited lines of metals dominate the UV, visible, and IR parts of the spectrum. In the far-IR and submillimeter, rotational transitions of molecules are present (not shown). For the continuum (shown in solid line), 2-γ radiation dominates in the UV, b–f emission in the visible and near IR, thermal emission from hot gas in the X-ray (not shown), dust emission in the IR, and f–f emission in the radio (not shown). Some of the broad emission features above the continuum in the IR are due to the stretching and bending modes of aromatic compounds.

1) The luminosity of a star is given by $L_* = 4\pi R_{\mathrm{photo}}^2 T_{\mathrm{eff}}^4$ where R_{photo} is the size of the photosphere and T_{eff} is the effective temperature. The photospheric radius is defined as the radius of the star where the optical depth reaches the value of 2/3.

spectra, recent observations of planetary nebulae in the infrared and millimeter wavelengths have shown that they also possess thick molecular envelopes mixed with solid-state particles. Mid- and far-infrared observations from the IRAS and Infrared Space Observatory (ISO) missions have found that dust emission represents a major fraction of energy output from planetary nebulae (Figure 6.4). The detection of molecular/dust envelopes in planetary nebulae clearly establishes the link to their progenitor AGB stars [283].

The existence of remnant AGB dust envelopes enables the search for protoplanetary nebulae, as they are expected to have infrared colors between those of evolved AGB stars and young planetary nebulae [284]. The first protoplanetary nebulae (AFGL 618 and AFGL 2688) were discovered as the result of ground-based follow-up of the Air Force Geophysical Laboratory infrared sky survey. However, a comprehensive understanding of the protoplanetary nebulae phenomenon was possible only after a systematic search for these objects among cool IRAS sources [281]. Although protoplanetary nebulae are faint in the visible due to the absence of emission lines, their structure can be observed through scattered light from circumstellar dust. The optical images of four protoplanetary nebulae obtained from observations with the Hubble Space Telescope (HST) are shown in Figures 6.5–6.7.

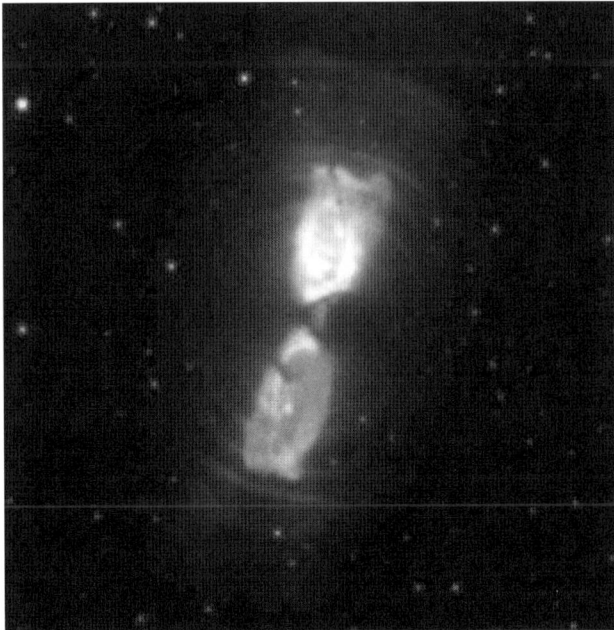

Figure 6.5 Color composite image of HST WFPC 2 observations of the protoplanetary nebula IRAS 17150-3224 (Cotton Candy Nebula) [551]. For a color version of this figure, please see the Color Plates at the beginning of the book.

Figure 6.6 Color composite image of HST WFPC observations of the protoplanetary nebula IRAS 17441-2411 (the Silkworm Nebula) [552]. For a color version of this figure, please see the Color Plates at the beginning of the book.

Figure 6.7 Color composite image of HST WFPC 2 I- and V-band observations of IRAS 16594-4656 (Water Lily Nebula) [553], a protoplanetary nebula that shows AIB emissions [554]. For a color version of this figure, please see the Color Plates at the beginning of the book.

6.2
Beyond the Asymptotic Giant Branch

The AIB features first make their appearance in the protoplanetary nebulae phase. To this date, no AIB feature has been seen in AGB stars, suggesting that they are synthesized in the circumstellar envelope during the post-AGB phase. Figure 6.8 shows the ISO SWS spectra of the protoplanetary nebula IRAS 07134+1005 and the young planetary nebula IRAS 21282+5050. The AIB features at 7.7 and 11.3 μm are clearly visible in the protoplanetary nebulae spectra. After the discovery of protoplanetary nebulae, ground-based observations have found that many C-rich protoplanetary nebulae possess strong emission features at 3.4 μm (Figure 6.9), which were later identified as being due to the symmetric and asymmetric C–H stretching modes associated with the methyl and methylene aliphatic groups [285, 286] (Figure 4.6). ISO observations have also discovered the 6.9 μm feature in protoplanetary nebulae due the aliphatic bending mode [287] (Figure 6.8). Furthermore, emission features at 11.3, 12.1, 12.4, and 13.3 μm, which can be identified as arising from out-of-plane vibrational modes of aromatic C–H bonds, have also been detected (Figures 6.8 and 6.10). These features correspond to 1 (solo), 2 (duo), 3 (trio), or

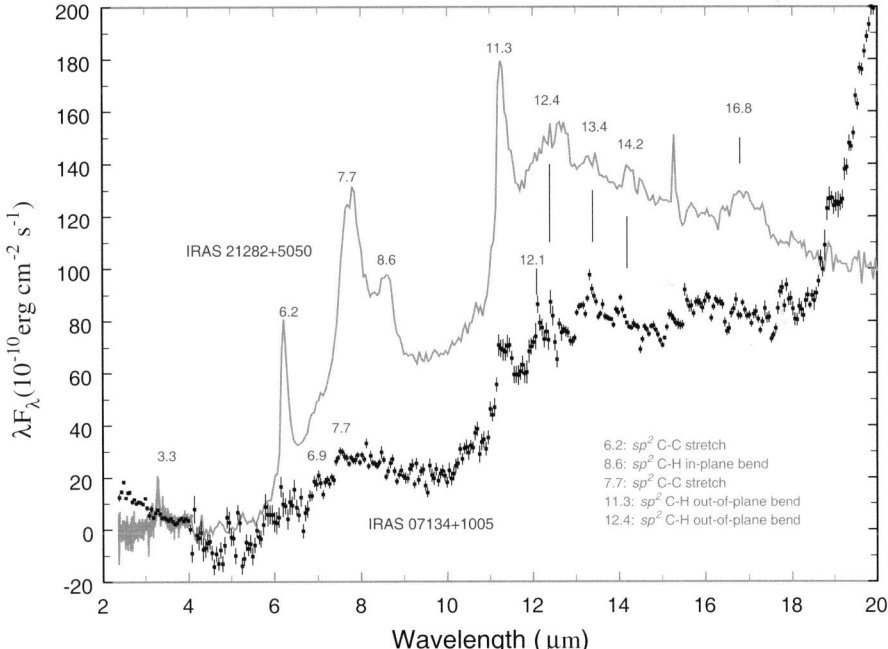

Figure 6.8 ISO SWS01 spectra of the young planetary nebula IRAS 21282+5050 and the protoplanetary nebula IRAS 07134+1005, showing various aromatic C–H and C–C stretching and bending modes at 3.3, 6.2, 7.7, 8.6, and 11.3 μm. Beyond the 11.3 μm feature are the 12.1, 12.4, 13.3 μm out-of-plane bending mode features from small aromatic units.

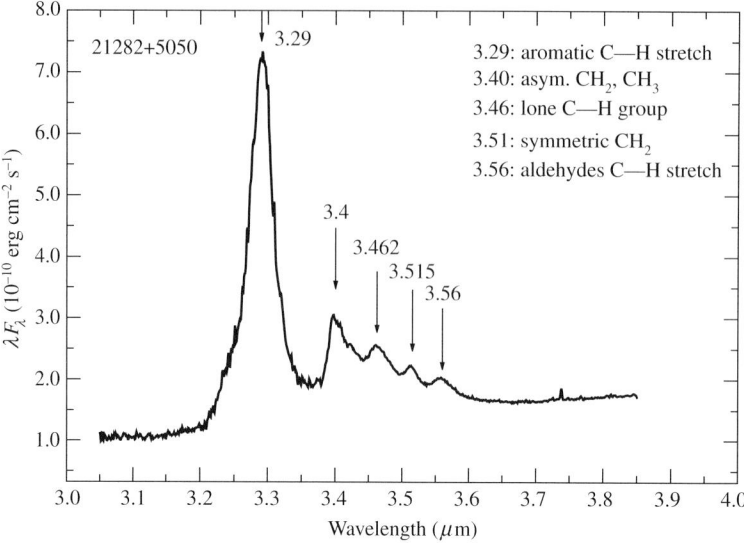

Figure 6.9 KECK spectrum of the young planetary nebula IRAS 21282+5050 showing the 3.56 μm feature possibly attributed to the aldehyde group, in addition to the 3.4 μm aliphatic C–H stretch features. Figure adapted from [286].

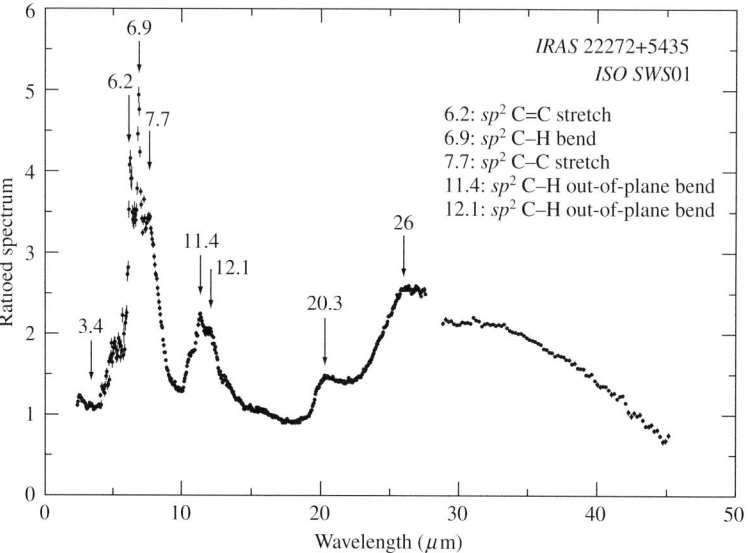

Figure 6.10 The ISO SWS01 spectrum of the protoplanetary nebula IRAS 22272+5435 after the removal of a continuum. The detected narrow emission features and their peak wavelengths are marked on the spectrum. The identification of some of these features are listed in the legend. Figure adapted from [289].

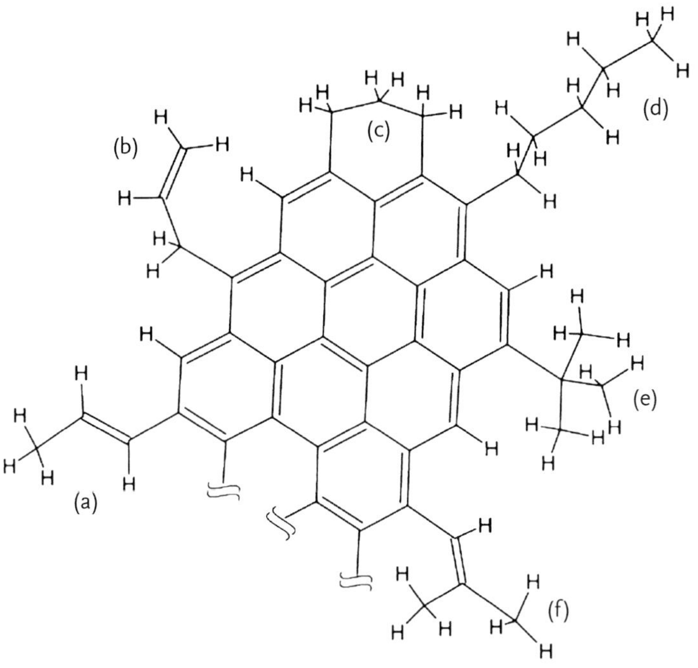

Figure 6.11 (a–f) Schematic chemical diagram illustrating the various possible side groups attached to aromatic rings that contribute to the plateau emissions. Figure adapted from [289].

4 (quarto) adjacent CH groups in each aromatic ring (Figure 3.24), and their out-of-plane bending mode frequencies can be slightly different [288]. Generally, the duo-CH bending mode is in the 11.6–12.5 μm range, the trio-CH mode is in the 12.4–13.3 μm, and the quarto-CH mode in the 13–13.6 μm range. While the 11.3 μm solo-CH mode is dominant in PNe, the other modes are seen in PPNe [287].

In addition to the infrared emission bands attributed to aromatic and aliphatic structures, very broad emission features at 8 and 12 μm are seen in the spectra of protoplanetary nebulae (Figure 6.10). Since the 6.9 μm band is known to originate from a mixture of $-CH_2$ and $-CH_3$ deformation modes [274], associated bending modes of other side groups can also be present. Examples include the $-C(CH_3)_3$ bending modes at 8.16 μm (Figure 6.11, site (e)) and the $=C(CH_3)_2$ (site (f)) bending mode at 8.6 μm. These features together can form a quasicontinuum similar to the observed broad feature.

Similarly, the 11.3 μm aromatic out-of-plane bending mode can be accompanied by a complex set of features due to out-of-plane vibrations of alkenes. Such groups can be connected directly to the aromatic rings (e.g., the $-CH=CHCH_3$ group at site (a) of Figure 6.11), or indirectly ($-CH_2CH=CH_2$; Figure 6.11, site (b)) through an alkyl $(CH_2)_n$ linkage to aromatic rings. Cyclic alkanes ($-CH_2CH_2CH_2-$, site (c)) may also contribute in the short wavelength part (9.5–11.5 μm) of this band and long chains of four or more $-CH_2-$ groups ($-(CH_2)_4CH_3$, site (d)) may contribute to the long wavelength end (13.9 μm).

The existence of the plateau emission features therefore suggests that the structure of these carbonaceous grains is complex and probably includes a variety of alkane and alkene side groups attached to aromatic rings. Comparison between the astronomical plateau features seen in Figure 6.10 and the infrared spectrum of semianthracite coal shows a lot of similarity, suggesting that the carbonaceous grains in protoplanetary nebulae may have a chemical structure similar to coal (Section 9.8).

6.3
Chemical Evolution

The detection of organic compounds in the ejecta of evolved stars gives us important information on how these species are formed. The evolution from AGB to protoplanetary nebulae to planetary nebulae is very short ($\sim 10^3$ yr [280]), and this gives us precise knowledge about the time scale of chemical synthesis. Since the AIB features first emerge in the protoplanetary nebulae phase, what are the steps leading to formation of ring molecules? Acetylene (C_2H_2), believed to be the first building block of benzene, is commonly detected in evolved carbon stars through its ν_5 fundamental band at 13.7 μm (Figure 6.12). Polymerization of C_2H_2 leads

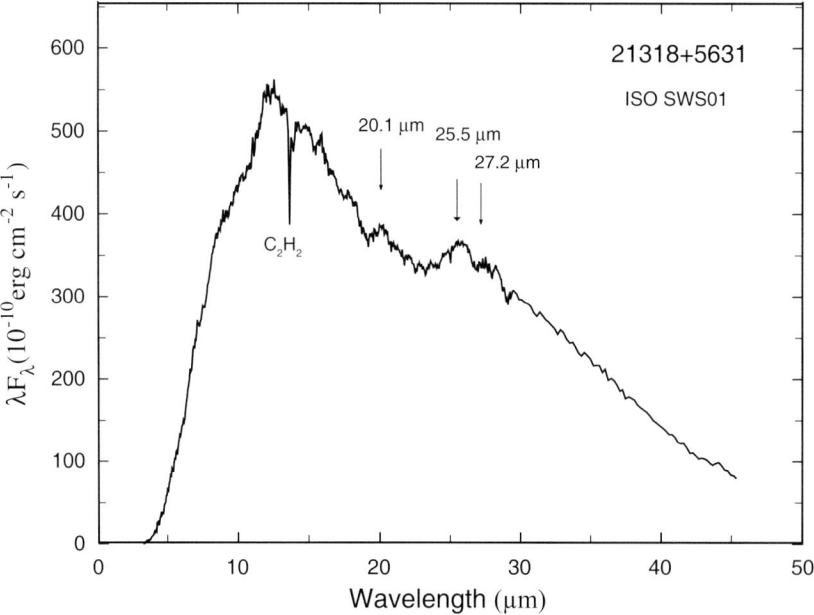

Figure 6.12 ISO SWS01 spectrum of the carbon stars IRAS 21318+5631. The strong absorption feature at 13.7 μm is due to the ν_5 bending mode of acetylene (Section 3.2). The 21 and 30 μm features have also made their appearance in emission.

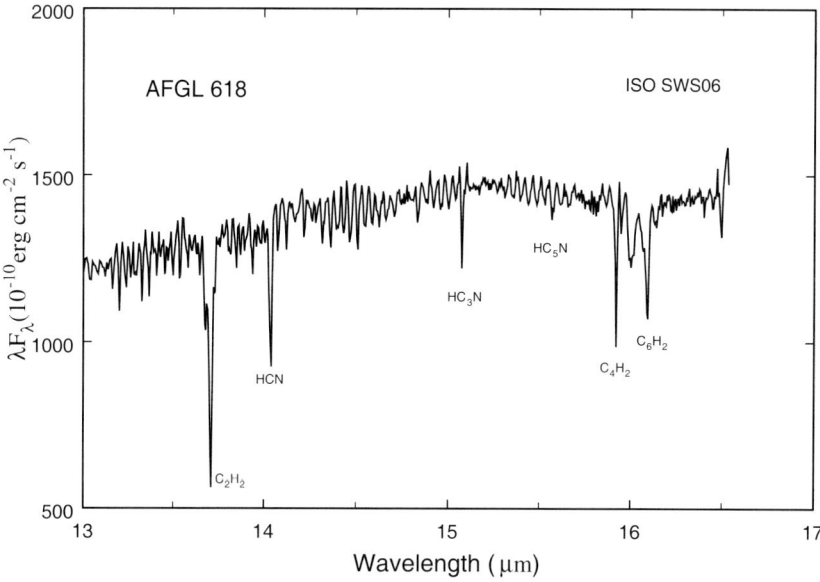

Figure 6.13 ISO SWS06 spectrum of the protoplanetary nebula AFGL 618. In addition to the 13.7 μm acetylene (C$_2$H$_2$) feature, vibrational features of diacetylene (C$_4$H$_2$) and triacetylene (C$_6$H$_2$) can also be seen.

to the formation of diacetylene (C$_4$H$_2$) and triacetylene (C$_6$H$_2$) in protoplanetary nebulae (Figure 6.13), culminating in the formation of benzene [290].

A summary of the changes in the relative strengths of the infrared emission features as stars evolve from AGB to planetary nebulae are given in Table 6.1. The weakening of the 3.4 and 6.9 μm from protoplanetary nebulae to planetary nebulae suggests a change from aliphatic to aromatic structures. This could be the result of photochemistry where the onset of UV radiation modifies the aliphatic side groups through isomerizations, bond migrations, cyclization and ring closures and transform them into ring systems [289]. Hydrogen loss can also result in fully aromatic rings that are more stable than alkanes or alkenes. Evidence for such H loss can also be found in the weakening of the 12.1, 12.4, and 13.3 μm features and the strengthening of the 11.3 μm feature from protoplanetary nebulae to planetary nebulae [287].

There are many advantages to using circumstellar envelopes over interstellar clouds to study the process of chemical synthesis. First, there is only one energy source – the central star, whose temperature and luminosity are well-known. The envelope often has a well-defined symmetry, making the geometry of the system simple. By using molecular lines and infrared continuum radiation as probes, the physical conditions of the envelope such as density, temperature, and radiation background ($\rho(r)$, $T(r)$, $I(r)$, respectively) are well determined. Most importantly, the chemical reaction times are constrained by the dynamical and stellar evolution times, which are 10^4 yr for AGB stars, 10^3 yr for protoplanetary nebulae, and 10^4 yr

Table 6.1 The evolution of emission features from AGB stars to protoplanetary nebulae and then to planetary nebulae.

IR features (μm)	Origin	Carbon stars	Protoplanetary nebulae	Planetary nebulae
Primarily features: 3.3, 6.2, 7.7, 11.3	Aromatic stretch and bending modes	no	yes	strong
secondary features: 3.4, 6.9	C–H aliphatic stretch and bend	no	yes	weak
12.1, 12.4, 13.3	C–H out-of-plane bend with 2, 3, and 4 adjacent H atoms	no	yes	weak
broad 8, 12	Bending modes from aliphatic sidegroups	no	yes	weak
broad 21	–	weak	strong	no
broad 30	–	yes	yes	yes

for planetary nebulae. These time scales give us precise knowledge of the chemical time needed to form one species from another and rigorously constrain the chemical models.

6.4
Enrichment of the Interstellar Medium

It has been known for some time that AGB stars, planetary nebulae, and supernovae are the major sources of heavy elements in the Galaxy. Heavy elements produced by nucleosynthesis are ejected by stellar winds from AGB stars, planetary nebulae, and supernovae into the diffuse interstellar medium, and the abundance of heavy elements increases with time as successive generations of stars enrich the chemical content of the Galaxy. We now realize that these objects not only distribute atomic nuclei, but also molecules and solids. The ability of AGB stars and planetary nebulae to synthesize molecules and solids, both organic and inorganic, has potential significant implications for the chemical enrichment of star formation sites as well as planetary systems in the Galaxy [291].

7
Organic Compounds in the Solar System

Besides the Sun, the Solar System consists of the planets and their satellites, asteroids, comets, and minor bodies in the outer Solar System. The Sun is a gaseous self-luminous body whose radiative energy output is fueled by nuclear fusion in the core. The Sun is an ordinary star, similar to the other one hundred billion stars in the Milky Way Galaxy. The four inner planets (Mercury, Venus, Earth and Mars) are rocky (called terrestrial planets), and the outer four (Jupiter, Saturn, Uranus and Neptune) are gaseous or icy (called Jovian planets). Most of the planets have satellites orbiting around them; the Earth has one (the Moon), and Mars has two (Phobos and Deimos) but Mercury and Venus have none. The Jovian planets have tens of satellites, and some are so large that they could qualify as planets if they were orbiting the Sun. The major satellites of Jupiter are Io, Europa, Ganymede and Callisto. The largest satellite of Saturn is Titan and the largest satellite of Neptune is Triton. Beyond Neptune are tens of thousands of minor bodies, with Pluto being the largest one.

Since the Jovian planets are massive, their gravities are able to retain all the gases in their atmospheres. The terrestrial planets, however, are unable to keep the lighter elements and therefore have lost most of their original H content. So while molecular hydrogen makes up most of the atmosphere of Jovian planets, the dominant gases in the terrestrial planets are CO_2, N_2, and O_2. The satellite Titan also has a substantial atmosphere, being made up of primarily N_2 and CH_4.

The general perception of the chemical composition of the terrestrial planets, comets and asteroids is that they are made up of metals, minerals, and ices. However, in recent years, organics have increasingly been recognized as another major component. Among these bodies, Earth is the most well studied, but organics on Earth are dominated by the results of life and are unique in the Solar System. Asteroids and comets can be physically and chemically studied through meteorites and micrometeorites that fall on Earth. Our newly acquired ability in the last several decades to send space probes to the planets, satellites, and comets has opened a new avenue of close spectroscopic observation and direct sample collection. Consequently, our appreciation and understanding of the organic components of the Solar System have increased dramatically.

Organic Matter in the Universe, First Edition. Sun Kwok.
© 2012 WILEY-VCH Verlag GmbH & Co. KGaA. Published 2012 by WILEY-VCH Verlag GmbH & Co. KGaA.

7.1
Techniques

Most objects in the Solar System, planets and their satellites, comets, asteroids, and so on, are necessarily studied by telescopic observations using the techniques of remote sensing. Spectroscopy is the most sensitive technique for determining the compositions of atmospheres and surfaces, and while direct detection of organic materials by remote sensing is difficult, a number of successful detections and identifications have been made, notably in the atmospheres of the giant planets. Especially in recent years, close-up measurements by infrared spectroscopy and mass spectroscopy have been enabled by spacecraft flying by or orbiting planets and their satellites, comets, and asteroids.

The organic content of Solar System objects can also be studied by the analysis of samples collected in Earth's stratosphere, meteorites, and material returned to Earth from sampling missions to comets and asteroids. Techniques that analyze the chemical contents of Solar System material samples can be divided into two categories: one based on the ability to extract partial contents of the sample (and therefore destroying part of the sample in the process), and the other relying on imaging and spectroscopic analysis. An example of the first technique is pyrolysis gas chromatography, which uses heating of the sample to release smaller molecules so that they can be separated by gas chromatography and identified by mass spectrometry. Since chemical modification can occur during the pyrolytic liberation of molecules, it is not straightforward to reconstruct an accurate structure of a given macromolecule from the molecular constituents. The technique of secondary ion mass spectroscopy (SIMS) uses a beam of ions to sputter several atomic layers of the sample, and subject them to mass analysis, allowing the isotopic ratios to be determined.

The most basic nondestructive technique is imaging. Scanning electron microscope observations reveal the overall structures and morphology of the sample. Examples of the spectroscopic technique include microscope-based Fourier transform infrared (μ-FTIR) spectroscopy which uses the infrared absorption spectrum to identify the functional groups, and solid state nuclear magnetic resonance (NMR) spectroscopy.

The development of synchrotron scanning transmission X-ray microscopes (STXM) provides a focused monochromatic X-ray beam of 30–40 nm which can yield X-ray spectra spanning the C, N, and O 1s X-ray absorption edges. This technique of X-ray absorption near-edge structure (XANES) spectroscopy of the submicron domain can indicate the types of organic functionality present by detecting the characteristic absorption features in the preionization region of the X-ray absorption spectrum. Based on the photoelectric effect, the absorption edge refers to the sharp rise in the cross section of photon absorption when the energy of the incident photon is equal to the binding energy of the inner electron. The excitation of an inner electron to a higher unoccupied electronic state in a molecular band will leave an X-ray signature of an atom in a solid. For example, the absorption of low-energy ($\sim$285.0 eV) X-ray results in the photoexcitation of carbon

Table 7.1 Various C-, N-, and O-XANES transitions and associated functional groups. Table adapted from [292].

Photon energy (eV)	Transition	Functional group[a]	
C-XANES assignments			
283.7	1s–π^*	$C=C^*-C=O$	Quinone
285.2	1s–π^*	$C=C^*-H, C$	Aromatic and olefinic carbon
286.1–286.3	1s–π^*	$C=C^*-C=O$	Aryl, vinyl-keto
286.5	1s–π^*	$C=C-C^*=O$	Vinyl-keto
286.7–286.9	1s–π^*	$C^*\equiv N$	Nitrile
287.2	1s–π^*	$C=C^*-OR$	Enol
287.3–288.1	1s–$3p/\sigma^*$	CH_x-C, H	Aliphatics
287.9–288.2	1s–π^*	$NH_x(C^*=O)C$	Amidyl
288.4–288.7	1s–π^*	$OR(C^*=O)C$	Carboxyl
288.9–289.8	1s–π^*	$NH_x(C^*=O)NH_x$	Urea
289.3–289.5	1s–$3p/\sigma^*$	CH_x-OR	Alcohol, ether
290	1s–π^*	$NH_x(C^*=O)OR$	Carbamoyl
290.3–296	1s–π^*	$RO(C^*=O)OR$	Carbonate
N-XANES assignments			
398.8	1s–π^*	$C=N^*$	Imine
399.8	1s–π^*	$C\equiv N^*$	Nitrile
401.9	1s–π^*	$N^*H_x(C=O)C$	Amidyl
402.1–402.3	1s–$3p/\sigma^*$	$C-N^*H_x$	Amine, pyrrole
402.5–402.6	1s–$3p/\sigma^*$	$C-N^*H_x$	Amino
403	1s–$3p/\sigma^*$	$CO-N^*H_x$	Urea
O-XANES assignments			
531.2	1s–π^*	$C=O^*$	Ketone
532.0	1s–π^*	$O-C=O^*$	Carboxyl
534.4	1s–$3p/\sigma^*$	CH_xO	Alcohol, ether
534.9	1s–$3p/\sigma^*$	$C=C-O^*$	Enol

[a] The asterisk indicates C, N, or O atom that is being excited.

1s electrons to unoccupied π^* orbitals of alkenyl and aromatic species, whereas a carboxyl group requires higher energy to cause the 1s–π^* transition. The methyl (–CH_3) and methylene (–CH_2) groups, which both involve the 1s–$3p/\sigma^*$ transition, can be distinguished as the intensity of the absorption is stronger in the methyl group [292]. Table 7.1 lists some of the functional groups and their associated photon energies that can be studied with the technique of C-, N-, and O-XANES spectroscopy.

7.2
The Sun

Since the Sun has a surface temperature of ~ 5800 K and molecules usually dissociate at high temperatures, one would not expect that molecules would be abundantly present. Indeed, the photospheric spectrum of the Sun is dominated by atomic absorption lines. However, several diatomic molecules have been detected through their electronic transitions in the solar photosphere, and they include MgH (in the $A^2\Pi - X^2\Sigma^+$ band), C_2 ($d^3\Pi_g - a^3\Pi_u$ Swan band and the $A^1\Pi_u - X^1\Sigma_g^+$ Phillips band), CH ($B^2\Sigma^- - X^2\Pi$ and $A^2\Delta - X^2\Pi$ bands), CN ($B^2\Sigma^+ - X^2\Sigma^+$ and $A^2\Pi - X^2\Sigma^+$ bands). In the infrared, vibrational-rotational lines of CO, CH, NH, and OH can be detected.

In sunspots, the temperature can be much lower, and can be as cool as 3300 K in large sunspots. In addition to the molecules in the photosphere, SiO, HF, HCl, and H_2O have also been detected.

7.3
The Earth

Organic matter on Earth is primarily of biological origin. Although the first that comes to mind is the biomass in plants and animals, in fact, the most common form of organic matter is kerogen (Table 7.2), the remnants of living matter (including bacteria) from the past. Under pressure and thermal processing, kerogen is gradually transformed into other more stable forms of carbon, for example, graphite and diamond (Section 2.2). Although kerogen and its by-products, coal and oil, all contain biosignatures in their structures, these signatures are completely lost in the end products of graphite and diamond.

The atmosphere of the Earth is different from those of other planets in the sense that the chemical composition of the Earth's atmosphere has been significantly altered from its primordial composition by biological processes. The high abundance of O_2 and CH_4 in the present atmosphere originates from biological sources (photosynthesis, bacteria, etc.). Nitrous oxide (N_2O) is produced by bacteria and algae. The atmosphere of the Earth therefore contains spectroscopic signatures that signifies that it is not primordial and it is possible that remote spectroscopic observations of the Earth's atmosphere will allow the observer to deduce the presence of life on Earth [294].

Table 7.2 Carbon pools in the major reservoirs on Earth. Table adapted from [293].

Pools	Fractional quantity (Gt)
Atmosphere	720
Oceans	38 400
Total inorganic	37 400
Surface layer	670
Deep layer	36 730
Total organic	1000
Lithosphere	
Sedimentary carbonates	>60 000 000
Kerogens	15 000 000
Terrestrial biosphere (total)	2000
Living biomass	600–1000
Dead biomass	1200
Aquatic biosphere	1–2
Fossil fuels	4130
Coal	3510
Oil	230
Gas	140
Other (peat)	250

7.4
Planets and Planetary Satellites

7.4.1
Planetary Atmospheres

Gas-phase organic molecules are commonly found in the atmospheres of the planets Jupiter, Saturn, Uranus, Neptune as well as satellites such as Titan and Triton. Although the first molecular absorption bands have been recorded in the spectra of Jupiter and Saturn as early as 1905, they were not identified as methane and ammonia until 1932. In 1944, the 619 and 725 nm methane bands were discovered in the spectra of Uranus, Neptune, and Titan, establishing that Titan has an atmosphere [295]. This limited inventory of organic molecules lasted over three decades until the development of infrared spectroscopy in the 1970s. Spectroscopic observations of the planets by ground-based telescopes led to the discovery of CH_3D, CO, H_2O, C_2H_2, C_2H_6, and so on in the spectra of Jupiter [296].

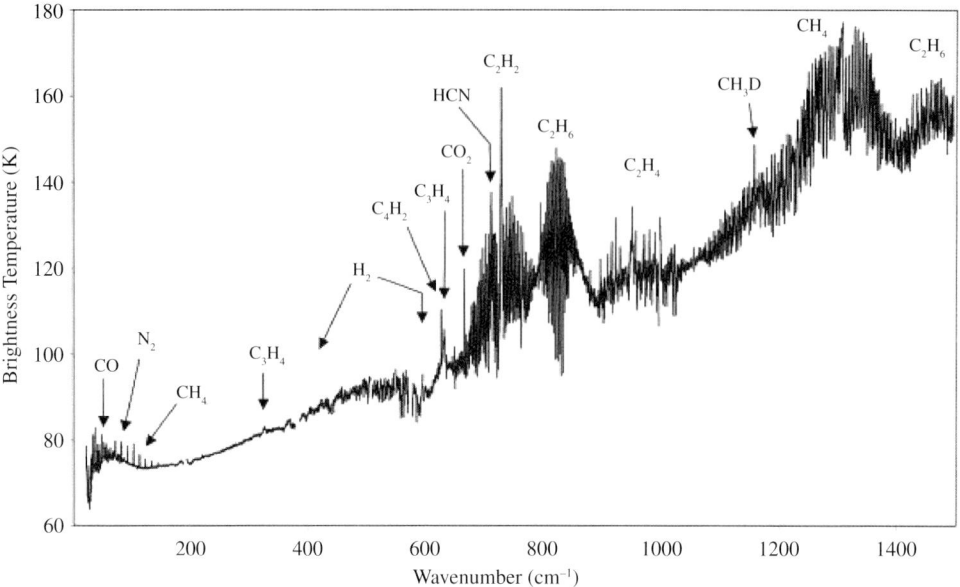

Figure 7.1 A Cassini CIRS spectrum of the atmosphere of Titan. Figure adapted from [299].

The Galileo Orbiter and Galileo Entry Probe have extensively studied the chemical composition of Jupiter. Methane can be formed by equilibrium reactions between H_2 and CO, and photodissociation of methane yields ethane, acetylene, and polyacetylenes ($C_{2n}H_2$). In Jupiter and Saturn, organic molecules can form clouds and haze observed in their atmospheres.

Since its original discovery in 1655 by Christiaan Huygens, Saturn's large moon Titan has attracted a lot of attention from planetary scientists. In recent years, Titan has been recognized as to being rich in organics, both on the surface and in the atmosphere. Infrared observations of the atmosphere of Titan began with the Voyager missions in 1980. Observations by the Infrared Radiometer Interferometer and Spectrometer (IRIS) on the Voyager spacecraft and ISO have revealed a large variety of organic molecules in Titan, including C_2H_6, C_3H_8, C_2H_2, C_2H_4, CH_3C_2H, C_4H_2, HCN, HC_3N, CH_3CN, and so on [297, 298]. Ground-based observations offer higher spatial resolution and are able to map the distribution of molecules such as HCN, CO, and HC_3N. On July 2, 2004, the Cassini spacecraft arrived at the Saturnian system after a 6.7 year journey. The close approach to Titan allowed much higher spatial observations to be taken. Figure 7.1 shows the Cassini Composite Infrared Spectrometer (CIRS) spectrum of Titan at mid-latitudes. A variety of hydrocarbons (C_2H_2, C_2H_6, C_3H_8, C_2H_4, C_4H_2, C_3H_4) as well as benzene (C_6H_6) were detected and mapped.

The Ion Neutral Mass Spectrometer (INMS) on board Cassini observed the jet-like plumes on Enceladus and found that they are composed of N_2, CO_2, and methane [300]. This raises the possibility of a subsurface liquid reservoir made up of water and hydrocarbons.

7.4.2
Ices

Because of their distances from the Sun, surface temperatures of outer Solar System bodies are low and gas-phase molecules often condense into their respective ice forms. In addition to water ice, frozen hydrocarbons, for example, CH_4, CH_3OH, C_2H_6, are also believed to be present on the icy surfaces of many planetary satellites, including Triton, Ganymede, and Callisto. These organic ices have been identified by spectroscopy.

In addition to the crystalline form of ice that we are familiar with in everyday life, ice can also be in an amorphous form. Amorphous ice has varying O–H bond angles and has a structure similar to that of amorphous silicates. Water ice can be detected through the absorption bands at 1.5 and 2 μm as well as the OH stretching mode at 3.0 μm and the bending mode at 6.0 μm. Ice has been detected on satellites of Jupiter, Saturn, and Uranus as well as on Triton (satellite of Neptune), and Charon (satellite of Pluto) through the technique of infrared spectroscopy [301, 302]. CO ice was first detected on Triton and Pluto in the 2.35 μm overtone band, and CO_2 ice was detected on Triton, Ariel, Umbriel, and Titania. Other molecular ice forms include N_2, HCN, methane, acetylene, ethylene, ethane, and O_3. Also possibly present in small quantities are organic species like nitriles, ketones, and esters.

7.4.3
Organic Solids

Spectroscopic observations with the Visible-Infrared Mapping Spectrometer (VIMS) on board the Cassini spacecraft of Saturnian satellites Iapetus, Phoebe, and Hyperion have also found evidence for aromatic and aliphatic hydrocarbons [303]. The hemisphere centered on the orbital motion apex of Iapetus has a very low (2–6%) albedo and red color, which may indicate the presence of complex organics [304].

The atmosphere of Titan contains solid particles similar to haze which are the result of condensation of organics. A layer of haze is believed to exist in the lower stratosphere. Cassini–Huygens observations of Titan support the scenario that methane is being converted to complex hydrocarbon-nitrile compounds in the atmosphere of Titan [300]. The exact chemical composition of the haze particles is not known, although it has been suggested to be a complex organics similar to tholins (Section 9.10) [305]. These complex solids are formed from exposure of N_2 and CH_4 gaseous mixtures to electrical discharges under cold plasma conditions. The condensation of these nanoparticles (or macromolecules) on surface grains that are blown into dunes by wind, and in liquid lakes found by Cassini RADAR observations. The in-situ measurements provided by the Huygens probe during its descent onto the surface of Titan provide valuable information about the chemical composition and isotopic abundance of the Titan atmosphere from 140 km all the way to the surface. The probe found that nitrogen and methane are the main constituents of the Titan atmosphere and the polar regions of the satellite show numerous lakes of liquid methane and ethane, probably with many other organics dissolved in them. For example, liquid ethane was identified in Titan's northern lake Ontario Lacus by Cassini's VIMS instrument [306]. The total inventory of methane in Titan's atmosphere contains 360 000 Gt of carbon. The amount of carbon in ethane and methane in the several hundred lakes or seas is estimated to be 16 000–160 000 Gt. Sand dunes cover ~20% of the Titan surface and contain 160 000–640 000 Gt of carbon [307]. The total amount of hydrocarbons on Titan is therefore larger than the oil and gas reserves on Earth (Table 7.2). The fact that there is more sand than liquid suggests that organics are primarily in solid form on Titan.

7.5
Meteorites

Meteorites are small remnants of asteroids that survive reentry through the Earth's atmosphere. About 10 tons of meteorites currently hit the Earth each year, although the rate was much higher in the past (Section 11.6). Meteorites are usually made of iron, nickel, and silicates. Some meteorites (called chondrites) contain round, millimeter-size inclusions (called chondrules) which are original molten material crystallized during the formation of the meteorites. Chondrules are believed to be the first solids to have condensed out of the interstellar gas that formed the solar nebula. Of special interest are a small fraction of chondrites that contain a high content of carbon; they are classified as carbonaceous chondrites. They are considered to be the most primitive meteorites. The most famous among this class are the Murchison, Murray, and Orgueil meteorites.

Prebiotic organic matter have been known to be present in primitive carbonaceous chondrite meteorites since the 1970s [308, 309]. Through the use of different solvents, various components of the soluble fraction can be extracted and analyzed. Among the extractable organic matter are carboxylic acids [310], amino acids [311], aromatic hydrocarbons [312], heterocyclic compounds (adenine, guanine, hypoxanthine, xanthine, uracil, thymine, cytosine, etc.) [313], aliphatic hydrocarbons [314], amines, amides [315], alcohols, aldehydes, ketones and other sugars [64]. From the mass spectra of Murchison extracts, sugars, sugar alcohols, sugar acids, and dicarboxylic sugar acids, have been found (Figure 7.2). The general decrease in abundance with increasing carbon number within the same class of compounds suggests that they are of abiotic origin. Recent high resolution molecular analysis of the soluble component of the Murchison meteorite has identified 14 000 molecular compounds and likely millions of diverse structures [316]. This observed molecular diversity suggests interstellar chemistry is extremely active and rich.

It is fair to say that almost all biologically relevant organic compounds are present in carbonaceous meteorites, although our assumption is that these are all of nonbiological origin. In terms of abundance, aromatic hydrocarbons make up the largest fraction of the soluble organic compounds in meteorites, followed by carboxylic acids [317]. We should also note that over 90 individual amino acids have been

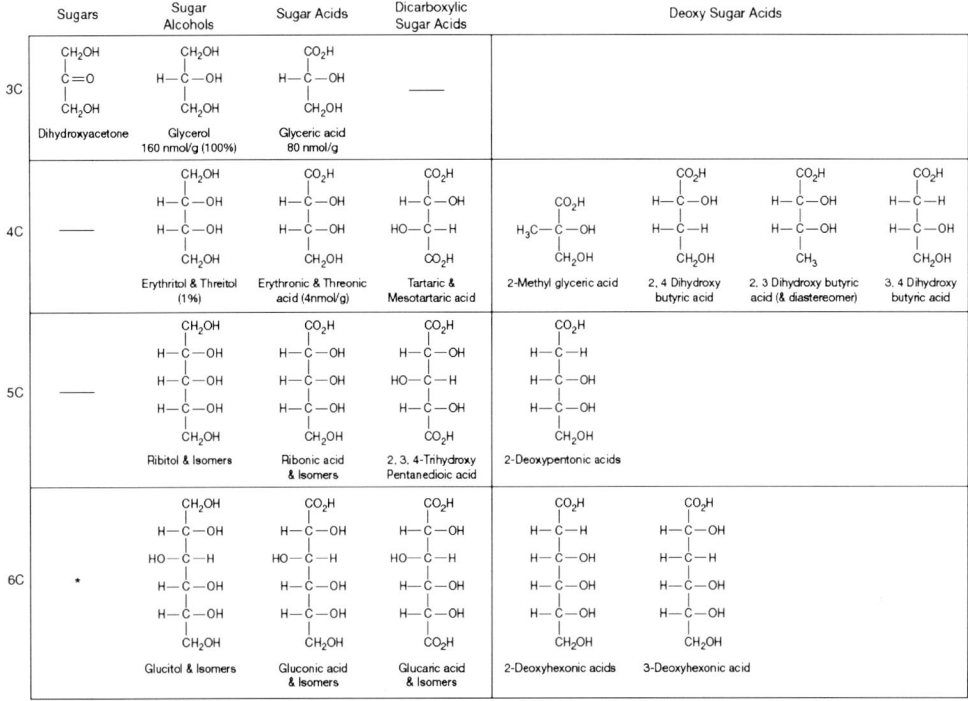

Figure 7.2 Examples of sugar-related organic compounds in the Murchison meteorite. Figure adapted from [64].

identified in meteorites. This number is far higher than the 20 amino acids found in life forms on Earth.

The remaining (and the majority, >70%) of the organic matter is in the form of insoluble macromolecular solids, which are now commonly referred to as Insoluble Organic Matter (IOM) [309]. Modern pyrolysis gas chromatography characterizes IOM as predominantly aromatic, with solid-state nuclear magnetic resonance spectroscopy finding various functional groups (Figure 7.3). Release of oxygen-containing molecules such as phenols, propanone, and nitrogen heterocyclics as a result of pyrolysis suggests that the IOM contains impurities. The structure of IOM can therefore be summarized as a complex organic solid composed of aromatic and aliphatic functional groups, as well as oxygen-containing functional groups.

Comparison of the IOM in the Murchison, Tagish Lake, Orgueil, and EET 92042 meteorites shows remarkable similarity [319]. Roughly speaking, the chemical structure of IOM resembles that of kerogen on Earth [320] (see Section 9.8). While the aromatic fraction increases from EET 92042, to Orgueil, to Murchison, to Tagish Lake, there is a corresponding decrease in aliphatic content [319].

The similarity between the 3.4 μm aliphatic feature observed in meteorites and the absorption feature seen in the galactic center has led to the suggestion that the

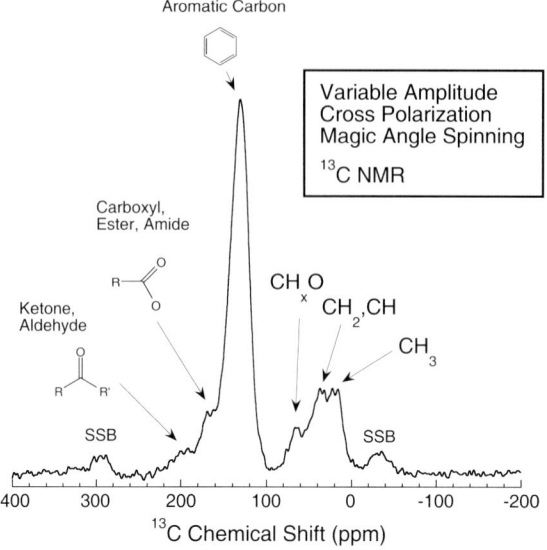

Figure 7.3 Functional groups identified in the Murchison IOM. The two features marked as SSB are spinning side bands of the main peak in this ^{13}C NMR spectrum. Figure adapted from [318].

diffuse ISM contains organic material similar to the IOM [321, 322]. The anomalous isotopic ratios of noble gases trapped in fullerenes (C_{60} to C_{400}) in the Allende and Murchison meteorites point to their extraterrestrial origin [323]. The carbon isotope ratios of nucleobases in the Murchison meteorite also suggests that they are interstellar [324]. The elevated ratios of ^{15}N/^{14}N and D/H in organic globules in the Tagish Lake meteorite provides additional evidence that interstellar organics are present in the Solar System [325]. Although early theories on the origin of meteoritic organics concentrated on synthesis in the solar nebula, the latest evidence suggests that the solar nebula is only involved in secondary processing of preexisting interstellar materials.

Organic carbon in chondrites may have been affected by parent body accretion and subsequent alteration, therefore affecting the determination of the extent of carbon of the most primitive origins.

7.6
Meteoroids and Interplanetary Dust Particles

Interplanetary space is filled with solid particles. Particles a centimeter or above are called meteoroids. Smaller (<0.1 cm) particles are called micrometeoroids. When meteoroids enter the Earth's atmosphere, the heat generated from friction can vaporize the meteoroid and the light it generates is seen as a meteor. If the object is large enough (>10 cm), some remnant will survive the entry and end up on Earth as meteorites.

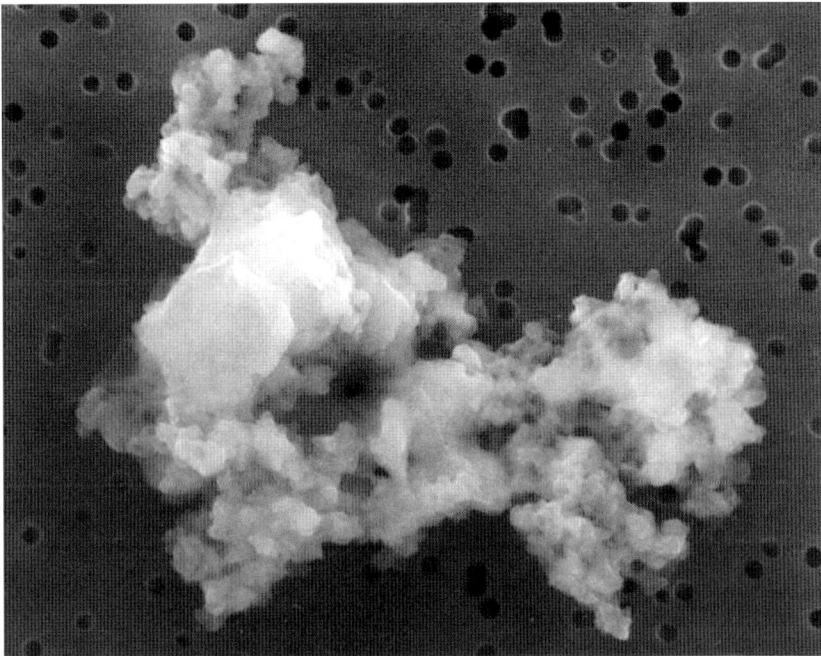

Figure 7.4 A scanning electron microscope image of the IDP U2012C11 collected by NASA from the Earth's stratosphere. The size of the particle is about 10 μm. Credit: NASA.

Interplanetary dust particles (IDPs) are micron-size particles from asteroids or comets that enter the Earth's atmosphere. From the impact scars on the Long Duration Exposure Facility Satellite, it is estimated that 20 000 to 60 000 tons of IDPs in the mass range of 10^{-9} to 10^{-4} g reach the Earth every year [326]. IDPs have also been actively collected in the stratosphere by NASA's U2 planes (Figure 7.4). IDPs are believed to originate from tails of comets and the collisions of asteroids with each other. Because asteroids's orbital plane is similar to that of the Earth, they have smaller relative velocities and asteroid IDPs therefore enter the Earth with lower velocities. Comets, however, have orbits that have high inclination angles with respect to the Earth's orbital plane and cometary IDPs enter with higher velocities. Basic constitutions of IDPs are silicates, oxides, and carbonaceous materials, including amorphous carbon, nanodiamonds, graphite, and macromolecular organics [327, 328].

In spite of their small size (<50 μm) and mass ($\sim 10^{-9}$ g), IDPs provide valuable samples of the primordial Solar System material for study in the laboratory. Laboratory analysis of IDPs collected from the stratosphere by high-flying aircraft with aerogel collectors has shown that they are rich in organic content. X-ray absorption near-edge structure spectroscopy and infrared spectroscopy have found aliphatic CH_2 and CH_3, carbonyl (C=O), and carbon ring structures in IDPs [329]. Figure 7.5 shows a comparison of the infrared spectra of IDPs with the Murchison meteorite. While both show the 3.4 μm aliphatic features, the Murchison has

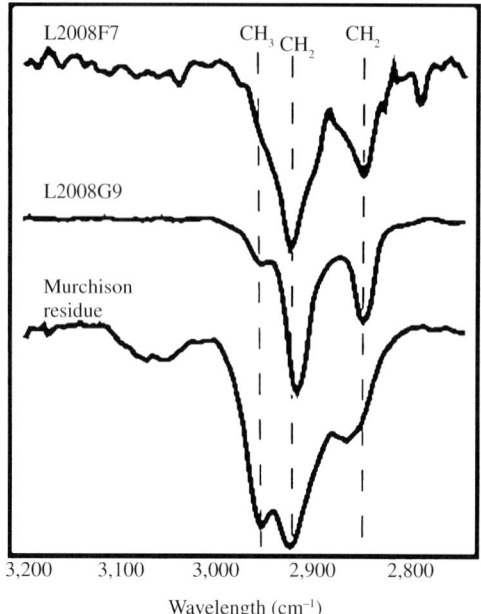

Figure 7.5 The infrared absorption spectra of IDP L2008F7 and L2008G9 and the Murchison meteorite showing the aliphatic stretching modes of CH_2 and CH_3. Note the similarity with the planetary nebula spectrum in Figure 6.9. Figure adapted from [329].

a much stronger 3.28 μm aromatic feature, suggesting that the Murchison is more aromatic than IDPs.

Using known X-ray cross sections, the elemental abundance can be inferred from X-ray spectra. For the IDP L2011*B2, the C : N : O ratio is found to be 100 : 10 : 50, which has higher O/C and N/C ratios than in the IOM extracted from the Murchison meteorite, suggesting that this IDP has been subjected to less processing than Murchison [330]. Accurate measurements of the elemental abundance are useful in determining the chemical makeup of the organic content.

Isotopic measurements of IDPs have shown evidence for deuterium enrichment, with D/H values as high as 50 times the observed Solar System values [331]. These results suggest that these IDPs contain remnants of interstellar materials that survived the formation of the Solar System.

7.7
Comets

Because comets are formed in the coldest regions of the Solar Nebula, they have not undergone significant thermal and chemical processing until they begin to venture into the inner Solar System. For this reason, they preserve a record of the chemical composition of the primitive Solar Nebula. Studies of the chemical contents of

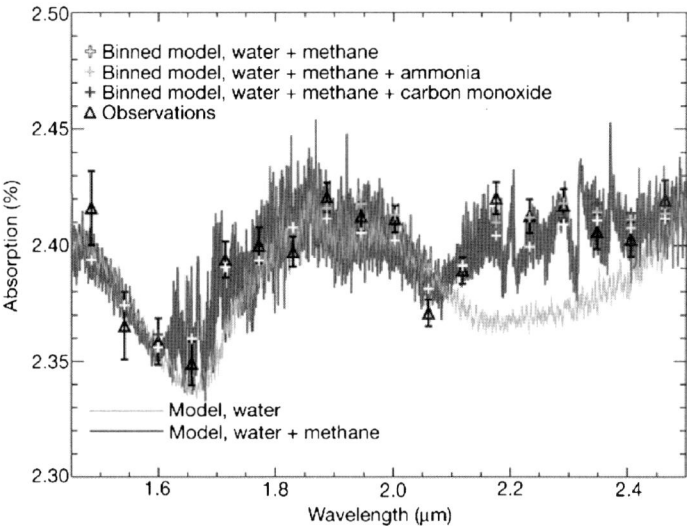

Figure 7.9 The observed spectrum of the transiting extrasolar planet HD 189733b (triangles) compared to the simulated spectra of $H_2 + H_2O$ (lower curve) and $H_2 + H_2O + CH_4$ (upper curve). The derived abundance is 5×10^{-4} for water and 5×10^{-5} for methane. Figure adapted from [355].

face areas of the planet and the star. If a planet possesses an atmosphere, then the "size" of the planet will be dependent on wavelength. By precisely measuring the amount of the dip in stellar flux during transit as a function of wavelength, one can obtain the spectrum of the planetary atmosphere. Another technique is to compare the spectrum of the star+planet system taken just before and after the planet disappears behind the star. The difference spectrum represents the spectrum of the planet. Molecules detected include H_2O, CO, CO_2, and CH_4. An example of the spectrum of extrasolar planets obtained with the first technique is shown in Figure 7.9. Two simulated planetary atmosphere spectra are compared to the observed data, showing that both water and methane are present in addition to hydrogen molecule. A small amount of ammonia could also be present.

7.11
Summary

From remote spectroscopic observations and laboratory analysis of actual collected samples, we know that the Solar System is rich in organics. Spectroscopic observations from ground-based telescopes, Earth-orbiting space telescopes, and instruments mounted on spacecrafts on fly-by missions have identified a large variety of organics in comets, planets, planetary satellites, and TNOs. Meteorites and IDPs are fragments of comets and asteroids that provide the physical samples we can analyze. The complex refractory organic solids in meteorites and IDPs give us a strong hint that these materials make up part of cometary nuclei and asteroids, a conclusion that is supported by comet fly-by missions and the red color of asteroids. These red colors are also common properties of outer Solar System objects, leading to the suggestion that they are likely to also be rich in complex organics [303].

Although the gas-phase hydrocarbons and other organic molecules are likely to be products of ongoing chemical processes in these bodies, the refractory complex organic solids are less obviously so. The most interesting question is how many of these organics were inherited from the parent molecular cloud from which the Solar System was condensed, and how many of them were subsequently synthesized after the formation of the Solar System. The IOMs in meteorites have mixed sp^2/sp^3 structures that are very similar to the carbonaceous solids produced by planetary nebulae and protoplanetary nebulae, so there may be a chemical structural link. However, if the planetary satellites organics are really tholin-like materials that contain a lot of N, then the link to stars is less strong as the element N is not abundantly produced in carbon stars. The key to the test of this link is of course in isotopic analysis: whether the D, C, N, isotopic ratios in Solar System organic solids are consistent with those expected in carbon stars.

8
Organic Compounds as Carriers
of Unsolved Astronomical Phenomena

Our ability to spectroscopically observe and identify ions, atoms, molecules, and solids from the Solar System to distant galaxies clearly demonstrates the universality of the laws of physics and chemistry throughout the Universe. The identification of forbidden lines of atomic ions (e.g., 500.7 nm line of O^{2+}) and rotational transitions of molecular radicals (e.g., HCO^+) not seen in the terrestrial environment further highlights the success of modern astrophysics and astrochemistry. While our understanding of quantum physics and chemistry led to the identification of unexplained spectroscopic features, the studies of these extraterrestrial atomic and molecular species have in turn enhanced our knowledge of atomic and molecular physics.

However, there are still a number of unsolved mysteries relating to spectroscopic features commonly seen around stars, in the interstellar medium, and in external galaxies. There are strong suspicions that some of these phenomena involve organic carriers whose spectroscopic properties are not known; or that they represent organic materials that do not exist or have not yet been artificially synthesized on Earth. Below are some mysterious astronomical phenomena that require further advances in chemistry to be explained.

8.1
Unidentified Infrared Emission Features

The term unidentified infrared emission (UIE) features refers to a family of emission features including the AIB at 3.3, 6.2, 7.7, 8.6, and 11.3 μm, aliphatic features at 3.4 and 6.9 μm, and broad emission plateaus at 8, 12, and 17 μm, as well as a host of weaker features that are too broad to be atomic or molecular lines. After the initial discovery of the UIE features, it was realized that they sit on top of a strong emission continuum which cannot be explained by free–free emission, reflected starlight, or thermal dust emission heated by stellar photons [243, 356]. For example, scattered starlight is expected to be strongly polarized, but the observed continuum emission is unpolarized. *The UIE phenomenon is therefore more than a collection of emission features, but an integrated phenomenon of emission bands, underlying continua, and broad emission plateaus.* These features were first classified by

Organic Matter in the Universe, First Edition. Sun Kwok.
© 2012 WILEY-VCH Verlag GmbH & Co. KGaA. Published 2012 by WILEY-VCH Verlag GmbH & Co. KGaA.

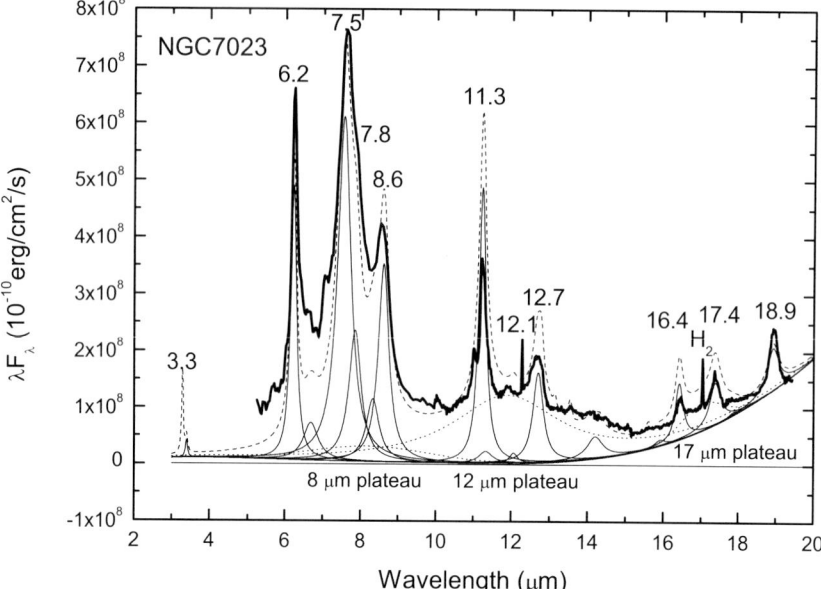

Figure 8.1 Spitzer IRS spectrum of the reflection nebula NGC 7023. Data in the 3–4 μm region is from ISOPHO. The underlying continuum is fitted by a modified blackbody ($\lambda^{-2}B_\lambda$ (60 K)). The marked features are UIE features and their respective contributions to the total spectrum are shown as spectral components under the total spectrum. At the base of these features are three broad plateau features (plotted as dotted lines) centered at 8, 12 and 17 μm. Because of the coolness of the central star ($T \sim 17\,000$ K), there are no atomic emission lines in this spectrum. The only line present is the $v = 0-0\ J = 3-1$ 17.03 μm rotational line of molecular hydrogen. The features at 17.4 and 18.9 μm have been suggested to originate from the C_{60} molecule [206]. IRS data courtesy of K. Sellgren.

Alan Tokunaga [357] and Tom Geballe [358], and the classification scheme was later refined by Els Peeters [359]. According to the notation of Peeters, sources of type A show a pronounced feature at 6.19–6.23 μm together with bands at ~7.6/7.8 μm and ~8.6 μm. Type B sources have bands at 6.24–6.28, 7.6/7.9 and 8.7 μm. A few type C objects (e.g., IRS 13416) exhibit features at ~6.3 and 8.2 μm, but do not have a band near 7.7 μm.

The UIE features are seen in very different radiation environments. The energy source responsible for the excitation of the features range from tens of thousands of degrees in planetary nebulae (the central star temperature of NGC 7027 is 200 000 K), to ~30 000 K in Hɪɪ regions, and to only thousands of degrees in reflection nebulae and protoplanetary nebulae. The UIE features seen in the reflection nebula NGC 7023 (Figure 8.1) are very similar to those seen in the planetary nebula NGC 7027 (Figure 1.6) in spite of the very different intensities of UV background in the two nebulae.

Some of the UIE features can be positively identified as vibrational modes of aromatic compounds. These bands can be designated as aromatic infrared bands

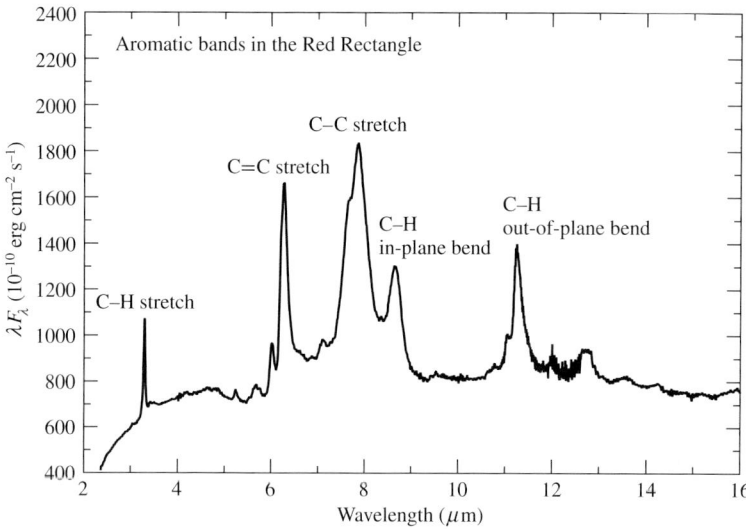

Figure 8.2 The aromatic infrared bands (AIB) are prominent in HD 44179 (the Red Rectangle), a reflection nebula surrounding a B8-A0 central star in the late stages of stellar evolution (Figure 1.5). The vibrational modes of the aromatic units are identified in this Infrared Space Observatory (ISO) SWS01 and SWS06 spectrum.

(AIB) and some examples of the AIBs are shown in Figure 8.2. In the 11–15 µm region, the C–H out-of-plane bending modes of aromatic units with a different number of exposed corners can give rise to features of different wavelengths. For example, the 11.3, 12.1, 12.4, and 13.3 µm features can be assigned to a lone (solo) C–H group, or two (duo), three (trio), or four (quarto) adjacent C–H groups, respectively [288]. From ISO observations of HR 4049, IRAS 21282+5050 and NGC 7027, a broad feature centered around 17 µm was identified [360]. Features at 15.8 and 16.4 µm have been found in the ISO spectra of galaxies [269]. Other features at 17.4, 17.8, and 18.9 µm are also seen in protoplanetary nebulae [287] and reflection nebulae [361]. From a model fit to the observed Spitzer IRS spectra of galaxies (Figure 8.3), a list of the detected UIE features from 5.27 to 33.10 µm is given in Table 8.1. Also included are the central wavelengths and profiles of the features derived from the spectral fit.

In addition to the bands, broad emission features up to several microns wide are seen together with the features. The 8 and 12 µm plateau features (Section 6.2) and a broad feature covering the 15–20 µm have been detected in young stellar objects, compact H_{II} regions, and planetary nebulae [363]. This feature is particularly strong in protoplanetary nebulae [364] (Figure 8.4) and has been suggested to arise from C–C–C in-plane bending of aromatic rings [363].

In addition to hydrogen, methyl ($-CH_3$) or methylene ($-CH_2$) side groups, it is also possible that other side groups such as carbonyl (C=O), aldehydic ($-HCO$), phenolic (OH), and amine (NH_2) can be attached to the aromatic units. These could give rise to the other UIE features and the broad plateau features [289].

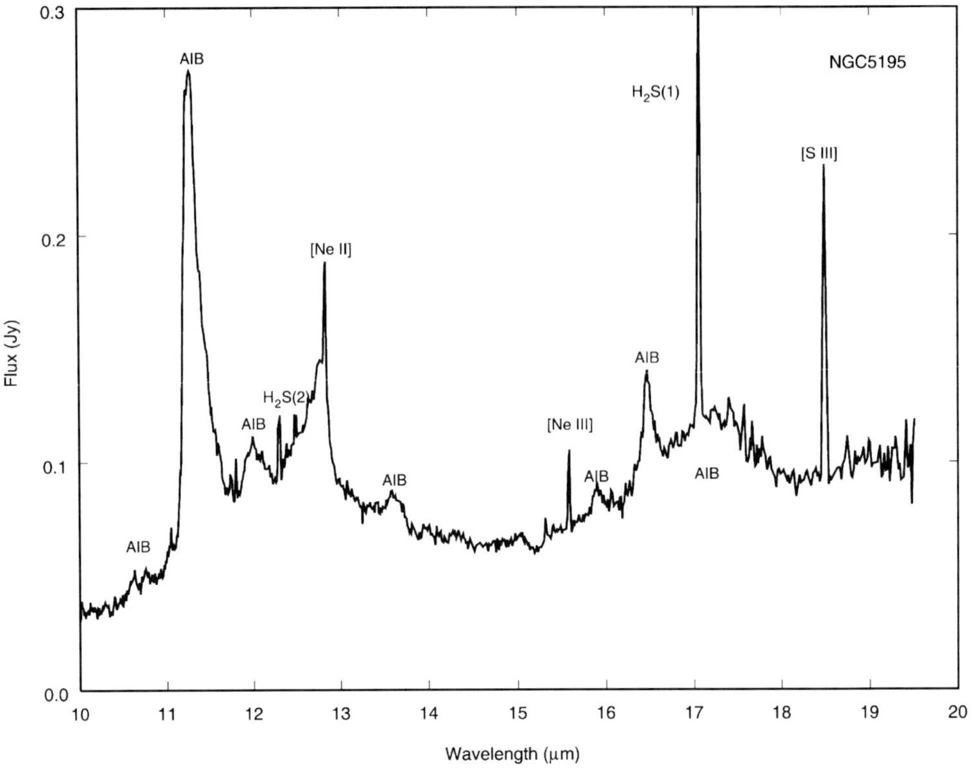

Figure 8.3 The UIE features are common-ly observed in star-forming galaxies as can be seen in this Spitzer IRS spectrum of NGC 5195. The amount of energy emitted in the bands can be as high as 20% of total infrared fluxes observed, suggesting that a signifi-cant amount of organic matter is produced in these galaxies. The narrow features are due to atomic or molecular (H_2) lines. Figure adapt-ed from [362].

The peak and profiles of the UIE features are known to vary from source to source. A correlation between the peak wavelengths of the 7.7 and 11.3 μm fea-tures with the temperatures of the exciting stars is observed (Figure 8.5), which has been interpreted to be the result of change in the ratio of aliphatic to aromatic content [365].

8.2
Diffuse Interstellar Bands

The diffuse interstellar bands (DIB) are absorption bands formed in interstellar clouds seen against the spectra of stars. Since their initial discovery in 1922 [366] and the identification of their interstellar origin in the 1930s [367], approximately 400 bands from near UV to near IR have been detected. Among the strongest DIBs are those at wavelengths of 443.0, 578.0, 579.7 and 628.4 nm. In the Milky Way

Table 8.1 UIE features seen in galaxies. Table adapted from [362].

Central wavelength λ_0 (μm)	Fractional FWHM γ [e]	Full-width at half maximum FWHM (μm)
5.27	0.034	0.179
5.70	0.035	0.200
6.22	0.030	0.187
6.69	0.070	0.468
7.42 [a]	0.126	0.935
7.60 [a]	0.044	0.334
7.85 [a]	0.053	0.416
8.33	0.050	0.417
8.61	0.039	0.336
10.68	0.020	0.214
11.23 [b]	0.012	0.135
11.33 [b]	0.032	0.363
11.99	0.045	0.540
12.62 [c]	0.042	0.530
12.69 [c]	0.013	0.165
13.48	0.040	0.539
14.04	0.016	0.225
14.19	0.025	0.355
15.90	0.020	0.318
16.45 [d]	0.014	0.230
17.04 [d]	0.065	1.108
17.375 [d]	0.012	0.209
17.87 [d]	0.016	0.286
18.92	0.019	0.359
33.10	0.050	1.655

[a] 7.7 μm complex
[b] 11.3 μm complex
[c] 12.7 μm complex
[d] 17 μm complex
[e] The features are assumed to have the profile $I_\lambda \sim \gamma^2/((\lambda/\lambda_0 - \lambda_0/\lambda)^2 + \gamma^2)$.

Galaxy, DIBs have been seen along the line of sight of over one hundred stars. In the Magellanic Clouds, DIBs have been seen in the spectrum of SN 1987A as well as in the spectra of reddened stars [368]. The interstellar medium of external galaxies can be probed by supernovae [369], and DIBs have been detected in external galaxies with redshifts up to 0.5 [370].

A survey of 160 stars with spectral resolution $\lambda/\Delta\lambda \sim 37\,500$ covering the spectral range of 360–1020 nm has been carried out at Apache Point Observatory in New Mexico in order to obtain a uniform set of high-quality DIB data. It

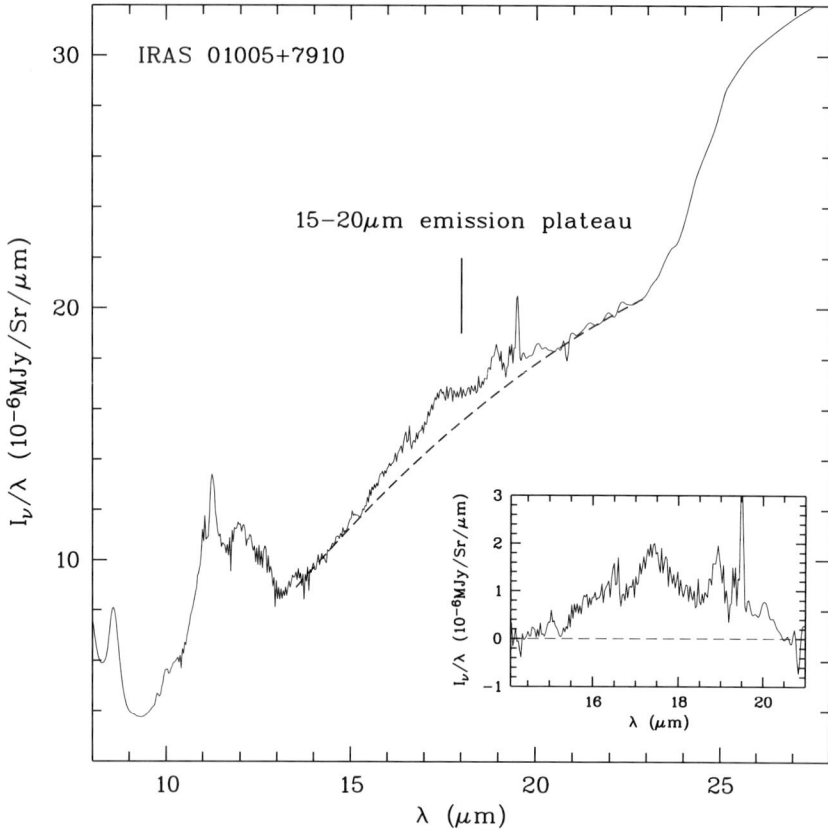

Figure 8.4 Spitzer IRS spectrum of the protoplanetary nebula IRAS 01005+7910. The 15–20 μm feature can be seen above the continuum (shown as a dashed line). The profile of the feature is shown in the continuum-subtracted spectrum in the insert.

has been found that the 619.6 nm and the 661.3 nm lines are strongly correlated with each other, suggesting that the two lines are due to two closely related molecules.

It is interesting to note that in spite of their widespread presence, none of the DIBs has been positively identified. Although the bands show a variety of widths, strengths, and profiles, they share the common property that they are all too broad (FWHM $\sim$ 0.06–4 nm) to be attributed to atomic lines. Consequently, molecular and solid-state origins have been extensively studied. The invariance of the wavelengths and the narrowness of some bands suggest that the carriers are gas-phase molecules, and the bands arise from electronic transitions of these molecules. An example of the gas-phase molecule origin of DIB is the recent proposal that the 488.1 and 545.0 nm DIBs are due to the $B^1B_1 - X^1A_1$ transition of propadienylidene (l-C_3H_2, Figure 3.18) [371].

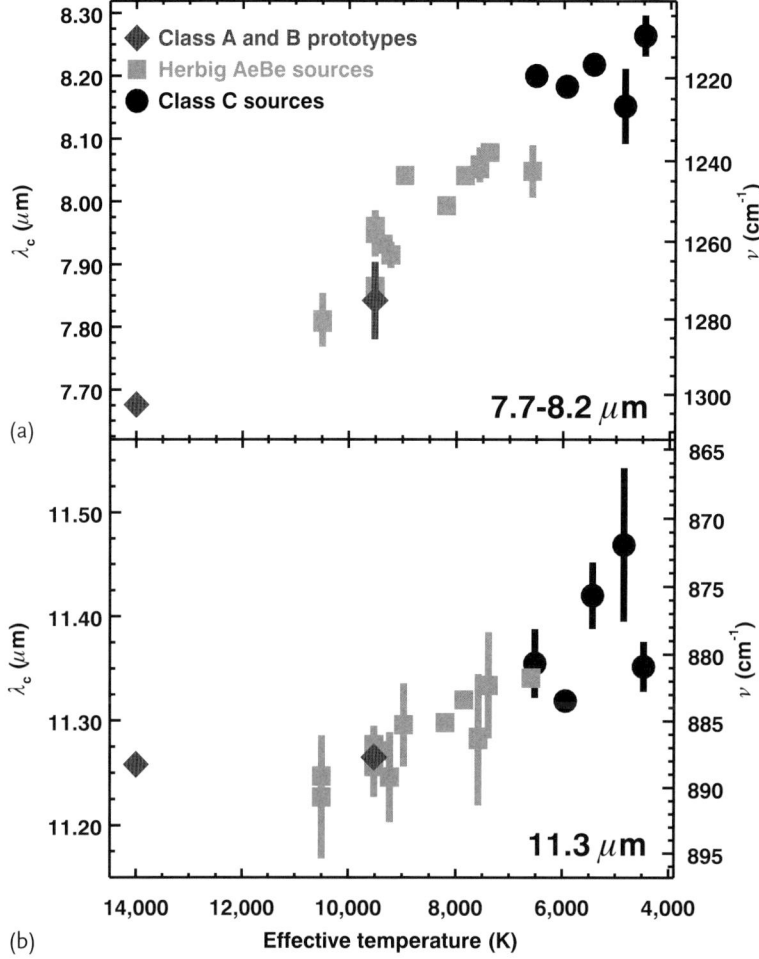

Figure 8.5 The correlation between the peak wavelengths of the 7.7 (a) and 11.3 μm (b) features with the temperatures of a group of red giants, post-AGB stars, T Tauri stars, Herbig Ae/Be stars, and a reflection nebula. Figure adapted from [365].

No matter what the carrier of the DIBs are, they must be made of abundant elements because the total amount of interstellar absorption due to all known DIBs is very large. Organic compounds are favored because only carbon atoms have the rich chemistry necessary to create the large variety of molecules to account for the DIBs [372]. Candidates that have been suggested include carbon chains, PAHs, fullerenes or nanotubes, including their variations in ionization and hydrogenation states. The carriers have not yet been identified because the carrier molecules are probably unstable under laboratory conditions. Although no solution exists at this time, the ubiquitous nature of DIBs seems to imply that complex organic compounds are widespread in the ISM.

8.3
The 217 nm Feature

The prominent feature in the interstellar extinction curve at $\lambda \sim 217.5$ nm is widely observed in the Galaxy (Figure 8.6), suggesting that its carrier is a common constituent of the diffuse ISM [373, 374]. It was extensively observed by the International Ultraviolet Explorer (IUE) satellite and is found to have remarkable constancy in its peak wavelength (217.5 nm $\pm \leq 25\%$, corresponding to 5.7 eV or 4.6 μm^{-1}). This is not just a local phenomenon, as the feature has been detected in galaxies as distant as redshift $z > 2$ [375]. A correlation between the strengths of the 217 nm feature and the λ 443 nm DIB has been noted [376], but it is not clear whether the correlation just reflects the degree of reddening (i.e., the column density of dust) along the line of sight [377].

The strength of the feature requires that the carrier be made of abundant elements such as C, Mg, Si, or Fe. It has been attributed to absorption by small interstellar graphite grains arising from the $\pi-\pi^*$ transition of sp^2 carbon. Another resonance due to $\sigma-\sigma^*$ transition occurs at 80 nm which could be responsible for the rise of the extinction curve toward the UV. However, a good model fit of observed data with graphite can only be achieved with fine-tuning of the optical constants and particle sizes and shapes. It has been suggested that a complete sur-

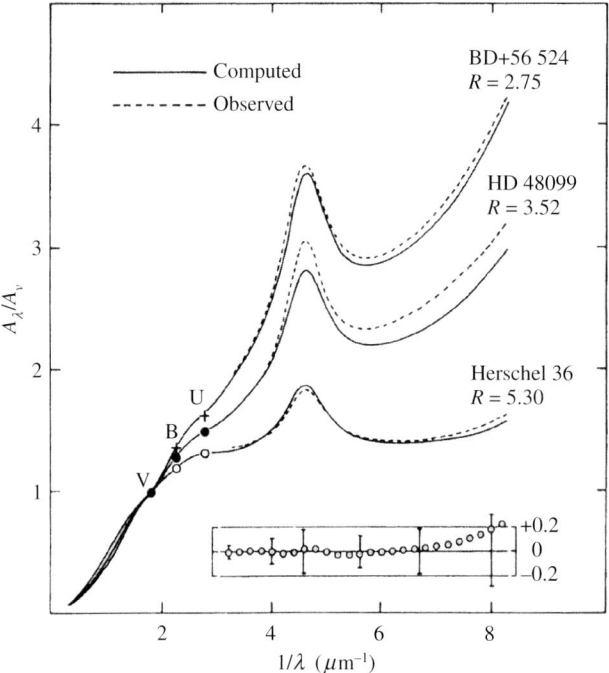

Figure 8.6 Interstellar extinction curves along three different lines of sight. The 217 nm feature is the most prominent feature in the ultraviolet part of the spectrum. Figure adapted from [378].

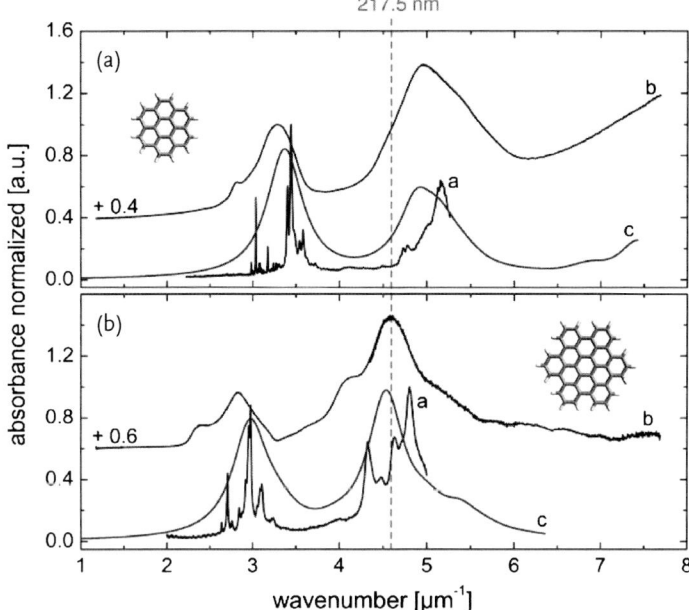

Figure 8.7 Electronic spectra of coronene ($C_{24}H_{12}$) (a) and hexa-peri-hexabenzocoronene ($C_{42}H_{18}$) (b). The position of the astronomical 217.5 nm feature is marked. The curves *a, b, c* are measurements done under different conditions. Figure adapted from [382].

face reconstruction to the sp^2 carbon in diamond can delocalize the π electrons and create a surface mode analogous to that of graphite. The strong absorption coefficient of diamond in the UV can also make diamond a contributor to extinction in the UV.

PAH molecules have been extensively discussed as a possible carrier of the 217 nm feature [379] as well as strongly dehydrogenated PAHs [380, 381]. Absorption spectra of mixtures of large PAH molecules show strong resemblance to the observed profiles of the 217 nm feature [382] (Figure 8.7). A model for the interstellar 217 nm feature based on a population of nanosized hydrogenated amorphous carbon grains processed by UV radiation field has been suggested [383]. This hypothesis is based on the observed structural evolution of carbonaceous grains under UV and thermal processing in the laboratory. Such structural changes can lead to changes in electronic structures and therefore the UV/optical properties of the material, leading to a shift of the absorption band to the 217 nm feature.

Other recent models include carbon onions [384, 385] and polycrystalline graphite [386]. Carbon onions are concentric shells of fullerenes made by electron beam irradiation of carbon soot in a transmission electron microscope, or by heat treatment and electron beam irradiation of nanodiamond. Polycrystalline graphite is composed of randomly oriented arrays of microscopic sp^2 carbon formed into a macroscopic solid. It is suggested that they formed in the ISM when diffuse starlight gradually strip away the peripheral atoms (primarily H) and make the carbonaceous materials increasingly aromatic. This results in sheets of sp^2 stacking together with random orientation inside a single grain. Calculated optical properties of such spherical grains show peak wavelength and width similar to those of the 217 nm feature.

8.4
Extended Red Emission

Extended red emission (ERE) is a broad ($\Delta\lambda \sim 80$ nm) emission band with a peak wavelength between 650 and 800 nm. ERE was first detected in the spectrum of HD 44179 (the Red Rectangle, Figure 1.5) [387, 388] and is commonly seen in reflection nebulae [389, 390]. ERE has also been detected in dark nebulae, cirrus clouds, planetary nebulae, HII regions, the diffuse interstellar medium, and haloes of galaxies. The central wavelength of the emission shifts from object to object, and even between locations within the same object. Other unidentified optical emissions include a set of bright visible bands seen alongside ERE in the Red Rectangle nebula [391] and blue luminescence which is seen both in this object and a number of other sources [392].

Since many solids emit visible luminescence when exposed to UV light, it is assumed that ERE is a photoluminescence process powered by far UV photons. In the diffuse ISM, approximately 4% of the energy absorbed by dust at wavelengths below 0.55 μm is emitted in the form of ERE, suggesting that the carrier of ERE must be a major component of the interstellar grains. It has been estimated that the intrinsic quantum yield of ERE is as high as 50%, and the ERE carrier intercepts ~20% of the photons absorbed by interstellar dust in the 90–550 nm range. This limits the chemical composition of the ERE carrier to a few abundant and highly depleted elements, for example, C, Fe, Si, and Mg [393]. Since metals do not undergo photoluminescence, the most likely candidates are C and Si.

Proposed carriers of the ERE include hydrogenated amorphous carbon (HAC, Section 9.5) [394], quenched carbonaceous composites (QCC, Section 9.7) [395], C_{60} [396], silicon nanoparticles [397, 398], and nanodiamonds [399]. In a material with mixed aromatic and aliphatic composition, for example, HAC, the tunneling of excitation energy from one aromatic ring to another is inhibited by the presence of sp^3 material, resulting in wider bandgaps [400]. Orange-red fluorescent emissions peaking from 670 to 725 nm are found in the spectra of QCC. Since this fluorescence rapidly decays upon exposure to air, the fluorescence may be caused by radicals and highly unsaturated molecules in QCC [395]. Crystalline silicon nanoparticles with 1.5–5 nm diameters have been suggested to possess the optical properties to satisfy the spectral and quantum efficiency requirements [397, 398]. Experimental studies have shown that the silicon nanoparticles can have quantum efficiencies near 100% and absorption coefficients ten times higher than average interstellar dust. Because the number of surface atoms relative to the number of volume atoms is high for nanoparticles, in order to avoid nonradiative recombination, all the dangling Si bonds at the surface of the nanocrystal are passivated by H or O atoms.

Many organic compounds are known to undergo photoluminescence. While it may be an extreme position to associate the origin of ERE to biological materials [401], there is almost no doubt that the ERE comes from a carbonaceous solid as the presence of ERE is correlated with carbon-rich objects. A key to distinguishing diamonds as a possible carrier for ERE from other carbonaceous compounds may lie in the identification of the narrow features seen above the continuum [388, 402]. An early attempt to make this identification has been made by Walter Duley [403], who pointed out the striking similarity between several sharp lines in the red luminescence spectra of brown diamonds and those observed in the luminescence spectra in the Red Rectangle.

8.5
The 21 and 30 Micron Emission Features

The strong emission feature at 21 μm was first discovered in four protoplanetary nebulae from observations by the Infrared Astronomical Satellite (IRAS) Low Resolution Spectrometer (LRS) [404] (Figure 8.8). This feature is very strong and the reasons they were not detected earlier are that the features lies in the mid-infrared, not easily accessible by ground-based observations and their presence is restricted to a limited class of sources. High resolution ($\lambda/\Delta\lambda = 2000$) ISO observations have found that all features have the same intrinsic profile and peak wavelength (20.1 μm) [405]. A comparison of the normalized profiles of six 21 μm sources observed with Spitzer IRS is shown in Figure 8.9. There is no evidence for any discrete substructure due to molecular bands in the observed spectra, suggesting that

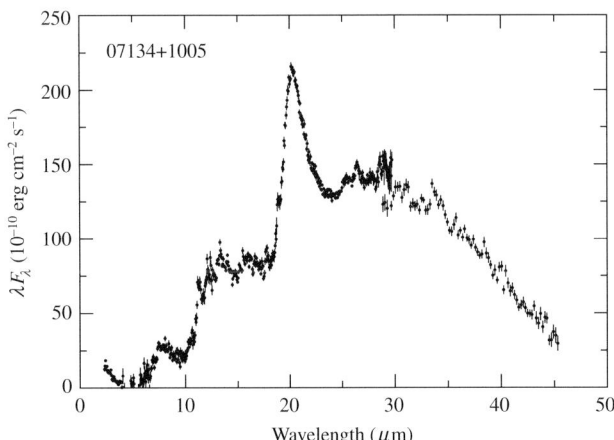

Figure 8.8 ISO SWS spectrum of the protoplanetary nebula IRAS 07134+1005 showing the strong unidentified 21 μm emission feature. Figure adapted from [291].

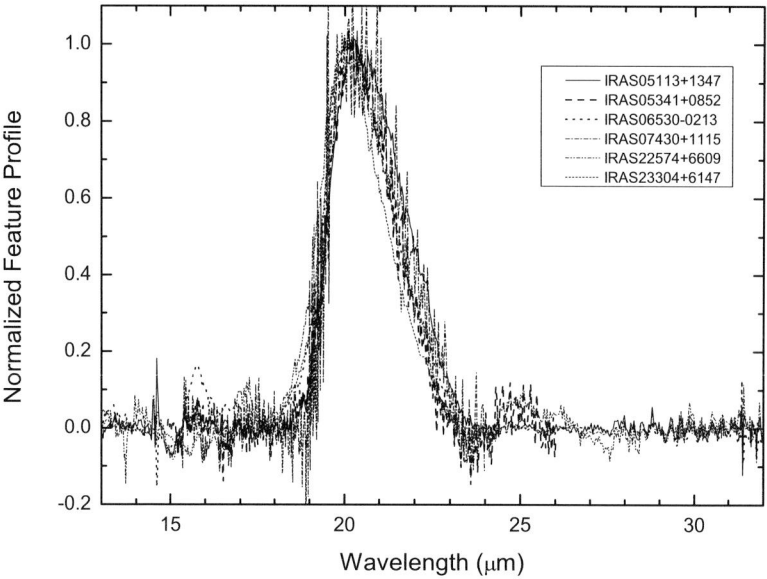

Figure 8.9 Comparison of the normalized 21 μm feature profiles in six protoplanetary nebulae. The spectra are obtained by Spitzer IRS Short-High (SH) ($R \sim 600$) observations. Figure adapted from [408].

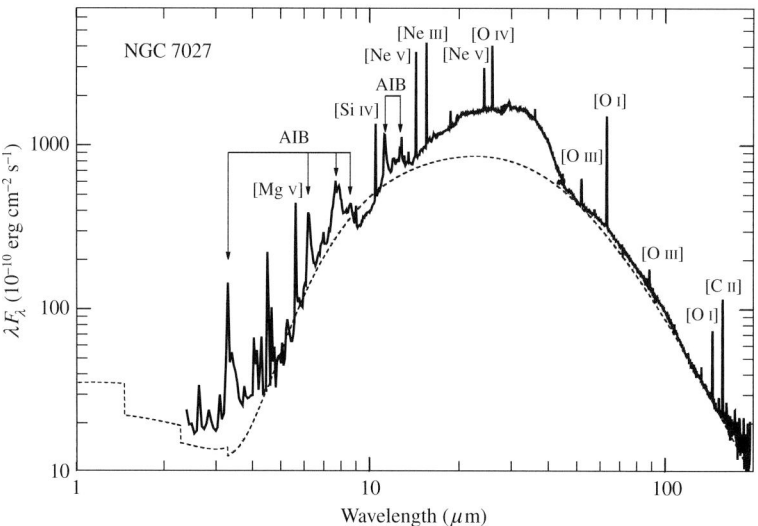

Figure 8.10 The ISO SWS/LWS spectrum of the planetary nebula NGC 7027 (Figure 1.4) between 1 and 200 μm. Some of the stronger atomic lines and AIB features are marked. The dashed line is a model fit composed of a sum of dust emission (with an emissivity of λ^{-1}), f–b and f–f gas emission. The strong 30 μm feature can be clearly seen above the dust continuum.

Table 8.2 Summary of the Spectral Features of Carbon-Rich protoplanetary nebulae and 21 μm Sources. Note 1: Colon indicates a marginal or uncertain detection, blank indicates lack of information, and "…" indicates that the object has not been observed in this spectral region. Table adapted from [406].

Object	SpT	C/O	C₂, C₃	3.3	3.4	6.2	6.9	7.7	8.6	8br	11.3	12.3	Class[a]	C₂H₂	15.8	21	30 μm
02229+6208	G8 Ia	…	Y,Y	Y	Y:*	Y:	Y	N	N	Y	Y	Y	A	…		Y	Y
20000+3239	G8 Ia	…	Y,…	Y	Y*	Y	Y	N	N	Y	Y	Y	A	…		Y	Y
05113+1347	G8 Ia	2.4	Y,Y	Y:	Y:	…	…	…	…	…	Y	Y	…	N:*	Y:*	Y	Y
22272+5435	G5 Ia	1.6	Y,Y	Y	Y	Y	Y	Y	N	Y	Y	Y	B		Y:	Y	Y
07430+1115	G5 Ia	…	Y,Y	Y	Y	…	…	…	…	Y:	Y:	…	A	…	…	Y*	Y
23304+6147	G2 Ia	2.8	Y,Y	Y*	Y:*	Y	Y	Y	Y	Y	Y	Y	A	Y:*	Y*	Y	Y
05341+0852	G2 Ia	1.6	Y,Y	Y	Y	Y	Y	N	N	Y	Y	Y*	B	Y*	Y*	Y	Y
22223+4327	G0 Ia	1.2	Y,Y	Y	N						Y		A			Y	Y
04296+3429	G0 Ia	…	Y,Y	Y	Y			Y			Y	Y	B			Y	Y
AFGL 2688	F5 Iae	1.0	Y,Y	Y	Y	Y	Y:	N	N	Y	Y	N:	A	Y		Y:	Y
06530−0230	F5 I	2.8	Y,Y	Y*	N	…	…	…	…	…	Y*	Y*	A	Y*	Y*	Y*	Y*
07134+1005	F5 I	1.0	Y,N	Y	Y:	…	Y	Y	N	Y:	Y	Y	A		Y:	Y	Y
19500−1709	F3 I	1.0	N,N	N	N	Y	N	Y	Y	Y:	Y	Y:	…		Y:	Y	Y
16594−4656	B7	…	N,N	Y	N	Y	N	Y	Y	Y:	Y	Y:	A			Y	Y
01005+7910	B0 I	1.2	N,N	Y	Y*	Y	Y	Y	N	N	Y	N	A			N:	Y
22574+6609	…	…	…,…,…	…	…	…	…	…	…	…	…	Y*	…	Y*	N*		Y
19477+2401	…	…	…,…,…	…	…						…	…	…	…	…	Y*	Y

[a] Classification scheme of Geballe [358] at 3.3, 3.4 μm.

the 21 μm feature is either due to a solid substance or a mixture of many similarly structured large molecules. To date, the number of known 21 μm sources is very small, ~16 in the Milky Way Galaxy [406] and ~8 in the Large and Small Magellanic Clouds [407]. A list of the currently discovered Milky Way Galactic 21 μm sources is listed in Table 8.2.

The fact that all of the 21 μm sources are carbon-rich strongly suggests that the carrier of this feature is carbon-based. Since the AIB emission features are also seen in the 21 μm sources (see Table 8.2), it is likely that the carrier of the 21 μm feature is related to aromatic compounds. Possible candidates that have been proposed include large PAH clusters, HAC grains [409], hydrogenated fullerenes [410], nanodiamonds [411], TiC nanoclusters [412], O-substituted 5-member carbon rings [413], nano-SiC grains with carbon impurities, and cold SiC grains with amorphous SiO_2 mantles [414, 415].

The unidentified emission feature around 30 μm was discovered from KAO observations [416]. It was first seen in carbon-rich AGB stars (e.g., IRC+10216 and AFGL 2688) and planetary nebulae (e.g., IC 418 and NGC 6572). More recently, the 30 μm feature was found to be common in carbon-rich protoplanetary nebulae, especially those showing the 21 μm emission feature. The prominence of the 30 μm feature can be seen in the spectral energy distribution of the planetary nebula NGC

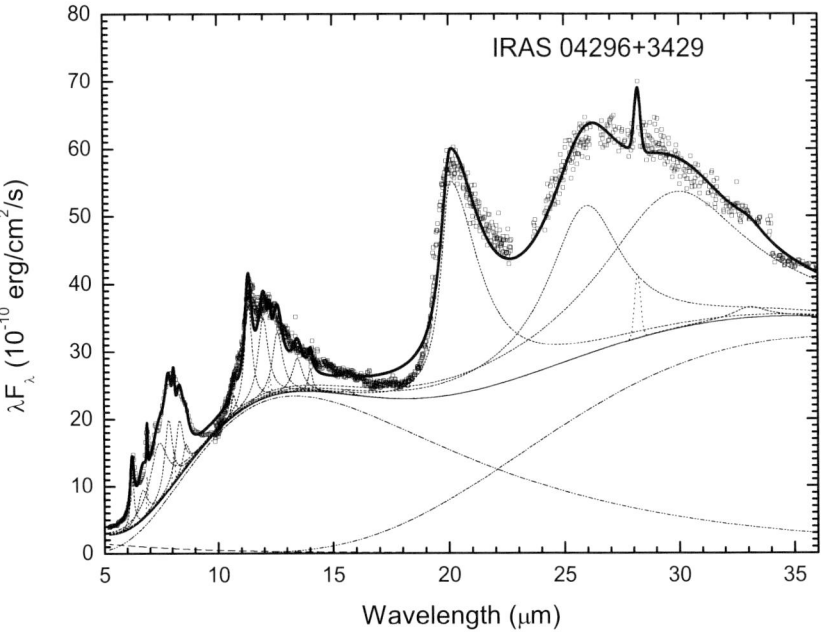

Figure 8.11 The Spitzer IRS spectrum (shown as squares) of the protoplanetary nebula IRAS 04296+3429 fitted by a combination of emission features. The general underlying dust continuum consists of two modified blackbod- ies ($\lambda^{-2}B_\lambda(T)$ with $T = 180$ and 60 K). The high point in the spectrum at 28 μm is due to the 28.22 μm $\nu = 0-0$ $J = 2-0$ line of H_2. A series of UIE, 21 and 30 μm features (see text) are used to fit the spectrum.

7027 (Figure 8.10). Among planetary nebulae in the Magellanic Clouds, about half of them possess the 30 μm feature [270].

The origin of the 30 μm feature is not known. The fact that a significant fraction (~20%) of the total luminosity of the object is emitted in this feature suggests that the carrier must be composed of abundant elements [417]. The first suggested identification was solid MgS based on a comparison with laboratory measurements. The alternative suggestion that the carrier is a carbonaceous material continues to be popular because the feature is only seen in carbon-rich objects.

The fact that the UIE, 21 μm, and 30 μm features are seen together in protoplanetary nebulae has been used to suggest that the carriers of these three phenomena are chemically related. Figure 8.11 shows the mid-infrared spectrum of the protoplanetary nebula IRAS 04296+3429. The spectrum consists of several broad features at 8, 12, 21 and 30 μm on top of a very broad continuum. The 8 and 12 μm features can be fitted by a superposition of UIE features at 6.2, 6.7, 7.5, 7.8, 8.3, 8.6, 10.7, 11.3, 12.0, 12.6, 13.5, and 15.9 μm. The 30 μm is best fitted as a combination of features peaking at 26 and 30 μm. The Spitzer IRS spectra of 10 protoplanetary nebulae have been successfully fitted by this superposition technique [364]. In general, the fraction of the energy emitted in the 30 μm feature is ~20%, in the 21 μm feature ~5%, and slightly smaller in the UIE features. It is interesting to note that the objects with the highest 21 μm flux ratios (IRAS 07134+1005, IRAS 06530−0213, IRAS 04296+3429, IRAS 23304+6147) all have stellar temperatures ~7000 K. If the flux ratio reflects the abundance of the carrier, then this implies that the carrier begins to be synthesized after the AGB at $T_* \sim 5000$ K, reaches maximum abundance at 7000 K, and is gradually destroyed when the star evolves to higher temperatures.

9
Chemical Structures of Organic Matter in Space

The wealth of unexplained astronomical phenomena as discussed in the last chapter has led to a large variety of proposals on the possible candidates of their carriers. In general, they should be either molecules or solids made up of commonly available elements in the Universe due to the ubiquitous nature of the phenomena. The most likely candidates are organic molecules and carbonaceous solids. The quantum processes leading to the emission of photons by a molecule are briefly described in Chapter 3.

The presence of solids in interstellar space has been known since the beginning of the twentieth century through their effects on the extinction of starlight [418]. The distribution of solid particles (also referred to as "dust" in astronomical nomenclature, although it has a different meaning from our ordinary meaning of household dust) can be seen as dark patches in a background of stars [419]. Since micronsize solid particles preferentially absorb and scatter blue light more than red light, their presence can be inferred by their effect on the colors of background stars. Many stars on the galactic plane appear redder in color than expected from their respective spectral types based on their photospheric spectra. The degree of extinction can be quantified by the difference between the observed and expected intrinsic colors of the stars and the amount of extinction as a function of wavelength is known as the extinction curve. The observed extinction curve along different lines of sight in the Galaxy all show a distinctive "bump" around 217.5 nm, which is a spectroscopic signature of the solid particles in the diffuse ISM. Although the carrier of this "bump" is widely attributed to graphite, its exact chemical composition has not been firmly established (Section 8.3).

Interstellar solid particles are heated through the absorption of visible star light, and reemit the energy in the form of infrared light. With the development of infrared astronomy, their existence can be determined by direct imaging and spectroscopy in the infrared wavelengths. Thermal emission from micron-size solid grains have been detected in the form of mid-infrared continuum radiation in the circumstellar envelopes of stars, interstellar clouds, and galaxies. In Figure 9.1, we show an infrared image of M 16 due to thermal emission by interstellar solid particles. In addition to stars and interstellar gas, solid particles are now recognized as a common constituents of the Universe.

Organic Matter in the Universe, First Edition. Sun Kwok.
© 2012 WILEY-VCH Verlag GmbH & Co. KGaA. Published 2012 by WILEY-VCH Verlag GmbH & Co. KGaA.

Figure 9.1 Spitzer IRAC 5.8 μm image of the star formation region M16 (the Eagle Nebula). The bright (light color) areas are due to thermal emission from solid particles. The points are background stars.

Early theories on the chemical composition of interstellar solids include metals (e.g., iron), ice, graphite, and diamond, and so on. These conjectures continued for several decades until the development of infrared spectroscopy allowed the observation of specific features, leading to the identification of amorphous silicates, silicate carbide, and other refractory oxides as carriers of these features (Section 6.1). A mixture of silicate and graphite materials has become the commonly assumed model of interstellar dust [420].

The search for the identification of carbonaceous solids has been more difficult. Some carbon-rich objects have featureless spectra, which have been suggested to be due to emission from amorphous carbon grains. From the observations described in the previous chapters, we learned that some of the unidentified features are likely to originate from some form of organic matter. However, there are still a large variety of natural or synthetic substances that are potential candidates. In this chapter, we will discuss some of these possibilities.

9.1
Optical Properties and Colors of Solids

Since gas-phase atoms and molecules emit radiation at specific frequencies, they have unique spectral signatures and can be identified exactly. Atoms and ions undergo electronic transitions and molecules can have electronic, vibrational and rotational transitions, all giving out spectral lines or bands that can be detected and measured spectroscopically. In a solid, rotational motions are not possible and the vibrational-rotational transitions seen in gas-phase molecular spectra are replaced by a broad, continuous band at the vibrational frequencies. A crystalline solid has a highly ordered lattice structure, with constant bond lengths and angles between atoms. Such a structure has translational symmetry and lattice vibrations create plane waves which are referred to as phonons. Due to the symmetry of the structure, only a few of the possible lattice vibrational modes are optically active. Therefore, crystalline solids have only a few sharp features in the infrared.

In contrast to crystalline solids, amorphous solids have their atoms arranged in a disordered manner. While the bond lengths in amorphous solids are nearly the same, the bond angles can have large variations. This lesser degree of symmetry results in most modes being optically active. The variation of bond lengths and angles also mean that there is a wider range of vibrational frequencies, leading to broader features. By comparing the absorption spectra of terrestrial minerals to observed astronomical spectra, we can identify the chemical composition of solids in space. The best known example is the identification of amorphous silicates through their Si–O stretching mode at 9.7 μm and the Si–O–Si bending mode at 18 μm. Crystalline silicates (olivines and pyroxenes) as well as several refractory oxides have been similarly identified. The ability to identify minerals and solids in space has led to the establishment of the field of astromineralogy [421].

Interstellar solids can also be identified through their absorption of visible and UV light. As in atomic and molecular transitions, the absorption of UV and visible light by a solid is due to changes in the electronic structure. Electrons in crystals are arranged in energy bands separated by band gaps where no electronic states exist. The size of the band gaps determines the optical properties of crystals. A solid with a large bandgap (e.g., diamond) will be transparent because no electronic transition can occur across the gap. For a crystal to have a strong color, its bandgap has to be between 1.7 to 3.5 eV, corresponding to wavelengths in the visible range. A bandgap of ~ 2.5 eV will cause most of the blue light to be absorbed, making the crystal a yellow-orange color. Many crystals, although transparent when pure, can have bright colors if impurities are present. For example, ruby (red) and sapphire (blue) are due to the presence of Cr^{3+} and Al^{3+} in Al_2O_3 (colorless), respectively. Many of the transition elements have electronic transitions in the visible, and their presence can bring colors to a crystal even the crystal itself has no bandgap in the visible region. The strong 217 nm absorption feature in the ISM (Section 8.3) is probably due to an electronic transition of an unidentified solid.

Large organic molecules have very many electronic states and their ultraviolet absorption bands are very broad. Macromolecular organic solids have their absorption edges in the red and often appear brown or black in the visible as their reflectance increases to the red. Remote identification of organic solids therefore has to rely on infrared spectroscopy, their visible colors, and albedo.

Photoluminescence can also occur in space when a solid particle responds to irradiation by higher energy photons (e.g., UV starlight) by emitting a lower frequency photon (e.g., visible light). Photoluminescence occurs when a photon of energy greater than the bandgap (separation between valence and conduction bands) is absorbed by a semiconductor, leading to the creation of an electron-hole pair. These pairs quickly thermalize to the limits of the conduction and valence bands. When they recombine, they emit a photon of energy close to that of the bandgap. Metals do not undergo photoluminescence because there is no bandgap.

9.2
Carbon Chains and Rings

Several members of the carbon chains family (C_2, C_3, C_5) have been detected in the Solar System, in the atmosphere of cool stars, in circumstellar envelopes of evolved stars, and in the diffuse ISM. Could the larger ($n > 5$) members of this family (in either chain or ring form) be widely present in the Galaxy and represent a reservoir of carbon? Carbon chains have long been suspected to be the carriers of DIBs [422]. If this is the case, the strengths of the DIBs require that the carbon chains be present in large quantities. Given their unstable nature, it has been difficult to manufacture and study them in the laboratory.

Recent measurements of carbon chain spectra have indicated that it is unlikely that smaller carbon chains (C_n, $n = 5$–10) and their ions can be responsible for the DIBs [139]. The question remains that for longer C_{2n+1} chains, they have electronic transitions in the visible that have large oscillator strengths, and therefore could explain the stronger DIBs. Figure 9.2 shows a mass spectrum of the products produced by laser ablation of graphite rod ionized by a focused 157 nm laser. Although the species C_{15}, C_{17}, C_{19}, and so on are produced in abundance, their lifetimes are too short for their spectra to be obtained. For ring systems, the electronic spectra of the carbon rings C_{14}, C_{18}, and C_{22} have been observed. However, there is no good match between the observed carbon-ring bands and the DIBs.

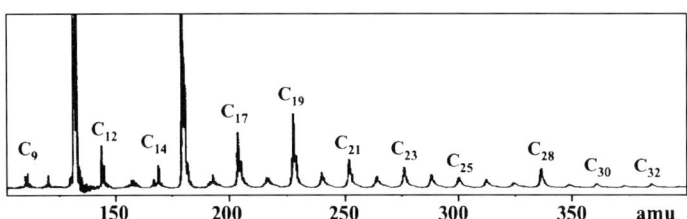

Figure 9.2 A mass spectrum obtained from laser ablation of a graphite rod. Peaks corresponding to carbon chains are marked. Figure adapted from [423].

9.3
Polycyclic Aromatic Hydrocarbon

The interest in PAH molecules as a possible constituent of the ISM began with the suggestion that they could be responsible for the UIE bands [171, 424]. Small, neutral PAHs are primarily excited by UV light, as laboratory measurements show a sharp cutoff in the UV with little absorption in the visible. In order to better match the observations, studies of PAHs have been extended to include PAH ions and clusters. From quantum calculations as well as laboratory measurements, it has been shown that the central wavelengths and the relative intensities of PAH bands can change markedly upon ionization. If PAH molecules in the ISM are indeed partially ionized, the resulting complexity of the spectrum would make comparison between theoretical or laboratory predictions with observations almost impossible. Also relevant are nitrogen-substituted PAH molecules. Since the N atom can replace a CH group in an aromatic ring without significant change to the molecule's chemical structure, such heterocyclic aromatic compounds are of particular interest because of their biochemical implications (Section 2.3).

If present, interstellar PAHs could contribute to the energy balance of the ISM through the absorption of diffuse starlight. Absorption of UV photons results in a rapid redistribution of the photon energy among the vibrational modes of the molecule in the ground electronic state. In a low-density environment where collisional deexcitation is not possible, the molecules undergo spontaneous deexcitation via infrared fluorescence, leading to the emission of infrared features. The absorption of diffuse UV light can also lead to the ejection of electrons through the photoelectric effect, which can contribute to the heating of the interstellar gas.

The PAH hypothesis suggests that the AIB features are due to infrared fluorescence of far-ultraviolet-pumped PAH gas-phase molecules [425]. In spite of the popularity of this hypothesis in the astronomical community, it is not clear that PAH molecules are the solution to the entire UIE mystery. We now know that the AIB features are seen in protoplanetary nebulae [287] and reflection nebulae with low-temperature central stars [246] where the UV radiation background is negligible. In order to extend the absorption cross section to the visible, the PAH molecules have to be in large clusters. The only difference between such PAH clusters and solid-state grains is that the PAH clusters are pure aromatics with regular structures. It is expected that the ionization state (positively, neutral, or negatively charged) of PAH is also related to the cluster size. A larger cluster may require a stronger UV radiation field to affect the molecules' ionization state.

Furthermore, the observed AIB features are not sharp, as it would be the case if they are due to molecular emission. The widths of the features are much broader than can be expected from Doppler effects. The dominant observed bands at 6.2 and 7.7 μm do not appear at these wavelengths in laboratory or theoretical emission spectra of PAH molecules, and the proponents of PAH have to appeal to a complex mixture of molecular PAHs of different sizes, structures (compact, linear, or branched) and charged states (neutrals, positive and negative ions). Most of the fits to the observed UIE spectrum rely on linear combinations of a large number of laboratory or theoretical spectra of neutral, ions, and hydrogenated PAH molecules.

As of 2010, no specific PAH molecule has been identified by astronomical spectroscopy. The difficulty in detecting PAH molecules through rotational transitions could be due to a variety of factors, including small or zero dipole moments and population dilution due to large partition functions. The most promising PAH molecule is corannulene ($C_{20}H_{10}$, Figure 9.3), which is highly symmetric and has a large dipole moment (2.07 debye), allowing its rotational transitions to be measured

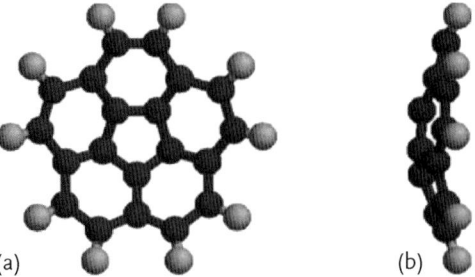

(a)   (b)

Figure 9.3 The top (a) and side view (b) of the PAH molecule corannulene ($C_{20}H_{10}$). Due to the presence of the central pentagonal carbon ring, the molecule is bent into the shape of a bowl. Because of this property, $C_{20}H_{10}$ can be considered as a transitional species between planar PAHs and fullerenes. Figure adapted from [427].

in the laboratory [426]. However, thus far, the results of astronomical searches have been negative [234, 427].

Finally, the AIB features often lie on top of strong plateau features or broad continuum, which have to originate from a solid or nanoparticles. It is therefore unlikely that pure PAH molecules can account for the entire UIE phenomenon.

9.4
Small Carbonaceous Molecules

Similar to the PAH hypothesis, the small carbonaceous molecules (SCM) hypothesis suggests that the UIE features are due to small, carbon-based molecules, in particular molecules with ring structures such as cyclopropenylidene (Section 3.11.2)

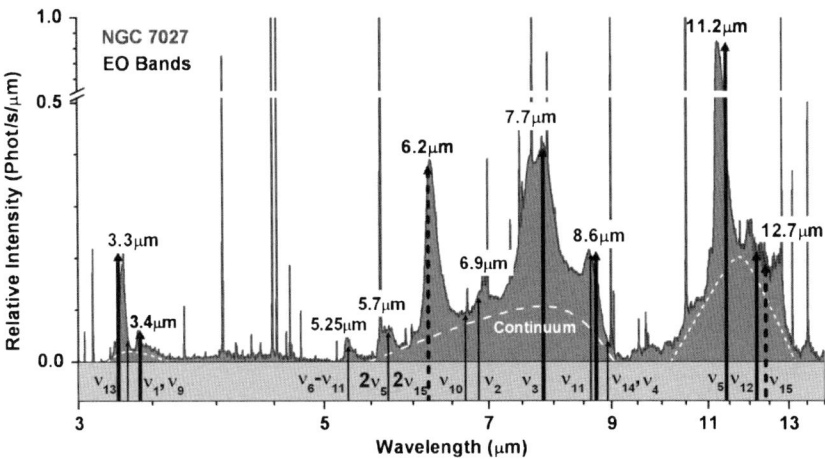

Figure 9.4 Comparison between the vibrational spectrum of ethylene oxide (arrows) with the UIE features in the continuum-removed ISO spectrum of NGC 7027. The spectral assignment of each of the vibrational modes are labeled below the arrows. Figure adapted from [151].

and ethylene oxide (Section 3.11.4) [151]. The strong vibrational modes of these molecules often coincide with the UIE features. For example, the 3.3 and 3.4 µm features are identified as the CH_2 symmetric and asymmetric stretching modes of ethylene oxide, whereas the 7.7 and the 11.3 µm features are identified with the v_3 and v_5 ring deformation modes respectively (Figure 9.4). One notable difference between the two hypotheses is that while the PAH model relies on anharmonic shifts to explain the observed UIE feature profiles, the SCM model is able to explain the profile with the rotational envelope of the vibrational modes.

The SCM model has the advantage that it is easily testable. If the UIE features are due to small ring molecules, then the abundance of these molecules can be determined from rotational transitions. While C_3H_2 has been widely detected in the ISM, ethylene oxide is much less widely distributed and in particular has not been detected in any source showing the UIE features.

9.5
Hydrogenated Amorphous Carbon

Hydrogenated amorphous carbon (HAC) is created by laser ablation of graphite in a hydrogen atmosphere. The structure of HAC contains islands of aromatic (sp^2) bonded C atoms joined together with a variety of peripheral sp^2 and sp^3 bonded hydrocarbons [428]. HAC has been proposed as being responsible for the AIBs as well as ERE, 21 µm feature, and so on.

Photochemical processing of HAC can create PAHs and fullerenes [429, 430]. The fact that fullerene has been detected in planetary nebulae has led to the suggestion that they are among the decomposition products of HAC [204].

9.6
Soot and Carbon Nanoparticles

Soot is formed by combustion of hydrocarbon molecules in a flame (Section 2.2.6). The fact that autoexhaust soot displays infrared features resembling the 6.2 and 7.7 µm UIE features was first noted by Lou Allamandola of NASA Ames Research Center [431]. Figure 9.5 shows a comparison between the spectrum of automobile exhaust and the astronomical UIE spectrum. Although the peak wavelengths do not match exactly, the profiles are similar. However, the interest of the NASA Ames group was quickly turned to the study of PAHs, which are gas-phased molecules rather than bulk materials as in soot.

The combustion of heavy hydrocarbons in a furnace yields carbon black, which is also a form of amorphous carbon similar to soot and HAC (Section 2.2.6). Chemically, soot and HAC are similar in that they are composed of H atoms and sp^2 and sp^3 carbon atoms (with $sp^3/sp^2 > 1$, as in coal). They are different primarily in the mode of production and the fact that HAC is deposited on to a substrate, whereas soot particles condense before they are deposited on the substrate. Once hydrocar-

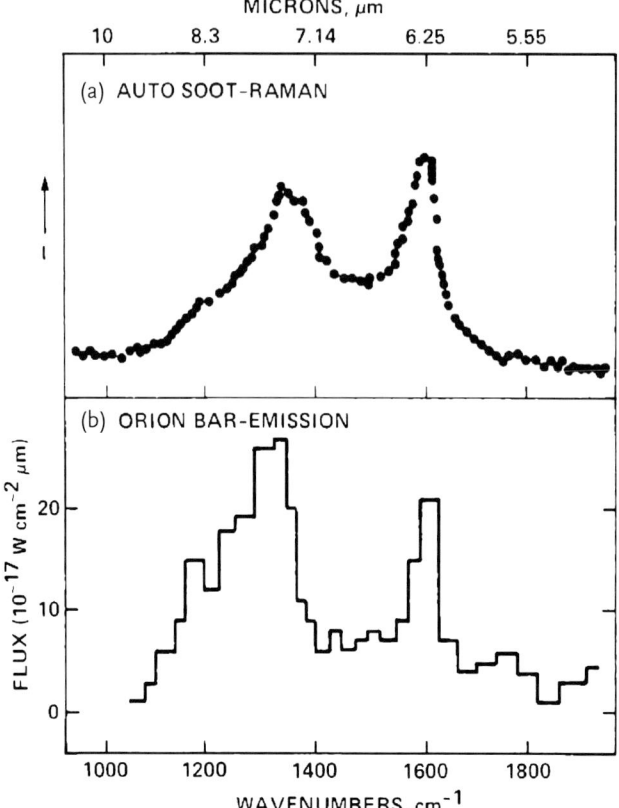

Figure 9.5 Raman spectrum between 5–10 μm of auto exhaust (a) in comparison with the astronomical spectrum of Orion Nebula (b) showing the 6.2 and 7.7 μm UIE features. Figure adapted from [431].

bons are made in a stellar atmosphere or a stellar wind, it is possible that soot and HAC can be produced in the circumstellar environment.

A variety of techniques have been used to simulate the production of carbon dust in the laboratory. Artificial soot has been produced by hydrogen flame or arc-discharge in a neutral or hydrogenated atmosphere [432]. Carbonaceous particles have been produced by pulsed laser ablation of carbon rods and condensation in 10 mbar argon atmosphere [433]. Laser-induced pyrolysis uses laser to dissociate gas-phase hydrocarbon such as ethylene, acetylene, and benzene, followed by condensation [434]. An example of the laboratory set up of laser ablation of graphite and condensation of carbonaceous matter in a quenching gas atmosphere is shown in Figure 9.6.

The end products of the condensation are often in the form of carbon nanoparticles (CNPs) which can be extracted from the flow reactor by a particle beam or by collection of condensate in a filter. The morphology and chemical makeup of the CNP are dependent on the techniques and conditions used to create them.

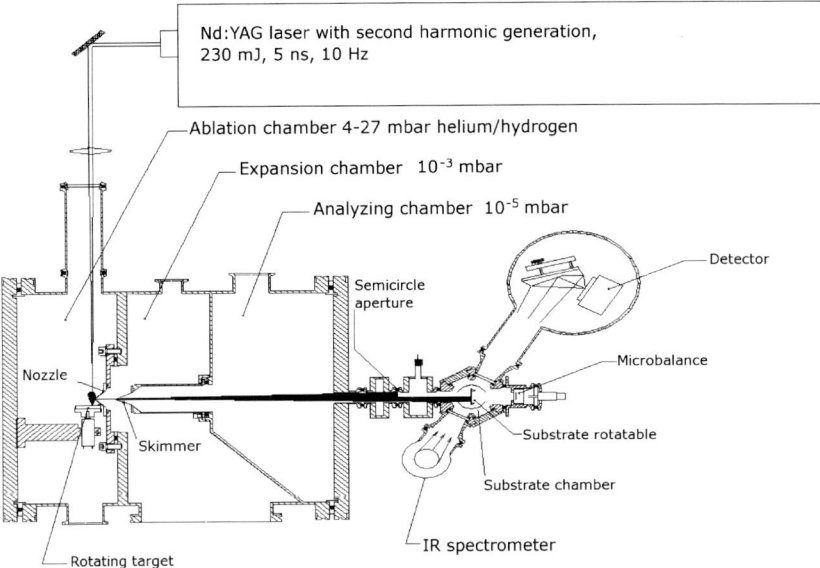

Figure 9.6 A schematic diagram illustrating the molecular beam apparatus containing a laser ablation source, particle extraction, deposition chamber, and infrared spectrometer. Figure adapted from [435].

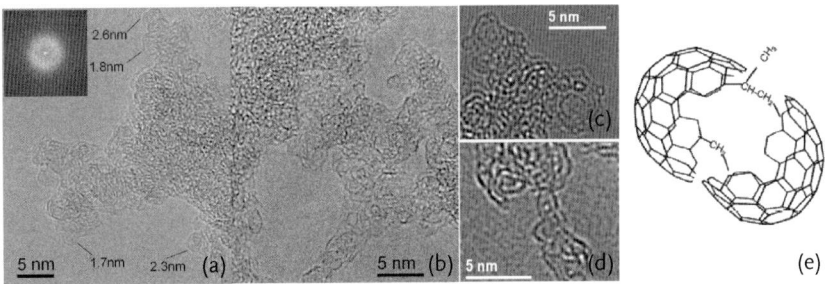

Figure 9.7 High resolution transmission electron microscopy (HRTEM) images of fullerene-like CNP produced in a high-temperature condensation process. Images (a) and (b) show grains condensed under low and high H contents, respective-ly, and (c) and (d) are images of elongated fullerene particles and their fragments. A schematic illustrating the possible links be-tween fullerene fragments is shown in im-age (e). Figure adapted from [438].

The CNPs created by the Waterloo group generally have sizes ranging from 1 to 10 nm and are made up of primarily sp^2 carbon rings connected by a network of aliphatic chains [436, 437]. These chains are dominated by sp^3-hybridized $-CH_2$ and $-CH_3$ groups. Fullerene-like particles have been found by the Jena group in high-temperature (>3500 K) condensates, including fragments of fullerene cages linked by aliphatic $-CH_x$ groups [438]. Images obtained with high resolution trans-mission electron microscopy (HRTEM) are shown in Figure 9.7.

9.7
Quenched Carbonaceous Composites

QCC are produced by the technique of hydrocarbon plasma deposition where methane gas is heated to 3000 K with a microwave generator, allowed to expand into a vacuum chamber and condensed on a room-temperature substrate. The resultant dark, granular material is shown by electron micrography to have an amorphous structure [439, 440]. In addition to this dark QCC, which forms in the plasma beam itself, there is also a light yellow-brown component (called filmy QCC) that forms on the walls of the vacuum tube.

Since QCC consists of only H and C and the conditions under which it is synthesized resemble the circumstellar environment of AGB stars, it could have

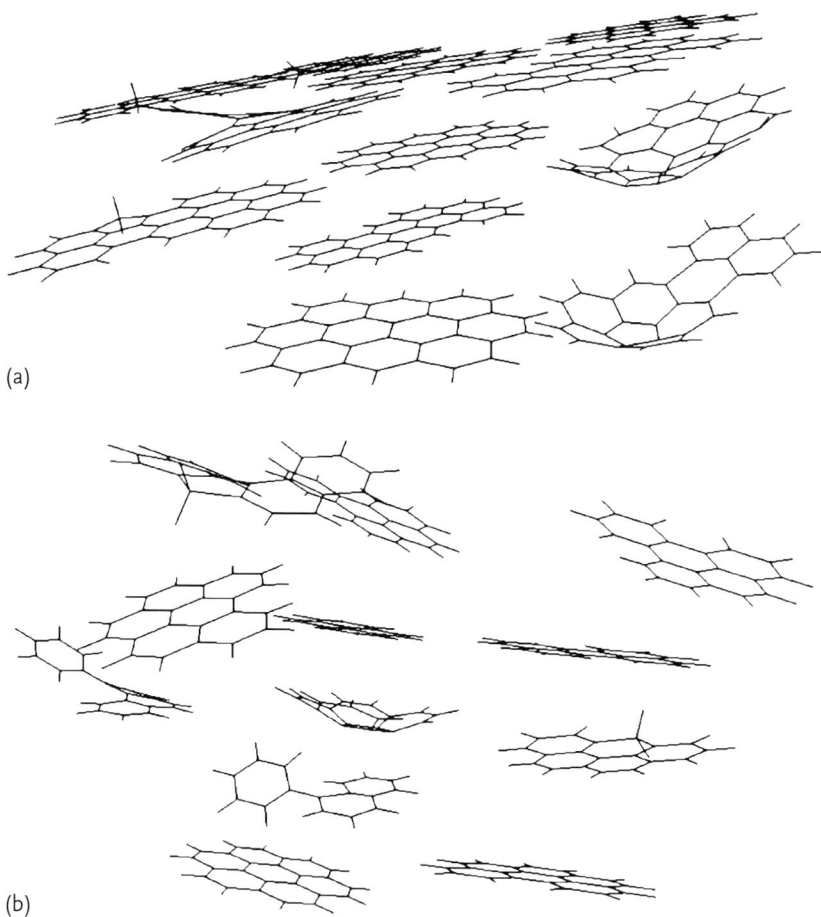

(a)

(b)

Figure 9.8 Simplified schematic diagram of the PAH components in the filmy-QCC. (a) Filmy-QCC collected close to the nozzle. (b) Filmy-QCC deposited far from the nozzle. Figure adapted from [442].

astrophysical significance. Mass spectrometry of QCC suggests that its aromatic component typically consists of one to four rings, and most have only one to two rings [441]. A schematic of the chemical structure of QCC is shown in Figure 9.8. Infrared spectroscopy of QCC reveals a mixture of sp, sp^2, and sp^3 sites. Upon heating, the strengths of the sp^2 features increase relative to the sp^3 features, suggesting that QCC can become graphitized at high temperatures.

9.8
Kerogen and Coal

Both coal and kerogen are amorphous carbonaceous solids widely present in the Earth's crust. Coal, formed from fossilized hydrocarbon materials, also contains a mixture of sp, sp^2, and sp^3 bonds. In addition to its aromatic character, coal also contains rich oxygen functional groups. Given the complexity of coal's origins, it is almost impossible to give it a molecular structure. In general, coal is a polymeric material containing clusters of aromatic rings connected by aliphatic bridges. These linked structures gives coal a high molecular weight, making it insoluble in all common solvents [47].

Kerogen is a solid sedimentary, insoluble, organic material found in the upper crust of the Earth [443]. In contrast to coal, which is found in bulk rocks, kerogen is usually found in sand-like dispersed form. Structurally, kerogen can be represented by random arrays of aromatic carbon sites, aliphatic chains $((-CH_2-)_n)$, and linear chains of benzenic rings with functional groups made up of H, O, N, and S attached. The O atoms can be in ester $(-O-C=O)$, ether $(C-O-C)$, or hydroxyl $(-OH)$ groups [444]. Kerogen is related to crude oil (Section 9.9) and natural gas as cycloalkanes (C_nH_{2n}) and alkanes (C_nH_{2n+2}), the ingredients of crude oil, are released from kerogens at depths of 2–3.5 km. At greater depths with temperatures > 150 °C, methane (CH_4) and ethane (C_2H_6) form the constituents of dry natural gas. In between the releases of oil and dry gas are the "wet gases" made up of propane (C_3H_8) and butane (C_4H_{10}).

In contrasts to pure C and H substances, for example, PAH, HAC, and QCC, kerogen contains impurities such as N, S, and O. The fractions of these elements relative to C are similar to those in lipids (fat and fatty acids, see Section 2.3.2), leading to the belief that kerogens on Earth are formed as a result of the decay of living organisms.

The idea that coal and kerogen can be the carrier of the UIE features was first raised by Renaud Papoular [445]. The laboratory spectra of coal and kerogen show a lot of similarity with astronomical spectra of protoplanetary nebulae [444, 446, 447]. The 3.4 μm feature and the broad emission plateaus at 8 and 12 μm seen in protoplanetary nebulae are difficult to explain by pure aromatic materials, but correspond well with mixed aromatic/aliphatic compounds such as coal and kerogen [321, 447] (Figure 9.9).

Objection to the coal model as an explanation to the interstellar UIE features is based on the inability of bulk coal materials to emit in near infrared wave-

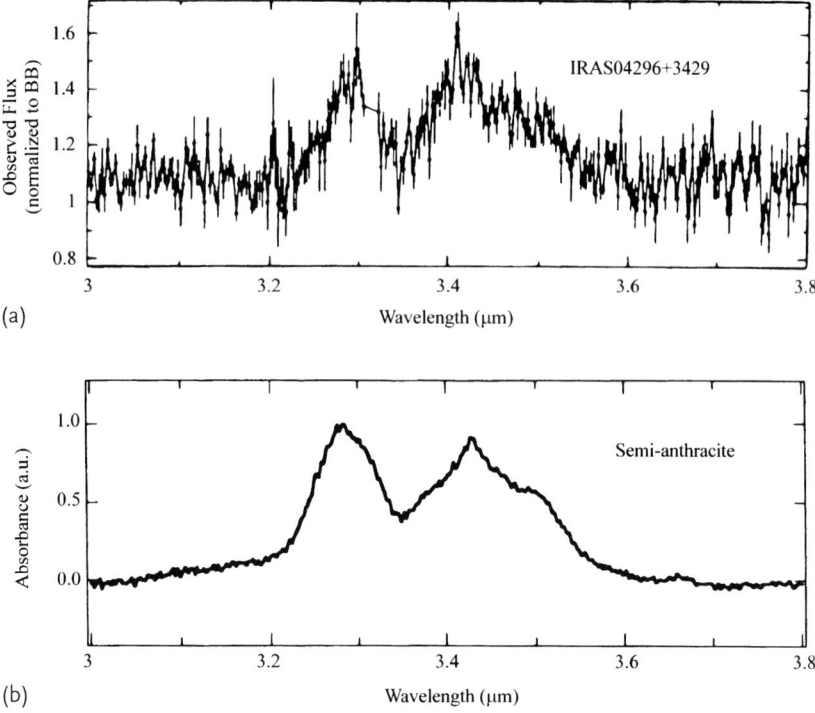

Figure 9.9 Comparison between the 3.3 and 3.4 μm emission features in protoplanetary nebula IRAS 04396+3429 (a) to the absorbance spectrum of semianthracite coal (b). Figure adapted from [447].

lengths [448]. The equilibrium temperatures of solid-state grains are too low to account for radiation in the 3 μm region under interstellar conditions, but the near IR continuum cannot be due to PAH either as molecules do not emit continuum radiation [449].

9.9
Petroleum Fractions

Petroleum is a mixture of different hydrocarbons, including alkanes, cycloalkanes, and different forms of aromatic hydrocarbons. The lighter alkanes (methane, ethane, propane, and butane) are in gaseous state and are usually called natural gas, whereas the heavier hydrocarbons are in the form of liquid or solids and are called crude oil. In addition to hydrocarbons, petroleum also contains traces of heavy elements such as N, O, and S as well as metals such as Fe, Ni, Cu, and so on.

As the case with coal, petroleum found on Earth is derived from the decomposition of living matter, hence the name fossil fuel. Remnants of living matter buried under high temperature and pressure was first transformed into kerogen, which under further heating produced petroleum [450, 451]. This theory of formation is based on the presence of biological markers, although there has been alternate theories of the origin of oil [452]. The abiological theory of oil was first developed in the Soviet Union in the 1950s and remained popular there for about 30 years. The theory proposes that petroleum hydrocarbons can be synthesized by inorganic means in the Earth's mantle [453]. Under high pressure and temperatures at depths >100 km, it is possible to convert methane to complex mixtures of alkanes and alkenes. The possibility of outgassing of hydrocarbons from the mantle along deep faults has also been discussed. This theory was expanded by Thomas Gold who suggested that oil and gas are converted from methane migrated from the interior of the Earth [454]. Gold based his theory on the observation that methane is abundantly present in the atmospheres of the Jovian planets (Section 7.4) and therefore could also be available on Earth.

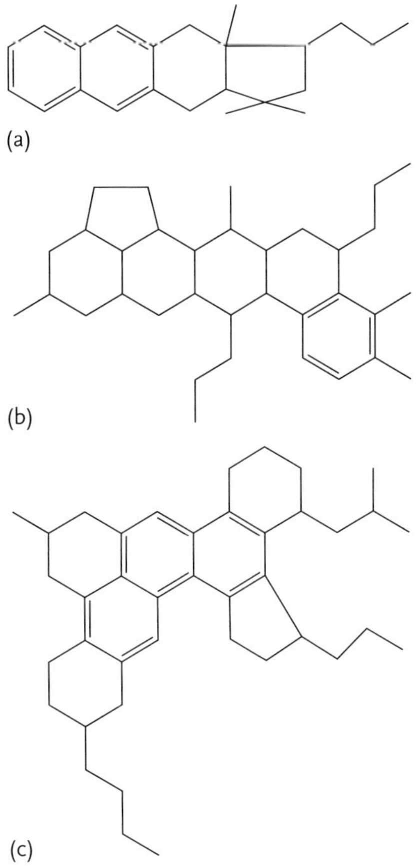

(a)

(b)

(c)

Figure 9.10 Schematic structures of petroleum fractions. (a) Distillate aromatic extract (DAE), (b) treated residual aromatic extract (T-RAE), (c) naphtenic oil. Figure adapted from [457].

Perhaps the most interesting theory relevant to our present discussion is the speculation by Fred Hoyle that primordial hydrocarbons were incorporated in the formation of the Earth, resulting in a vast reservoir of oil in the deep interior of the Earth [455]. If such primordial hydrocarbons did exist, then they must have been synthesized elsewhere and brought to the early Solar System.

Comparisons between the petroleum distillate extracts with the spectra of protoplanetary nebulae have led to the suggestion that they can be the carrier of the UIEs [456, 457]. These distillate aromatic extracts (DAE) are complex mixtures of hydrocarbons with molecular weights in the range of 400 to 1400 dalton. Typical structures are sketched in Figure 9.10. In contrast to the linked structure of coal

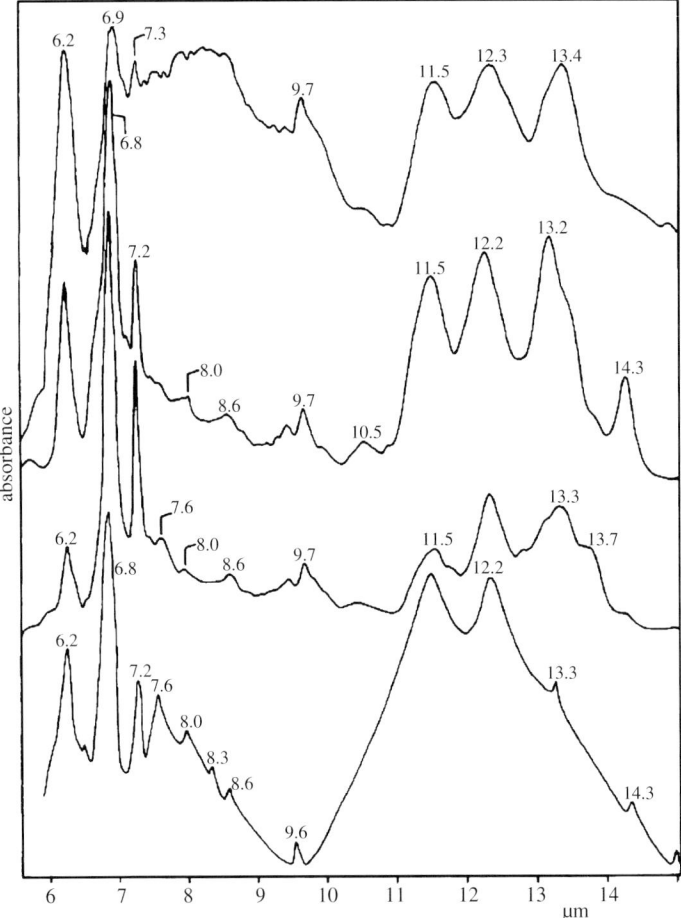

Figure 9.11 Comparison between the Fourier Transform Infrared Spectra of anthracite coal (top curve), modified fraccion 2 (second curve), distillate aromatic extract (third curve), and ISO spectrum of the protoplanetary nebula 22272+5435 (bottom curve, see Figure 6.10). Figure adapted from [457].

and kerogen, DAEs are simple but diverse polycyclic aromatic compounds with aliphatic and cycloaliphatic attachments and therefore have much lower molecular weights than coal. In other words, they are more complex than PAHs, though not as complicated as coal. Any natural synthesis pathway, for example, in the circumstellar environment during the late stages of stellar evolution (Section 6.2) will lead naturally to a mixture of undefined petroleum fractions, which together will have the spectral characteristics (bands and plateaus) as observed. In Figure 9.11, we can see that many of the features seen in the emission spectrum of the protoplanetary nebula IRAS 22272+5435 can be found in the laboratory absorption spectra of two petroleum fractions and coal.

9.10
Tholin and HCN Polymer

Hydrogen cyanide (H−C≡N) is highly reactive and is well known for its toxicity. In its neutral or ionized forms, it can react with itself to produce a variety of other compounds. The most interesting example is the nucleobase adenine (Section 2.3.4), which chemical composition ($C_5H_5N_5$) is just five times HCN.

Our current interest in HCN polymer as a component of complex extraterrestrial organic matter can be traced back to the early nineteenth century when Joseph Louis Proust found in 1807 that a complex polymer can emerge from a basic (pH > 7.0) solution of HCN [458]. In 1959, when John Oró injected HCN gas into an ammonia mixture solution, he found that its color changed to orange in a few minutes, and then to reddish color, and deep red, finally ending in black [459]. When this black HCN polymer is stirred with water, it dissolves into a solution that contains a number of amino acids [460] and nucleobases [461]. A pathway of amino acid synthesis from HCN was proposed by C.N. Matthews [462] (Figure 9.12).

Although the structure of HCN polymers is complex, its constituents can be determined by cleaving functional groups, methylating the resulting products and subjecting them to gas chromatography-mass spectrometry (GCMS) analysis (Section 7.1). From the mass spectra, a variety of products can be identified. In a sample spectrum shown in Figure 9.13, the identified products include methyl derivatives of glycine (8), alanine (10), urea (12), glycine dimer (17), a triazine derivative (18) and small amounts of the pyrimidine uracil (19) and the purines adenine (22) and xanthine (23) [463].

Tholins are refractory organic materials formed by UV photolysis of reduced gas mixtures (N_2, NH_3, CH_4, etc.) under cold plasma conditions [465]. Various ways of energy injection have been used, including spark discharge [466], current discharge [467], and UV light [468]. Instead of gaseous mixtures, the irradiation of ice mixtures has also been used. Tholins are generally poorly soluble in common solvents and fluoresce strongly when excited by visible radiation.

Depending on the ratio of gas mixtures, the bulk samples of tholins have colors ranging from yellow to dark brown [469]. The optical properties of tholins is depen-

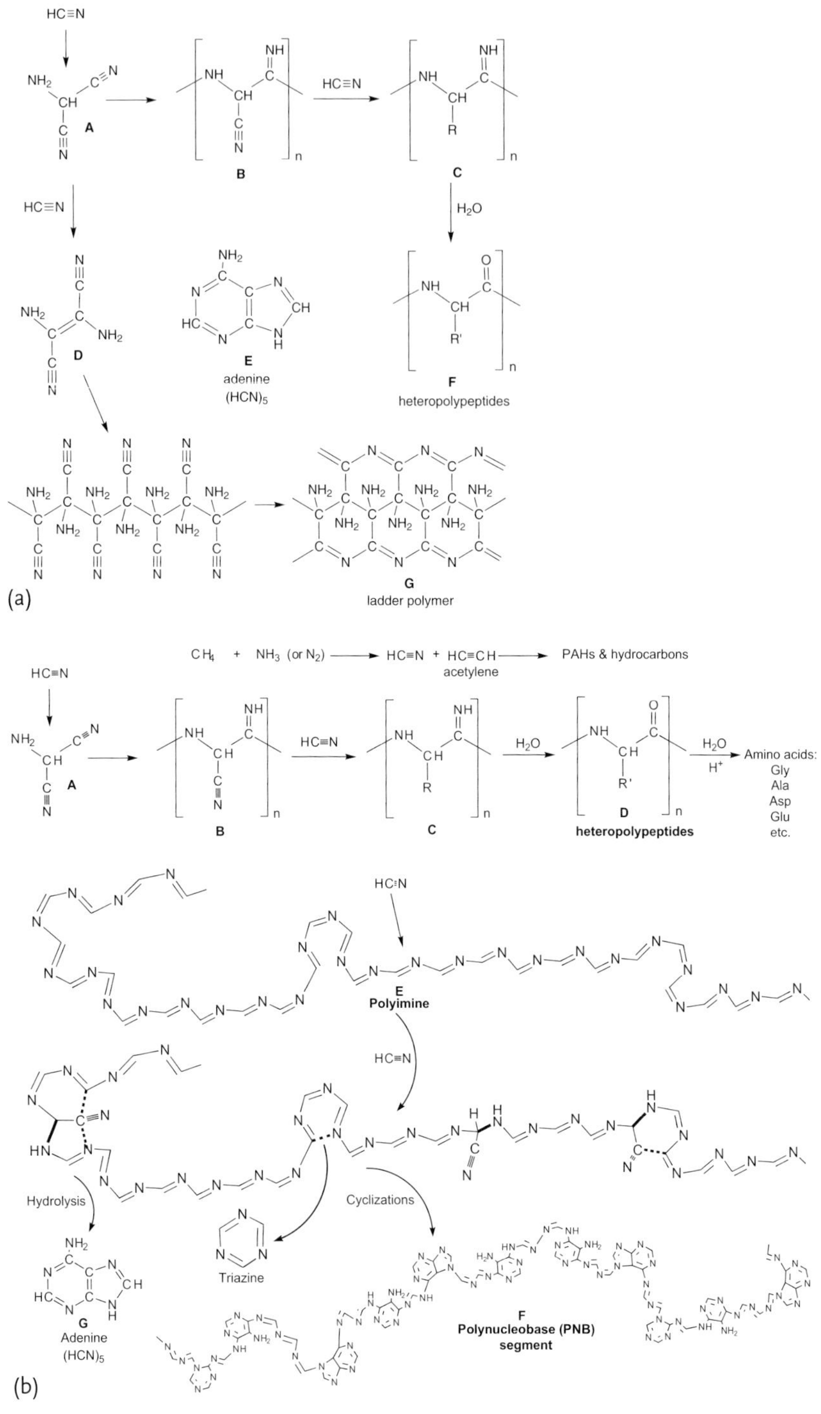

Figure 9.12 Pathways and structures for HCN polymerization proposed by C.N. Matthews. (a) shows polyamidine route to amino acids and (b) shows a chemical path to nucleobases via an imine structure of HCN polymer. Figures adapted from [463, 464].

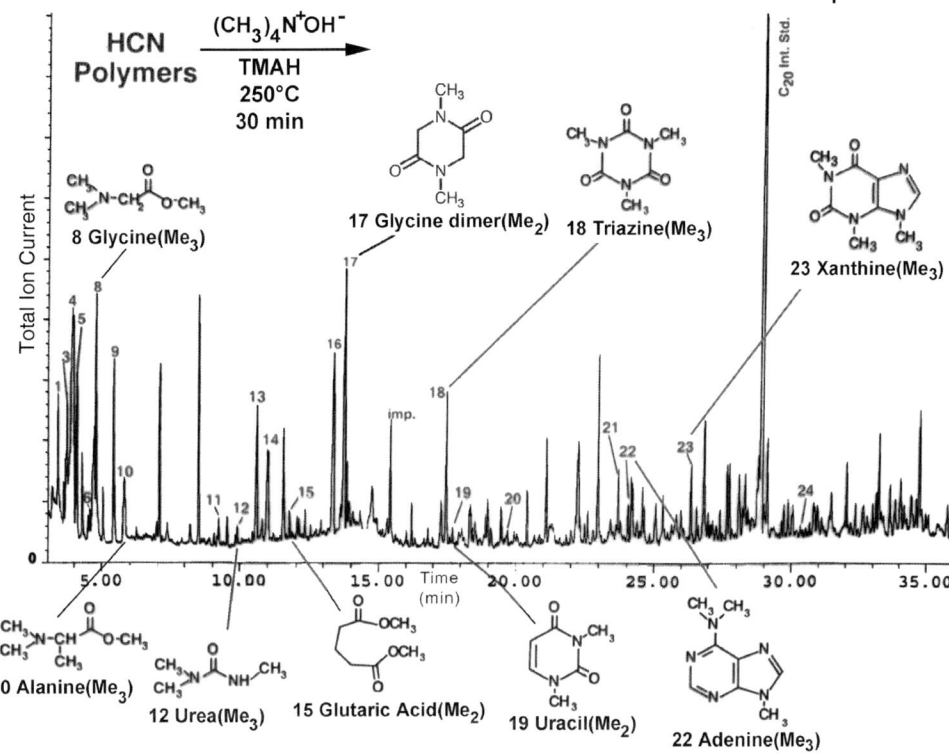

Figure 9.13 A GC-MS chromatogram showing products from the tetramethylammonium hydroxide thermochemolysis of HCN polymer. Figure adapted from [463].

dent on the chemical composition and the sp^2/sp^3 ratio. In terms of chemical structure, tholins can be described as amorphous hydrogenated carbon nitrides. High resolution transmission electron microscopy (HRTEM) images of tholins show a very disordered structure (Figure 9.14). This is confirmed by X-ray diffraction studies which show a broad continuum, suggesting a disordered or quasiamorphous structure (Figure 9.15).

Various functional groups have been identified through infrared spectroscopy, including cyanides ($-CN$), aliphatic CH_2 and CH_3, amines (NH_2) and C=N groups (Figure 9.16). UV Raman spectroscopy shows features at 690 and 980 cm^{-1}, pointing to the presence of C_3N_3 rings in the macromolecular network [469]. The reddish color of tholins's visible reflective spectra has led to the suggestion that tholins are abundantly present in the surface of Titan, TNOs, and asteroids [470].

Tholins and HCN polymers share many structural and spectroscopic similarities. In terms of differences, tholins are generally chemically and optically more homogeneous whereas HCN polymers can have a much wider range of chemical variations. The case for they being the analog of meteoritic and IDP organics is supported by the presence of the $-CN$ band (Figure 9.17).

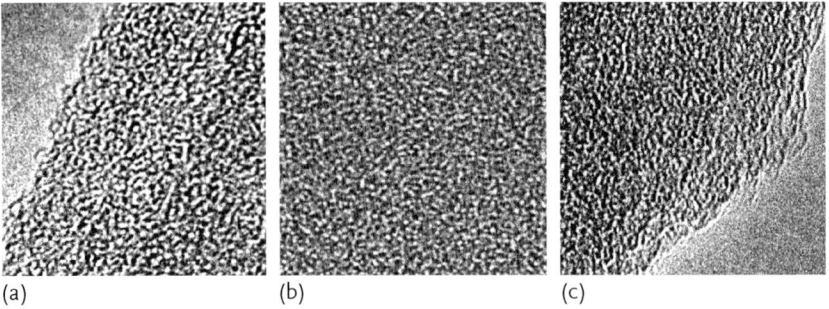

Figure 9.14 High resolution transmission electron microscopy images of an amorphous carbon film (b) and two samples (a), (c) of tholins made from different initial gas mixtures. Figure adapted from [469].

9.11
Biological Materials

Astronomers use the term "dust" to refer to solid-state particles of micron-size or smaller, which is assumed to be due to ice, metals, inorganic or organic compounds. However, the household "dust" that we encounter in our daily lives have a very different composition. In addition to organic particles such as soot and smoke, it also contains a large fraction of biological materials, namely, bacteria and fungal spores, skin flakes, lint, cotton fibers, and so on. Although some of these dust particles have natural origins, many are the result of human activities such as cooking, farming and industrial production.

Could cosmic "dust" also have a biological component? Just as minerals have specific vibrational bands, biological matter also has spectroscopic signatures. Many biological materials have absorption features in the infrared. There has been suggestion that polysaccharides can provide a good fit to the 10 μm astronomical absorption features. For example, cellulose, one of the most common organic compound on Earth, has a strong absorption feature in the 10 μm region [473]. Although there is no terrestrial mineral that provides an exact spectral match to the astronomical 10 μm feature, it is now commonly accepted that this feature is due to amorphous silicates.

The possibility that the 3.4 μm and other astronomical spectral features are due to biological materials (e.g., bacteria) has been made by Holye and Wickramasinghe [474]. Comparison of the laboratory spectrum of E. coli bacteria with high resolution astronomical spectra shows that they do not match well [322]. At present, there is no convincing evidence for the existence of biological materials in space.

Biological pigments have also been appealed to as the cause of the extreme red colors of asteroids and Kuiper Belt Objects. The red color of the surface of planetary surfaces such as those of Europa has been suggested to be due to microorganisms that are continuously generated and brought to the surface [475].

The ERE has also been attributed to chloroplasts and bacterial pigments [401]. Many biological pigments absorb blue light and fluoresce at longer wavelengths.

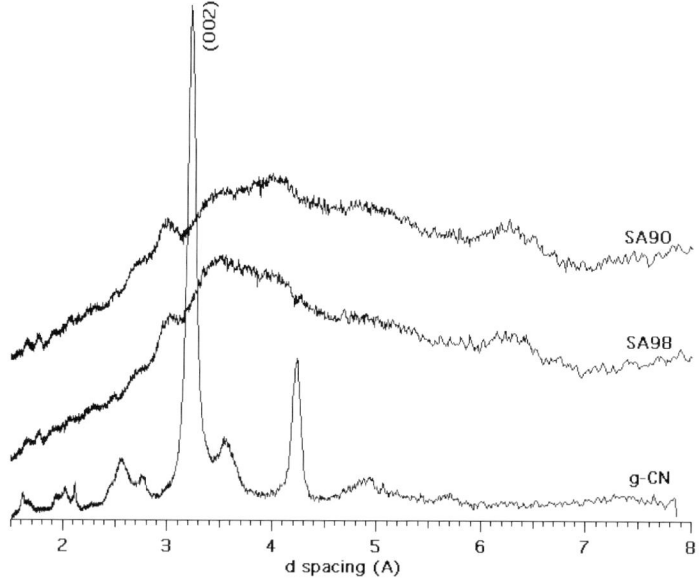

Figure 9.15 X-ray diffraction pattern of two tholins with different elemental compositions (top two curves) compared to that of a crystalline graphite carbon nitride (bottom curve). The diffraction pattern is clearly different from the highly crystalline nature of graphite-CN, suggesting a disordered structure of tholins. Figure adapted from [469].

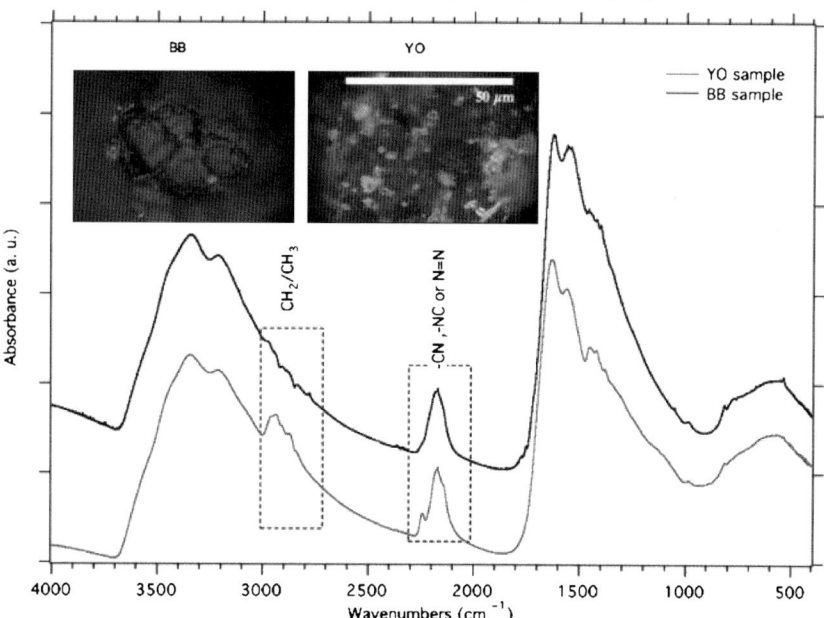

Figure 9.16 Laboratory spectra of two samples of tholins: black/brown (BB) and yellow/orange (YO). The 3.4 µm aliphatic CH_2/CH_3 and the 4.6 µm nitrile (−CN) and isocyanide (−NC) groups are identified. Figure adapted from [471].

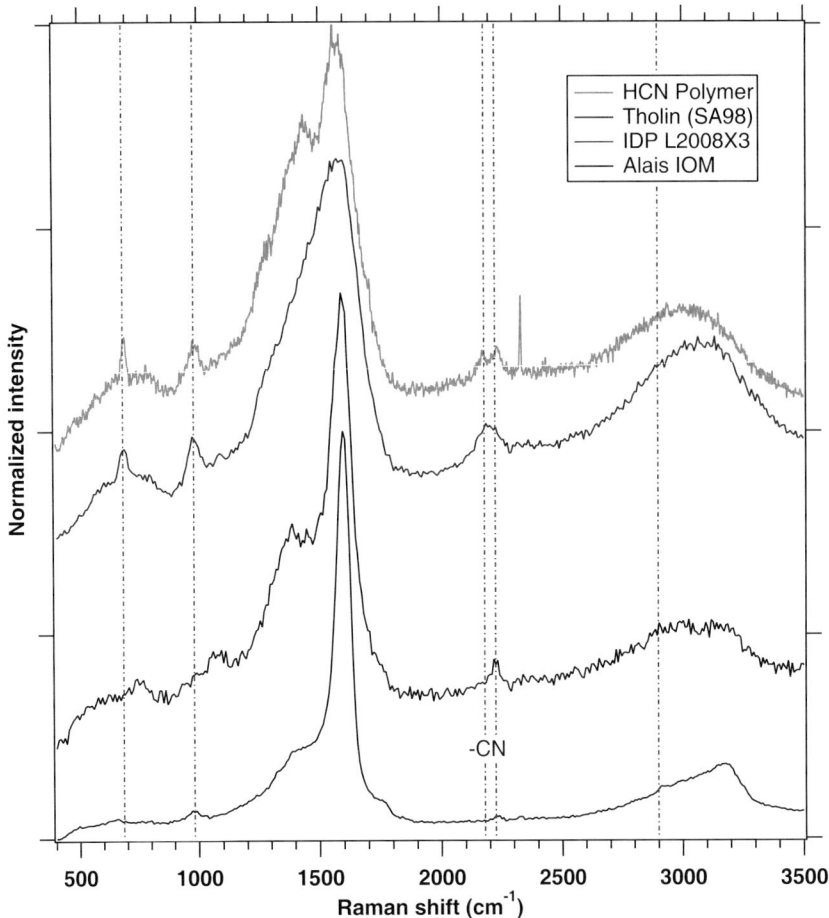

Figure 9.17 The Raman spectra of the IDP L2008X3 and the carbonaceous chondrites Alais (bottom two curves) in comparison with the spectra of HCN polymer and tholins (top two curves). Figure adapted from [472].

Both chlorophyll and many bioluminescent bacteria fluoresce in a broad spectrum centered in the red color. For the 227.5 nm feature, heterocyclic aromatic compounds have also been suggested as a carrier. The absorption properties of several isomers of $C_8H_6N_2$ (cinnoline, quinazoline, quinoxaline, phthalazine, etc.) have been calculated and were found to have absorption peaks around 220 nm [476]. Organic molecules with two or more multiple bonds such as $C=C-C=C$, $C=C-C\equiv C$, $C\equiv C-C\equiv C$, $C=C-C=O$, $C=C-C=N$, $C=C-C\equiv N$, and so on have electronic transitions in the 220 nm region [477].

The main issue of the biological hypothesis is not whether they exist, but whether they have the abundance to account for the strength of the observed features. For

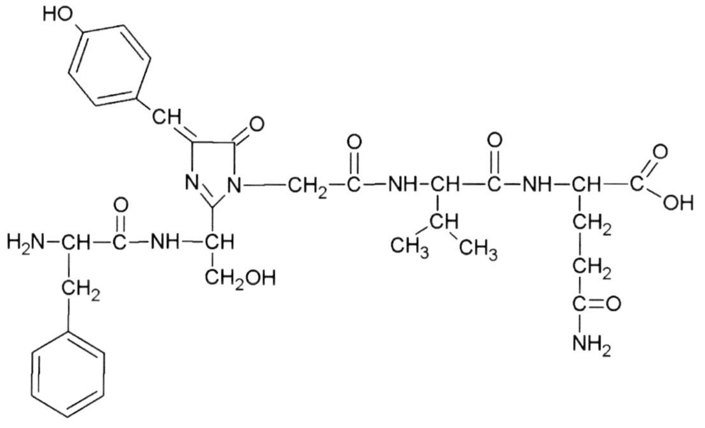

Figure 9.18 Chemical structure of the fluorescence green fluorescent protein.

example, the ERE and the 3.4 μm band are extremely common and wide-spread phenomena. A large amount of bacteria is needed in order to account for them.

In recent years, there has been an increasing degree of awareness of biolumi-nescence in nature. For example, the green fluorescent protein from the jelly fish Aequorea victoria has generated a significant amount of research to use it as a genetic tag to visualize spatial or temporal processes in living systems. The fluo-rescence displays an emission maximum at λ 509 nm upon exposure to blue light (absorption maximum at λ 400 nm and a weaker maximum at λ 470 nm). The green fluorescent contains 238 amino acids and the fluorescent activity has been traced to the Ser-Tyr-Gly peptide segment of the chromophore [478] (Figure 9.18). Genetic engineering has yielded other molecules that emit in the red part of the spectrum [479]. Since the fluorescence chromophore in the green fluorescent pro-tein contains 33 C, 9 O, and 7 N atoms, it would seem unlikely that such a complex biomolecule would be widely present in interstellar space. However, it is certainly possible that a variety of other similar fluorescent organic molecules collectively serve as the agent for the ERE, rather than having to exclusively rely on inorganic materials.

9.12
Summary

Our understanding of the chemical composition of astronomical "dust" has under-gone a significant evolution throughout the years. For the first part of the twentieth century, our knowledge of interstellar dust was derived from the extinction curve based on photometric measurements. The inference of them being metals, ice, or graphite is nothing more than conjecture. The actual observations of vibrational bands from infrared spectroscopic observations firmly established that amorphous silicates and silicon carbides are constituents of circumstellar dust. It was only after the IRAS spectroscopic survey in the 1980s that the astronomical community came to accept that silicates grains are widely present in the diffuse ISM and galaxies. The efforts toward carbonaceous grains of complex organic nature has taken even longer to be recognized. While it was easy to dismiss "coal in space" as being too fantastic, evidence has been gaining in favor of carbonaceous grains being a significant component of interstellar dust.

In this chapter, we discussed a number of organic compounds as possible carriers of unidentified astronomical phenomena, for example, DIB, UIE, ERE, and so on. While there is no clear consensus on their specific chemical origins, it seems likely that complex organic matter is needed to explain these phenomena. Actual identification of the specific substance has to rely on further laboratory simulated production of the substances and the spectroscopic studies of the products.

10
Laboratory Simulations of Molecular Synthesis

From the previous chapters, we have learned that nature can synthesize gas-phase organic molecules and complex organic solids in circumstellar envelopes of stars over very short time scales. Although we do not know the time scale of formation of molecules in the interstellar environment, over 160 molecules are detected in interstellar clouds. Various models for the synthesis of molecules under interstellar conditions through gas-phase chemical networks have been extensively explored. Under the interstellar radiation field, molecules undergo photoionization and photodissociation, and reform under biparticle neutral–neutral and ion–molecule reactions. Molecules can also collide with electrons through radiative recombination and dissociative recombination [480, 481]. Surface reactions on solid-state grains is increasingly recognized to play a significant role of the synthesis of molecules, from the simple H_2 molecule to complex organics.

We also learned that biologically relevant organic molecules, such as, amino acids and nucleic acid bases, are already present in the early Solar System through their detections in meteorites. The presence of IOM in meteorites also suggests that these complex organics are interstellar or stellar in origin and were incorporated in primordial materials of the presolar nebula. The chemical pathways leading to the synthesis of these organics in circumstellar and interstellar environments are therefore interesting topics in astrochemistry.

Our knowledge of chemical synthesis is limited to our experience in the laboratory. Although significant progress has been made in our understanding of biological systems through chemical structures and networks, the experiments that we conducted were performed under high-density conditions in thermodynamical equilibrium. The physical and radiation conditions of the circumstellar and interstellar environments are very different from the laboratory, and specifically they often correspond to nonthermodynamical equilibrium conditions.

Since the original Millar–Urey experiment [482], there have been many efforts in simulating the chemical reactions under conditions of the early Earth. Most of these experiments begin with a mixture of simple inorganics in a liquid solution which are subjected to energy injection in the forms of radiation [483], heat [484] or electric charges. The early experiments used strongly reducing gas mixtures which were thought to correspond to primitive Earth atmospheres. Later experiments use a less reducing mixtures, for example, CO_2, CO, N_2 and H_2O, which are

Organic Matter in the Universe, First Edition. Sun Kwok.
© 2012 WILEY-VCH Verlag GmbH & Co. KGaA. Published 2012 by WILEY-VCH Verlag GmbH & Co. KGaA.

now believed to be closer to the chemical makeup of the early atmosphere. Generally speaking, the amount of biologically relevant compounds produced is less than those obtained in a strongly reducing mixture.

10.1
Laboratory Simulation of Chemical Processes under Interstellar Conditions

As the number of organic molecules detected in the ISM increases, there have been an increasing degree of interest to simulate the chemical processes under interstellar conditions. A possible site of organic synthesis is on grain surfaces. In interstellar space, ice is expected to accumulate on the surface of solid grains. With the injection of external energy, chemical reactions can take place on the ice mantles of grains. A number of experiments have been conducted on ice or grain surfaces, which are subjected to UV light, γ rays, or high-energy particles bombardments. Starting with simple ices, for example, H_2O, CO, CH_4, NH_3, CH_3OH, C_2H_2 and so on at low (~ 10 K) temperatures, an organic substance called "yellow stuff" can be produced by irradiating the sample by UV radiation followed by warming. When this substance was carried into space and subjected to four months of solar radiation, its color changed from yellow to brown, suggesting increased absorptivity as the result of increased carbonization, photodissociation and depletion of O, N, and H from the sample. The near infrared spectrum of the final sample shows the 3.4 μm feature due to the methyl ($-CH_3$) and methylene ($-CH_2$) aliphatic groups (Section 4.4) as well as broad plateaus around 3 and 6 μm suggestive of the OH stretch in alcohols and carboxylic acids (3 μm), and C=C, C=O, C–OH, C≡N, C–NH_2, and so on stretches, to CH, OH, and NH_2 deformations (beginning at 5.5 μm and peaking at 6 μm). These spectral properties resemble those of kerogens (Section 9.8) and IOM extracted from carbonaceous meteorites (Section 7.5), and this final sample was given the name "organic refractory matter" by Mayo Greenberg [485, 486].

Variations of the Greenberg experiment have been performed using mixtures of H_2O, CH_3OH, CO, NH_3 ices (in ratios consistent with ices in molecular clouds). Upon ultraviolet photolysis, a rich variety of moderately complex organic molecules can be produced. These include CH_3CH_2OH (ethanol), HC(=O)NH_2 (formamide), $CH_3C(=O)NH_2$ (acetamide), R–C≡N (nitriles), and hexamethylenetetramine (HMT, $C_6H_{12}N_4$) as well as more complex unidentified organic materials that include amides ($H_2NC(=O)$–R), ketones (R–C(=O)=R′), and polyoxymethylenes (POM, $(-CH_2O-)_n$), where R is used to represent a generalized alkyl group [487]. Among the major organic residues produced is HMT, it contains four N per molecule and is known to yield amino acids upon hydrolysis by concentrated acid. If this molecule can indeed be produced in interstellar space as this laboratory simulation suggests, it would represent a significant step in the synthesis of prebiotic materials.

Other experiments involving ultraviolet photolysis of amorphous water ice mixed with CH_3OH, HCN, and NH_3 followed by warming and ice sublimation are known

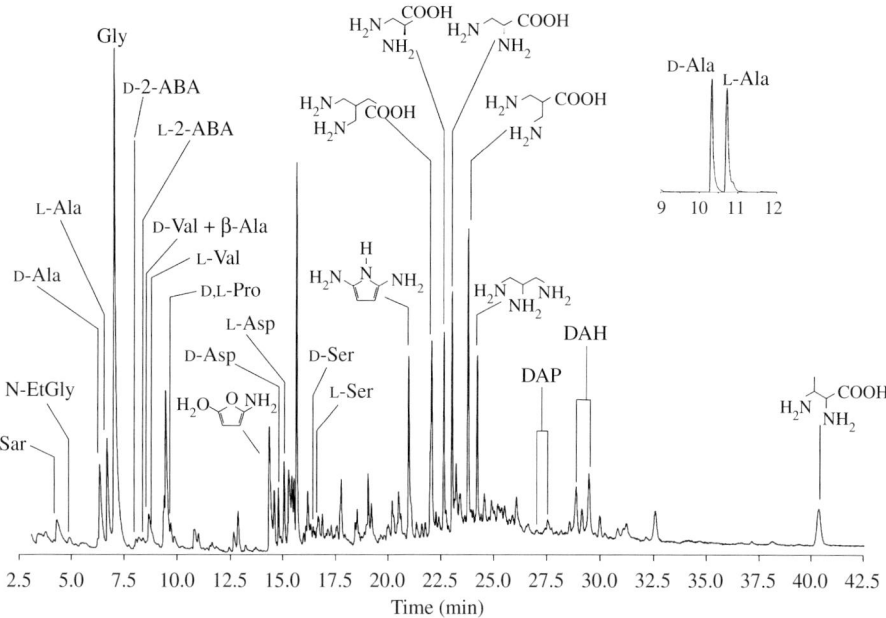

Figure 10.1 Gas chromatogram showing the amino acids (e.g., Gly, glycine; Ala, alanine; Pro, proline; Val, valine, etc.) and other compounds (e.g., DAP, diaminopentanoic acid; DAH, diaminohexanoic acid) generated in UV irradiated ice. Figure adapted from [489].

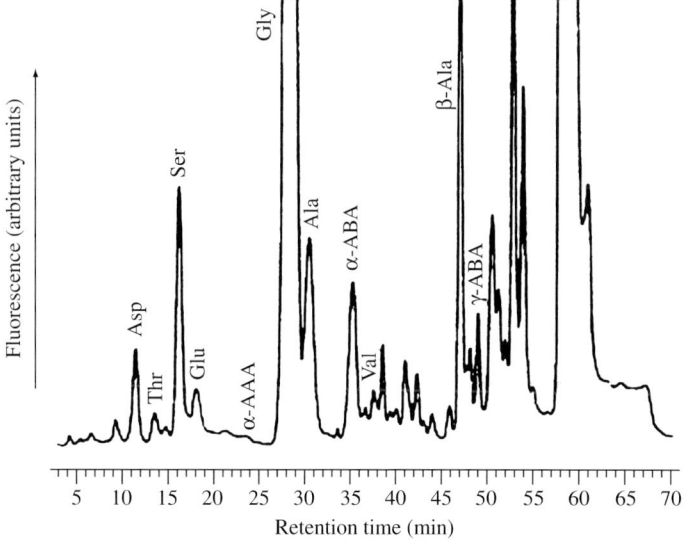

Figure 10.2 High-performance liquid chromatogram of amino acids produced by UV irradiation of icy mixture of methanol, ammonia, and water. The labeled products are: Asp: aspartic acid, Thr: threonine, Ser: serine, Glu: glutamic acid, AAA: aminoadipic acid, Gly: glycine, Ala: alanine, ABA: aminobutyric acid, Val:valine. Figure adapted from [490].

to produce organic residues containing *N*-formyl glycine, cycloserine, and glycerol. After hydrolysis, glycine, alanine, serine, glycerol, ethanolamine and glyceric acid are observed (Figure 10.1). These results suggest that spontaneous generation of amino acids in the ISM is possible [488, 489].

Similar experiments using different initial ingredients also yield many amino acids. When a gas mixture of carbon monoxide, ammonia, and water is irradiated with high-energy protons, a complex organic compound, which Kensei Kobayshi calls CAW, results. The molecular mass of this compound can be as high as 3000. When the same mixture is cooled in a nitrogen bath to turn into ice form and subjected to UV irradiation, a variety of proteinous amino acids (glycine, alanine, aspartic acid) and nonproteinous amino acids (β-alanine, α- and γ-aminobutyric acid) can be detected (Figure 10.2). By subjecting a mixture of water, methanol, ammonia and carbon monoxide ices to low temperature UV photolysis, it has been shown that a complex mixture of amphiphiles and fluorescent molecules can be created. Among the products are water insoluble droplets that show the properties of lipids [491].

10.2
Simulations of the Synthesis of Carbonaceous Nanoparticles in Space

The laboratory technique of laser vaporization of graphite followed by supersonic expansion into an inert gas is the most widely used method to create carbon clusters. Carbon clusters with sizes ranging from one to hundreds of atoms and in neutral, anionic and cationic states can be produced. The most spectacular success of this technique was the discovery of fullerene [32], although clusters of different masses and isomeric forms can be produced by changes in the laser power, gas pressure, and geometry of the nozzles. Variations of this method such as laser vaporization of thin diamond films and organic polymers have also proved to be successful in producing carbon clusters.

The realization that carbon clusters form spontaneously through energetic processing of carbon-rich materials has led to the speculation that similar processes are at work in the atmospheres of carbon stars. The carbonaceous compounds made in the circumstellar envelopes of protoplanetary nebulae are likely to be a solid-state material made up of islands of aromatic rings connected by aliphatic chains. In addition to hydrogen, other functional groups including other heavy elements O and N, such as methyl ($-CH_3$), methylene ($-CH_2$), carbonyl (C=O), aldehydic ($-HCO$), phenolic ($-OH$), and amino ($-NH_2$) groups can be attached to the peripherals of these aromatic units. These structures are subjected to radiation processing as the star evolves to the planetary nebulae stage with an accompanying increasing amount of UV radiation output. This could result in consolidation of the rings and aromatization of the structure. To what extent these chemical structures are further processed by interstellar radiation field once they are ejected into the interstellar medium is uncertain.

We now know that the synthesis of amorphous aromatic and aliphatic compounds from gas-phase molecules occurs in the circumstellar envelopes of evolved stars over time scales of less than 1000 years (Section 6.2). How such reaction can take place in the gas phase under extremely low density ($n_H \sim 10^6 \, cm^{-3}$) conditions is difficult to understand theoretically. The process may resemble that of soot production in flames where gas-phase molecules first nucleate into nanometer-size particles and then coalesce and grow into grains of thousands of atoms [492]. This scenario is supported by the fact that polyacetylenes first appear in the post-AGB phase of evolution. In the ISO spectrum of the protoplanetary nebula AFGL 618 (Figure 6.13), diacetylene (C_4H_2) and triacetylene (C_6H_2) are seen in addition to acetylene. These polyacetylenes probably contribute toward the size growth as well as the aromatization of the grains in the post-AGB phase of evolution.

Based on combustion experiments under typical terrestrial hydrocarbon pyrolysis conditions, acetylene in argon or air at pressures of $\sim$1 atmosphere, PAHs are formed at temperatures of $\sim$2000 K through hydrogen abstractions and acetylene additions. Model simulations including a network of chemical reactions under circumstellar conditions show that PAH production occurs in temperature interval of 900–1100 K [493]. An illustration of possible chemical pathways of circumstellar synthesis starting from acetylene to benzene to aromatic molecules is shown in Figure 10.3.

Using the formation process of soot as a guide to the condensation of carbonaceous grains in stellar atmospheres, we can assume that the following steps are necessary. First, there must be a step of homogeneous nucleation where polyacetylenes and small aliphatic molecules are grouped into ring-like structures, followed by aggregation of multiple aromatic rings. The transition from gas-phase molecules to solid-state particles probably occurs under non-LTE conditions and the details are unknown except for the observational fact that we know that it happens. Once the particles of molecular weights of 10^3–10^4 are formed, they can collide and coalescent with each other to form larger spherical particles [494]. There also exist possibilities of surface migration and the development of active surface sites for reaction with free-flowing gas molecules. Finally, particles may stick together to form chain-like structures, according to the rules of fractal geometry. For example, an ensemble of N units can proceed to form an ensemble of $N/2$ two-unit clusters, which in turn form an ensemble of $N/4$ four-unit clusters. The final shape of the grain can be quite irregular and will depend on the details of the growth process [495]. The processes of grain nucleation and growth have been extensively studied through empirical and physical models in the combustion science community [496].

The detection of organic hollow globules in carbonaceous chondrite meteorites [497] has led to laboratory simulations for the production of these globules in a He and benzene plasma [498]. It is suggested that these organic globules represent the final products in the evolution of carbonaceous matter in the circumstellar envelopes of evolved stars. Benzene grows to PAHs and the coagulates to form aromatic grains. The vacancies are produced in the grains by energetic plasma particles, eventually forming a hollow interior.

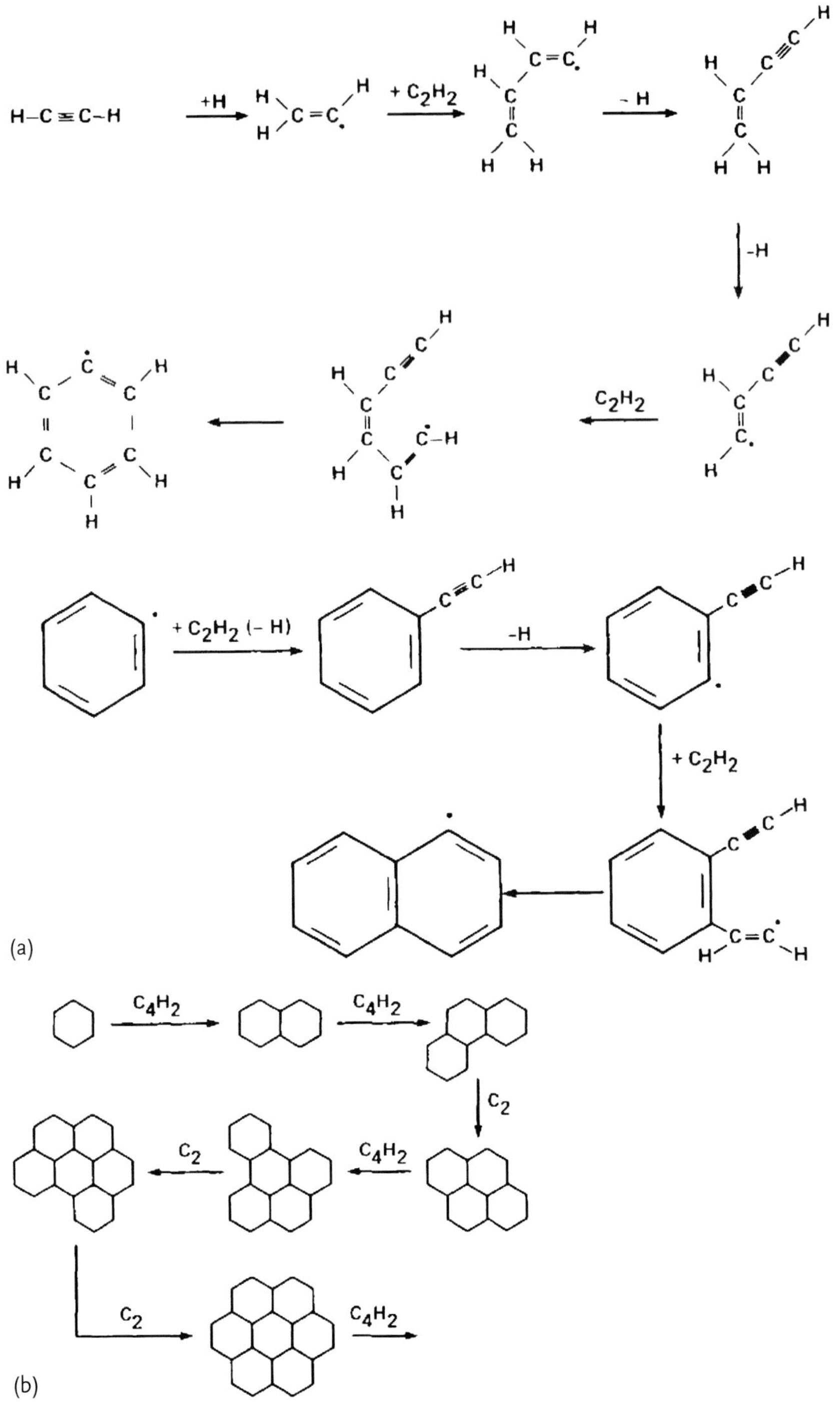

Figure 10.3 (a) and (b) represent two possible chemical pathways of chemical synthesis of aromatic compounds in circumstellar envelopes of evolved stars. Figure adapted from [171].

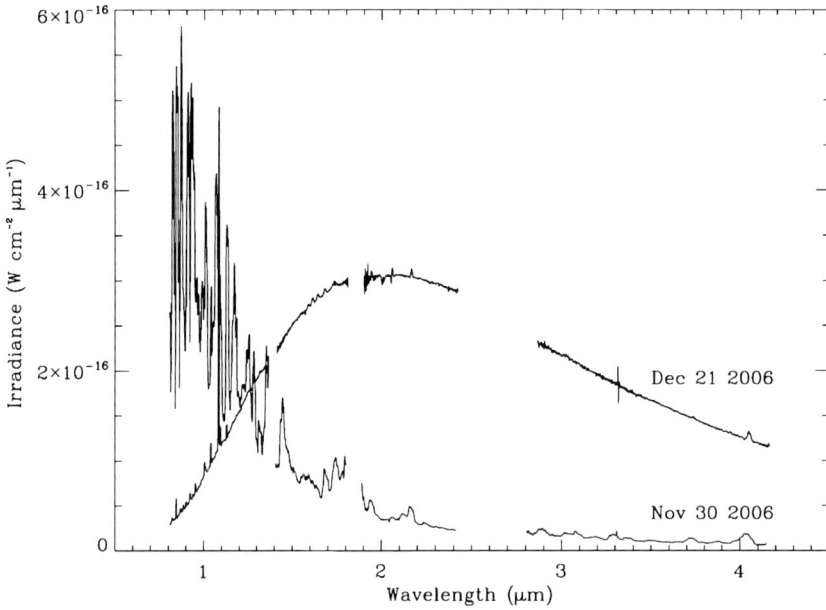

Figure 10.4 IRTF SpeX spectra of V2362 Cyg on November 30, 2006 and December 21, 2006, showing a dramatic change from an ionized gas spectrum to a dust dominated spectrum. The dust component has a color temperature of 1410 ± 15 K. Figure adapted from [500].

No matter what the theoretical difficulty in dust condensation is, nature manages to do it with no problem. Since the 1970s, it has been known that dust condenses in the outflow of nova ejecta in a matter of days, as soon as the condensation temperature is reached [499]. The nova V2362 Cyg, which had an outburst on April 2, 2006, was followed by multiple ground-based and space-based spectroscopic observations. Figure 10.4 shows that dust condensation took place within a period of less than one month [500]. Figure 10.5 shows the spectrum of the nova at 446 days after outburst. Above the dust continuum, one can see a prominent dust feature and excesses at 8, 12, and 17 μm. This shows that organic solid particles have condensed in the nova ejecta.

The UIE features are detected in the classical nova V705 Cas, with detected features at 3.28, 3.4, 8.1, 8.7, and 11.4 μm, where the 8.1 and 11.3 μm features are attributed by the observers as shifted 7.7 and 11.3 μm features respectively [501]. When the first infrared spectrum was taken at day 157 after the outburst, the 8.2 μm feature is already present. It continued to strengthen. The 11.4 μm feature was detected in day 251, with both fading with time (Figure 10.6).

The ease of formation of organic dust is not limited to novae. On May 14, 2008, an optical transient source was discovered in the galaxy NGC 300. The luminosity of the source was not as high as a supernova, and was interpreted as an energetic explosion of a 10 $M_\odot$ star [503]. A Spitzer IRS spectrum of the object obtained on August 14, 2008 shows strong, broad emission features at 8 and 12 μm (Figure 10.7) [504].

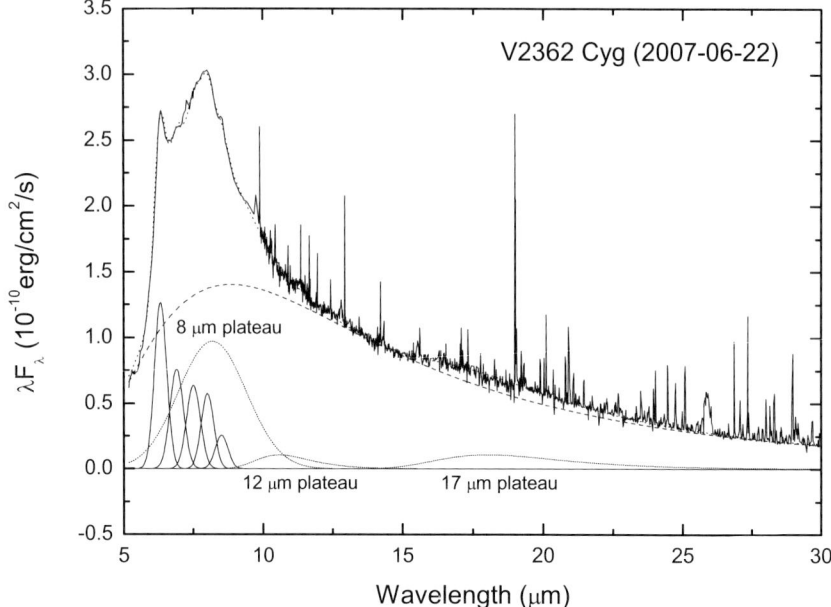

Figure 10.5 Spitzer IRS spectrum of V2362 Cyg at 446 days after outburst. The spectrum has been fitted by a modified blackbody ($\lambda^{-0.5} B_\lambda$ (365 K), shown as a dashed-line) plus plateau features at 8, 12, and 17 μm (dotted lines), and discrete features at 6.2, 6.9, 7.6, 8.3, 8.6 μm.

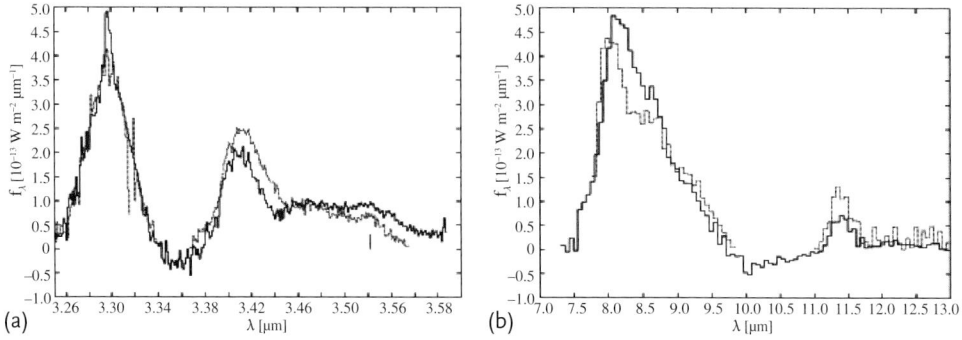

Figure 10.6 Profiles of the 3.3 and 3.4 μm features (a) and the 8.2 and 11.4 μm features (b) of Nova V705 Cas at two different epochs (253 and 320 days after outburst, shown as solid and broken lines respectively). Figure adapted from [502].

We note that both the classical novae and this supernova-like transient source first form dust, their spectra show strong, broad emission features around 8 and 12 μm, resembling the aliphatic in-plane and out-of-plane bending modes that are seen in protoplanetary nebulae (Section 6.2). It is reasonable to expect that when a carbonaceous solid first form suddenly condenses from the gas phase, its chemical structure must be heterogeneous. With a small number of aromatic rings formed

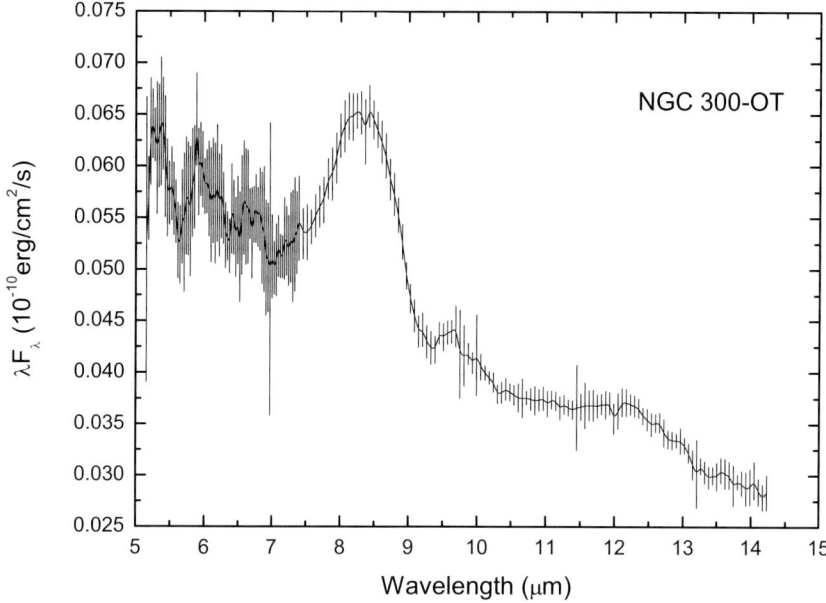

Figure 10.7 Spitzer IRS spectrum of the optical transient source NGC 300-OT 93 days after discovery. Broad emission features around 8 and 12 μm can be seen. Figure adapted from [504].

(probably through the soot condensation schemes discussed above), a diverse arrays of aliphatic branches can easily attach to the rings, and both their in-plane and out-of-plane vibrational modes create the broad features observed in novae and supernovae.

10.3
Chemical Pathways of the Synthesis of Biomolecules

As of yet, we do not know what is the most chemically complex organics that the early Earth inherited from the Solar System. If we assume that there were only nonbiologically materials present, then there must be chemical pathways that transformed these ingredients into the simplest forms of life. In order for a living system to be created from basic nonliving matter (Section 11.2), several building blocks of life are needed: (i) amino acids for building proteins; (ii) phosphate for the backbone of nucleic acids; (iii) purine and pyrimidine bases for the genetic code of DNA and RNA; (iv) sugars for the nucleic acids; and (v) fatty acids for cell membranes. The availability of these basic ingredients in the ISM has been discussed in the previous chapters (amino acids, Section 3.15.1; phosphate, Section 3.12; purine and pyrimidine, Section 3.15.3; sugars, Section 3.15.2). The detection of fatty acids and the amphiphilic molecules that make up cell membranes is more remote. All that one can say from astronomical observations is that we do observe carbon chains (Section 3.9) that could be combined into longer chains.

10.3.1
Synthesis of Amino Acids

Proteins (polypeptides) are formed by the joining of amino acids by peptide bonds. Two amino acids are linked together to form a dipeptide by eliminating an H atom from one and a hydroxyl group ($-OH$) from another, and releasing a H_2O molecule in the process. However, this process requires energy input to proceed in water. It is difficult to synthesize polymers to the lengths of 30–100 as required to get a self-replicating system going. A scenario to build up polymers on mineral surfaces was developed by Leslie Orgel [505]. In this scenario, clay minerals act as catalysts for polymerization reactions, allowing oligomers to indefinitely grow in length.

10.3.2
Synthesis of Nucleic Acid Bases

The bases found in nucleic acids have two (pyrimidines) or four (purines) N atoms in the aromatic rings (Figure 2.6), and how the N atoms can be incorporated into the carbon rings is a key question within the synthesis of nucleobases. The simplest aromatic heterocyclic molecule pyridine (C_5H_5N) can be formed by adding a HCN molecule to a phenyl radical (C_6H_5). The gradual addition of four HCN molecules lead to the synthesis of diaminomaleonitrile (DAMN, $(H_2N)_2-C=C-(C\equiv N)_2$), which upon photolysis forms diaminofumaronitrile and aminoimidazole carbonitrile [506]. The addition of another HCN forms adenine, one of the five nucleic acid bases (Figure 2.6). The introduction of NH_3 can react with diaminomaleonitrile to form guanine and adenine. Reactions between cyanoacetylene (HC_3N), a molecule widely detected in circumstellar and interstellar environments, with cyanate (NCO^-) can lead to the formation of cytosine and uracil.

10.3.3
Lipids

Protocells are made possible by the formation of membranes through the synthesis of lipids or fatty acids. In present life, the lipids used in the formation of cell membranes in archaea are different from those in bacteria and eukarya, and the pathways to their synthesis are different. While lipids in bacteria consist of ester bonds between glycerol and fatty acids, lipids in archaea consist of chains of twenty or forty carbon atoms. In the laboratory, straight chain fatty acids with five to twenty carbon atoms have been synthesized using nickel iron as a catalyst [507].

As possible precursors of lipids, there are cyanopolyynes that have been extensively detected in the circumstellar environment (Section 3.9.1), from which hydrogenation and hydrolysis of the cyano group can yield fatty acids. Straight carbon chains such as C_3, C_5 (Section 3.9.3) are known to be present in the circumstellar and interstellar medium, and the addition of HCN can yield aliphatic nitriles leading to fatty acids [507].

10.3.4
Synthesis of Sugars

Laboratory synthesis of sugars begins with two formaldehyde molecules condensing to form glycolaldehyde (CH_2OHCHO, 2 C atoms), which in turn reacts with another formaldehyde molecule to form glyceraldehyde ($CH_2OHCHOHCHO$, 3 C atoms), 1,3-dihydroxyacetone ($CO(CH_2OH)_2$, 3 C atoms), and other sugars. This process is called formose (short for formaldehyde and aldose) reaction [508]. Since sugars form the backbone of DNA and RNA, their syntheses in space are of great interest. Both formaldehyde (Section 3.5.1) and glycolaldehyde (Section 3.15.2) have been detected in the ISM, and the detections of longer monosaccharides are within current technical capabilities. The detection and abundance determination of a series of sugars will allow us to test the theories of synthesis pathway.

10.3.5
The RNA World

Assuming that the above products are made available, either delivered from space or synthesized on Earth, what will happen next is unclear. Nucleic acids contain the codes that instruct the biosynthesis of enzymes. However, enzymes are now known to catalyze the replication of the nucleic acids. So the question remains as to whether the formation of DNA preceded that of protein, or vice versa? Since RNA can serve both as an information carrier and as a biological catalyst, it has been suggested that RNA may catalyze the formation of other RNA molecules and thus can replicate itself. The RNA molecules are therefore the precursors to the present DNA protein world.

Development in molecular biology has convincingly demonstrated the remarkable similarity of the gene sequences among all living organisms, strongly pointing to the conclusion that all modern organisms are derived from a common progenitor (Last Universal Common Ancestor, or LUCA). Therefore, LUCA represents the "root" of the tree of life, before the branching of the bacteria on one side and archaea and eukaryotes on the other. This early microorganism probably used nucleic acids for information storage and proteins as catalysts, similar to modern bacteria.

11
Origin of Life on Earth

The study of the origin of life as a scientific discipline has to be seen in the historical context of (i) the Copernican revolution where Earth is no longer seen as the center of the Universe; (ii) the expansion of spatial scale, where the Sun is only one of a hundred billion stars in the Galaxy and our Milky Way is one of hundreds of billions of galaxies; and (iii) the Darwinian revolution where advanced life forms on Earth descended from simple life forms over billion-year time scales. The fact that the Earth is among more than 400 known extrasolar planets and is probably just a nondescript piece of rock among a billion others also reinforced our present idea of the Earth being far from unique. The main uncertainty is whether biological evolution is so complex and difficult that only one planet in our Galaxy has managed to create life so far. Do physical conditions such as temperature, chemical composition of the atmosphere, and the existence of water on the surface put such stringent constraints on life that it would never flourish anywhere else? To what extent did geological events such as continental drift and extraterrestrial impacts influence the development (and termination) of life on Earth?

It has been argued that given the old age of our Galaxy and the number of stars and planets inhabiting the Galaxy, any existing life forms elsewhere would have certainly visited us already. However, even if extraterrestrial life forms had visited us, we may not have recognized them. For example, if our young and relatively backward technological society had the ability to go back several hundred years to leave behind a DVD containing thousands of pictures, videos and music, our ancestors would not be able to see it as more than a piece of shining metal, or to decipher its contents. An artifact left behind by an alien advanced civilization is likely far too elusive or mysterious for us to notice and comprehend. The absence of evidence for visits by extraterrestrials is therefore no proof of their not having done so. If we were indeed visited, either by advanced life forms or by robots that they sent, they could have easily seeded life on Earth without our ever realizing it had happened.

It is clear that some hypotheses on the origin of life, although within the realm of possibility, are difficult to disprove. As scientists, all we can do is to use our present knowledge of astronomy, physics, chemistry, and biology to investigate whether theories of the origin of life stand up to observational and experimental tests.

Organic Matter in the Universe, First Edition. Sun Kwok.
© 2012 WILEY-VCH Verlag GmbH & Co. KGaA. Published 2012 by WILEY-VCH Verlag GmbH & Co. KGaA.

11.1
Design or Accident

The oldest theory of the origin of life is divine intervention by a supernatural being that created life. A variation of this hypothesis is that of seeding or experimentation by intelligent extraterrestrials. An advanced civilization elsewhere could have either directly seeded the Earth with life or provided basic ingredients with automatic instructions to create life. A specific example of this hypothesis is the release of bacteria from a spaceship [509].

A more fantastic scenario is that life on Earth was developed as an experiment or a zoo, similar to a bacterial colony in a medical laboratory or an ant hill in a sand box. Although both bacteria and ants possess a certain degree of intelligence, are they able to recognize the true state of their situation? The realization of such a situation would greatly hurt our pride, but it is probably no more than the shock and disappointment that human beings went through several hundred years ago when we learned that the Earth was not a chosen place specifically built for us, and the Sun, the Moon, and the stars were not created for our pleasure or convenience. The Darwinian revolution that occurred 150 years ago showed that humans, plants, animals, and bacteria all had the same common ancestor has probably humbled us more than any other event in our history.

11.2
Prebiotic Chemical Evolution: the Oparin–Haldane Hypothesis

The theory of the spontaneous creation of life, which posits that life spontaneously arises from nonliving matter in a short period of time, today as in the past, was popular for several hundred years. It was convincingly disproved by the experiment of Louis Pasteur, who in 1864 showed that these supposedly spontaneously created forms of life in fact are derived from preexisting microorganisms. The conclusion is that on Earth today, life only comes from life. If life indeed originated from nonliving matter, it could only have done it billions of years ago when conditions on Earth were very different from today's.

Aleksandr Ivanovich Oparin (1894–1980) [510] and John Burdon Sanderson Haldane (1892–1964) [511] first proposed the idea that the origin of life on early Earth can be explained using the laws of chemistry. This hypothesis is built upon the success of organic synthesis such as Friedrich Wöhler's 1928 artificial synthesis of urea, thereby dispensing the idea of the "vital force" (Chapter 1). After the Miller–Urey experiment in 1953 [482, 512] which demonstrated that complex organics such as amino acids and sugars can be synthesized from simple molecules such as hydrogen, methane, ammonia, and water by the injection of energy into an hospitable environment, the endogenous origin of organic matter and life has remained the dominant theory for the last 50 years.

The requirements for the Oparin–Haldane hypothesis are (i) a suitable environment (the "primordial soup") with correct physical conditions and chemical compositions; (ii) external injection of energy in the form of processes such as solar radiation, lightning, volcanic eruptions, geothermal, external impacts, and (iii) yet unknown mechanisms to transform organic molecules into life, creating the combined abilities to absorb energy, self assemble into forms increasing complexity, and evolve. The third requirement is particularly challenging for it requires the molecular entity to replicate itself – but not perfectly, so that the replica will have room to improve itself through natural selection.

Even if complex organic molecules can be synthesized from simple inorganic molecules with ease, there is still a long way to life. It is well known that if one applies energy to a mixture of organics, one is likely to obtain amorphous organic structures similar to tholins or tar (Sections 9.8, 9.10). In fact, in the original Miller–Uray experiment, tar was produced together with a small amount of amino acids [482]. A similar result was also observed in the synthesis of adenine from HCN [461]. Experiences in chemical laboratories have shown that organics do not tend to self-organize spontaneously.

Most chemical pathways lead to many different variations in chemical structures. Among the many three-dimensional structures possible, our known forms of life only utilize a small fraction of the possibilities available. Do the other pathways all lead to dead ends, or is there an infinite number of possible life forms? This problem is being tackled on two sides: beginning with experimental synthesis using inorganic ingredients (Chapter 10) and the search for the simplest life forms from geological fossil records.

In the extreme form of this reductionist philosophy (which Oparin himself subscribed to), life is nothing more than a form of matter and it is possible to create life artificially. Today's scientists are much more cautious. The majority would take the view that life developed on Earth from nonliving matter over a very long time and under special circumstances and there is no realistic chance of creating life in the laboratory of today.

11.3
Panspermia

Panspermia refers to the idea that life is common in the Universe and it is spread from place to place through interstellar travel of life forms. Although theories of panspermia have been around for some time [513], it has never found much support except in the works of Holye and Wickramsinghe [514–516]. These authors argue that if life can develop from inorganic matter on Earth, life must have been common in millions of other solar systems as the Galaxy is old ($\sim 10^{10}$ yr). Living organisms from those systems could just as easily have been transported to our Solar System and seeded life on Earth. They also cited the fact that microorganisms can survive and indeed thrive under extreme conditions as evidence that bacteria can endure long interstellar journeys. The analogs of bacteria revived from bees embedded in amber for 25–40 million years [517] and in 250 million year old salt crystals [518] have also been cited as evidence of the viability of panspermia.

11.4
Evidence for External Impacts in the Past

Results from recent planetary explorations have demonstrated unequivocally that terrestrial planetary surfaces have many craters as the result of meteorite, asteroid, and comet impacts. There are about half a million craters on the Moon with diameters larger than 1 km, and impact structures with a diameter >300 km (called impact basins) number about 40. The dark regions of the Moon (called maria, Latin for seas) are in fact craters of gigantic proportions, now filled with lava from later volcanic eruptions. An example of a lunar impact site is the Orientale Basin, which is 930 km in diameter and was formed by impact about 3.80–3.85 billion years ago (Figure 11.1). Between 4.5 and 3.8 billion years ago, external impacts created large basins covering the entire surface of the Moon (Color plate C2). The impact rate during that time period is estimated to be at least 100 times higher than the rate today.

Since the Earth has a much larger surface area than the Moon, the number of impacts suffered by the Earth must have been correspondingly higher. In addition to being geometrically larger, Earth also has a much larger gravitational cross-section to attract near Earth objects. Assuming current asteroidal approach velocities, the Earth should have suffered 2.5 times the lunar impact rate per km^2 [519]. It is estimated that there have been more than 22 000 impact craters created with a diameter larger than 20 km, with ∼40 impact basins ∼1000 km in diameter, and several with diameters of ∼5000 km [520]. Unfortunately the crater record of the early Earth has been completely erased by erosion, tectonics, and other geological events. Terrestrial impact sites are preserved only if they are young, large, located in a geologically stable region, or were buried by younger sediments which have since been removed.

Our appreciation of external impacts on Earth is relatively recent. The first terrestrial impact crater was recognized only in the early twentieth century [521], and the idea of external impact remained controversial for several more decades. The number of known terrestrial impact craters is about 160, although new examples have continued to be discovered [522]. The oldest crater on Earth is suggested to be Suavjarvi in Russia, which has been traced back to ∼2.4 Ga. Other large, old craters are Vredefort (2.02 Ga) and Sudbury (1.85 Ga). Although it is difficult to estimate the sizes of the original craters due to subsequent erosion, the suggested sizes are 200–300 km and 150–200 km respectively [523, 524]. The best-known crater is probably Chicxulub, the result of an impact that occurred at the Cretaceous-Tertiary boundary. This impact is believed to have caused the massive extinction event 65 million years ago [525–527]. The discovery of the K-T impact has led to an increased appreciation of terrestrial impacts as an agent for global change.

Where do these impactors come from? When the path of a meteorite is captured on film or tracked by other techniques, the pre-impact orbit of the parent body can be calculated. For example, the orbits of the Innisfree, Lost City, Peekskill, Pribram, and Tagish Lake meteorites can be traced back to the main asteroid belt.

Figure 11.1 The Orientale Basin on the Moon as imaged by the Lunar Orbiter IV spacecraft. The near side of the Moon (as well as the direction of the Earth and the Sun) is to the right of this image. An outline of the State of Texas is overlaid for size comparison. Illustration credit: Lunar and Planetary Institute and Rick Kline of Cornell University.

The size distribution of the impact craters on the ancient surfaces of the Moon, Mars, and Mercury suggests that the impactors originated from the main asteroid belt [528].

11.5
Impact-Origin of Life Hypothesis

It is commonly assumed that the impact rate gradually declined from a peak at 4.5 Ga to a rate at 3.8 Ga similar to the present date rate. However, some have argued for a more rapid decline followed by a cataclysmic period between 3.9– 4.0 Ga [529]. Such an impact period would have stripped away the existing atmosphere and vaporized the oceans of the early Earth. The dust created by vaporized rocks in the atmosphere would have blocked a substantial fraction of sunlight, making the creation of any life form based on photosynthesis difficult.

The time of the cataclysmic impact is nearly coincident with the emergence of life on Earth, estimated to have occurred around 3.85 Ga. Are these two events causally related? It has been suggested that impact events created subsurface hydrothermal systems that extended across the diameter of the craters and downward to depths of several kilometers. These subsurface hydrothermal systems had conditions conducive to prebiotic chemistry and eventually led to the first thermophilic organisms [530]. In this scenario, life originated as the result of conditions created by impacts.

11.6
Delivery of Organic Compounds in the Early History of the Earth

The crater record of the terrestrial planets provides the main constraints on the total mass delivered to Earth as well as the size distribution of the impactors. The total mass delivered to Earth from all accumulated impacts is estimated to be 13–500 times that of the Moon and the total energy imparted exceeds 10^{28} J, or 3×10^{12} megatons of TNT [520]. Given the amount of extraterrestrial matter delivered to Earth, there must have been a contribution to the organic content on Earth from external delivery. There are three categories of objects that contribute to the delivery of organic matter to Earth. One is the km-size asteroids and comets with masses between $10^{13}–10^{15}$ kg. The second group is meter-sized objects with masses $\sim 10^6$ kg. The third group is meteoric dust with submillimeter sizes and masses around 10^{-9} kg. Contributions from the first group of objects probably peaked during the Hadean eon 4.5–3.8 Gyr ago and peaked around 3.9 Gyr ago during the Late Heavy Bombardment (LHB) period. The LHB impactors primarily consist of asteroids ejected from the main asteroid belt by large-scale dynamical instability. Meteorites constitute the second group of objects. The third group of objects consists of micrometeorites and IDPs, which are distinguished only through their collection method. While particles collected in Antarctica and Greenland are referred to as micrometeorites, those collected in the atmosphere are called IDPs. The delivery rate of this group is estimated to be $\sim 4 \times 10^7$ kg yr^{-1} [531].

At least some of the micrometeorites and IDPs are remnants of comets. We know that meteor showers are caused by leftover dust particles of earlier comets that crossed the Earth's orbit. Comets that have exhausted their volatile components would still retain their solid nuclei made up of organic compounds. Collisions between these objects and Earth can deliver organics to Earth. Comets therefore have been suggested to have influenced the chemical makeup of the early Earth [532]. Since comets retain their original chemical composition, the chemical makeup of comets that we observe today is probably representative of the molecules and solids that were delivered to Earth by past impacts.

The current thinking of the history of the Solar System is that most of the mass of the original solar nebula was incorporated into the Sun, and the remaining planetesimals aggregated into planets or were ejected from the Solar System. However, a small fraction of the pristine materials resided in comets and asteroids. These bodies were not subjected to extensive processing as the rest of the Solar System was and therefore may retain much of the original nebular materials, including organic compounds. The possibility that the early Earth was enriched by organic compounds delivered by extraterrestrial sources has been under discussion for some time [533]. If organic matter in IDPs survived heating during the atmospheric deceleration, they could contribute a significant amount of prebiotic organic matter to the surface of the Earth. Based on the average fraction of organics found in meteorites and the observed IDP flux (Section 7.6), the current flux of extraterrestrial organics hitting the Earth is 300 ton yr^{-1} [252]. This possibility was considered by Edward Anders [534] as important to the origin of life (Color plate C 3).

Although most of the organics would be protected inside the meteorite and would likely survive intact through the passage of the atmosphere, only a small fraction of the meteorites on Earth today contains significant amounts of organics (mainly in carbonaceous chondrites, Section 7.5). For the third group, atmospheric entry heating may have chemically transformed the organic matter in IDPs, although particles with diameters < 10 μm are likely to have escaped extensive entry heating. It has also been suggested that robust polymers such as HCN polymers are much more likely to survive impacts, and the early Earth may have been covered with organics delivered by comets [463]. The peptides and nitrogen heterocycles in HCN polymers give rise to proteins after contact with water (Section 9.10).

The same argument for survival can also apply to extraterrestrial kerogen. One may argue whether extraterrestrial abiological kerogen-like materials have any relevance to the origin of life on Earth. Experiments have shown that membrane structures can be developed from lipid-like organic components from the Murchison meteorite [535]. This opens the possibility that cellular life can follow.

The most direct link between stars and the Earth is found in the presolar grains in meteorites, where diamonds [194], silicon carbide [536], corundum [537], and silicates [538] are found to have isotopic ratios identifiable with AGB stars [539]. This clearly demonstrates that inorganic minerals manufactured in AGB stars can survive through journeys over the ISM and reach the Solar System and Earth.

Although there is no corresponding strong evidence of organic compounds having arrived from outside the Solar System, there is evidence that the organic compound D/H ratios in some Solar System objects have nonsolar values [540]. The excess of D, ^{13}C, ^{15}N, and so on, found in some IOM also points to an interstellar origin [320, 541].

Whether such extraterrestrial delivery is responsible for or has accelerated the creation of life is a matter of conjecture. What we do know is that the ingredients for life are commonly produced by stars and are widely spread over the Galaxy by stellar winds. If these stellar materials played a role in the origin of life on Earth, they could have easily done the same elsewhere in the Galaxy. The chemical link between AGB stars, the ISM, the early Solar System, and Earth represents an area of study with great potential and significance.

12
Lessons from the Past and Outlook for the Future

In our everyday life, we are surrounded by organic matter. It comes in the form of plants, trees, animals, microbes, and other organisms. What these forms of organic matter all have in common is that they are part of living biological systems. Even the nonliving organics such as coal and oil, are remnants of living organisms from the past. The association of organic matter with biological systems probably has biased our view on the existence of organic matter in space. We believe that stars are too hot for complex molecules to survive and that interstellar space is too cold and not dense enough for complex molecules to form via chemical reactions.

The history of studying organic matter in space has been a history of continued denials based on the argument of theoretical impossibility, followed by gradual acceptance of organic molecules and compounds of increasing complexity after observational evidence became overwhelming. At each stage, the scientific community has moved reluctantly, being surprised at every turn that as soon as new tools became available, organic matter at the limit of instrumental capability was detected. The question we now face has changed from whether organic matter exists in space, to what degree of molecular complexity can be detected at the present state of technology.

12.1
Our Journey in Search of Organic Matter in Space

In the early twentieth century, astronomers were empowered by the possibility of using atomic lines in absorption in the photospheric spectra of stars, and atomic lines in emission in nebular spectra to derive the physical conditions and chemical abundance of atmospheres of stars and interstellar clouds. The use of spectroscopy as a means to determine density, temperature, and elemental abundance marked the beginning of the discipline of astrophysics [14]. The discipline of astrochemistry developed much later. Even after the detection of simple molecules like CN, CH and CH^+ in interstellar space, C_3 in comets, and C_2 and CN in evolved star atmospheres in the mid-twentieth century, no effort was made to extrapolate these discoveries into a comprehensive study of molecules in space. The search for answers to the mystery of the DIBs has caused some chemists such as G. Herzberg

Organic Matter in the Universe, First Edition. Sun Kwok.
© 2012 WILEY-VCH Verlag GmbH & Co. KGaA. Published 2012 by WILEY-VCH Verlag GmbH & Co. KGaA.

to believe that molecules are common in the ISM. It was only after the detection of widespread CO through its λ 2.6 mm $J = 1-0$ transition that the prevalence of molecules in our Galaxy was appreciated. The continued development of sensitive radio receivers operating at higher and higher frequencies has led to the discovery of more than 160 molecules in the ISM.

The next breakthrough came with the development of infrared spectroscopy. The extension of spectroscopic observing abilities to the near infrared allowed scientists to detect the stretching modes of molecules, and various molecular bending modes could be measured by mid-IR spectroscopic observations from space-based telescopes. Infrared spectroscopy is a long-established technique in identifying the structures of organic compounds as different functional groups have specific spectroscopic signatures [274]. The UIE features were discovered serendipitously. The 3.3 and 11.3 μm features were discovered using spectrophotometry through atmospheric windows with the modest 1.5 m Mount Lemmon Telescope [542, 543]. The complete family of UIE features at 3.3, 6.2, 7.7, 8.6, and 11.3 μm were found by the KAO [25]. The fact that these features are much broader than atomic lines had baffled the observers. The first suggestions that these features were due to aromatic compounds [544, 545] were generally ignored by the astronomy community. After several years, the community was more receptive to a halfway compromise, settling on PAH molecules as the popular explanation of the UIE features [171, 424]. After the launch of the ISO mission, the UIE features were observed with much better spectral resolution, in a much larger sample of sources, and covering a broader spectral range. It is now quite certain that the UIE features at 3.3, 6.2, 7.7, 8.6, and 11.3 μm are due to aromatic compounds, and it is therefore appropriate to name them as the AIB features. Aliphatic and plateau emission features were found, and a variety of larger and more complex carbonaceous compounds (HAC, QCC, coal, and so on, see Chapter 9) have been proposed as their carrier.

In the planetary science community, access to samples of organic solids in meteorites, IDPs, and comets allowed scientists to analyze the contents directly. It was realized that many kinds of organics are common in the Solar System, and more than 90 different kinds of amino acids and almost every kind of organics have been found in the soluble component of primitive carbonaceous meteorites (Section 7.5). In addition, there are insoluble organic compounds of an amorphous and complex nature. The IOM found that carbonaceous meteorites have chemical structures similar to kerogen, the most common form of organics on Earth formed from decayed living matter. The observations of organic haze in the atmosphere and on the surface of Titan also sparked interests in N-rich complex organics such as tholins.

Systematic observations of objects in the late stages of stellar evolution, from AGB to protoplanetary nebulae, to planetary nebulae, have shown that organic synthesis is taking place in the circumstellar environment. Carbonaceous materials with aliphatic and aromatic structures are shown to have developed in very low-density conditions in a very short time ($\sim 10^3$ yr). This is the first demonstration that stars can actually synthesize complex organics naturally and efficiently, without involving life as an intermediate step. The remarkable similarity between the spectra of protoplanetary nebulae and kerogen raises the possibility that the IOM in meteorites could be stellar in origin, although isotopic ratio confirmation is needed.

The link between evolved stars and the Solar System was established when meteoritic presolar grains were found to carry the unique isotopic signatures of evolved stars. The case is made stronger by the fact that the most well established presolar grain SiC is also known spectroscopically to be widely produced by carbon stars. This link demonstrates that star dust can actually survive the journey through the ISM, incorporated into the early Solar Nebula, and reached the surface of the Earth. Not only are the individual atoms in our bodies made in evolved stars, but the products of the stellar molecular factory are among us on this Earth.

Strong emissions from aromatic compounds are seen in molecular clouds that contain hot central stars as heating sources. Even in cooler surroundings, the AIBs are seen in emission in reflection nebulae powered by cool central stars. Although there are no energy sources to excite organic matter into emission in the diffuse ISM, their presence is inferred by the detection of absorption features (e.g., the 3.4 μm aliphatic feature) along the line of sight to bright stars. We have to remember that most of the volume of the Galaxy is cold ($T < 100$ K) and most of the baryonic mass is in the form of molecular hydrogen. How much complex organics matter exists in this cold component is unknown.

With the capability of infrared spectroscopy extended to external galaxies, we find that in certain classes of galaxies, the amount of energy output from the UIE features alone accounts for 20% of the total energy output in the infrared (Figure 8.3). Since we do not know the exact nature of the carrier of these features, it is difficult to translate this flux ratio into abundance. However, it is clear that the organic matter is widely present in galaxies, and its existence is not isolated into limited classes of stars or small regions of the ISM. Perhaps more significantly, the UIE features are seen in distant galaxies, implying that organic synthesis happened only a few billion years after the Big Bang.

At the same time, extensive amounts of observational data have built up on several unidentified phenomena, including the DIB, 220 nm feature, 21 and 30 μm features, and ERE, that point to the necessity of involving organic matter as an explanation. In this book, we have shown that organic matter is widely present in the Solar System, in stars, in the diffuse ISM, and in external galaxies. Is there a common origin of these forms of organic matter, or are they synthesized independently at different times? We know that life on Earth evolved from organic matter 3–4 billion years ago, and the organic matter is being deposited back to the Earth after the death of living organisms. The origins of organics in meteorites, comets, asteroids, and IDPs, are less clear, but they certainly reflect the chemical composition of the solar nebula. The most likely scenario is that the solar nebulae inherit the organics from the parent molecular clouds. The fact that organic molecules and solids are commonly observed in the spectra of star formation regions lends support to this hypothesis. However, star formation regions such as Orion are large, complex clouds with multiple components and central stars. Our present observational data do not provide enough information for us to determine where these organics are formed, and over what time scales that they are formed.

12.2
The Origin and Evolution of Organic Matter in Space

The only regions where we have concrete knowledge of molecular synthesis are the circumstellar envelopes of evolved stars. The molecules and solids form out of the outflows of stars, and the chemical reaction time is limited by the dynamical time of outflows, which is on the order of 10^4 yr. I would like to present the following scenario for the synthesis, evolution and delivery of organic matter in the Universe. The journey begins with the nucleosynthesis of carbon in AGB stars. The most basic ingredient of organic matter, carbon, is synthesized by nuclear reactions in the very last stage of stellar evolution during the AGB phase. These freshly synthesized elements are brought to the stellar surface by convection motions in the stellar envelope. Because of the large size of the AGB star, the surface temperature is very low. Near the end of the AGB, the stellar radius is about 1 AU and the surface temperatures are lower than 3000 K. At such low temperatures, simple molecules can form in the photosphere and solid-state particles can condense in the upper extended atmospheres. The newly formed solid particles efficiently intercept the stellar radiation and the radiation pressure can drive the grains into supersonic speeds. Collisions between the grains and the gas drag the gas with the grains, accelerating the gas to speeds above the escape velocities of the stars. This forms a steady stellar wind which gradually removes masses from the stellar envelope [278]. In this stellar outflow, new molecules are formed through gas-phase or grain-surface chemical reactions. The physical extent, chemical composition, and physical conditions of the stellar wind can be determined by millimeter-wave and infrared imaging and spectroscopy. From these measurements, we can determine that the typical mass loss rate of AGB stars is $\sim 10^{-5}$ $M_\odot$ yr^{-1}. In other words, the entire envelope of a solar-mass star can be depleted in a hundred thousand years.

Assuming a dust-to-gas ratio of 0.003, the ejection rate of solid particles per AGB star is 2×10^{15} kg s^{-1}. The number of heavy mass-losing AGB stars in the Galaxy is at least 100 000, giving a dust ejection rate of 6×10^{27} kg yr^{-1}. These solid particles aggregate into dense interstellar clouds which eventually lead to the formation of new stars over time scales of millions of years. It is difficult to estimate what fraction of the star dust remains in the diffuse ISM (we know that grain signatures such as the 220-nm and 3.4 μm features are seen in the diffuse ISM). Although we are uncertain what fraction of these grains eventually accumulates in star-forming regions, we do know that over the lifetime of the Galaxy, 6×10^{37} kg of dust has been produced by stars. This dust is spread throughout the Galaxy, some left in the diffuse ISM and some aggregated into nebular material for the formation of the next generation of stars and planets.

Although the Miller-type experiments are very interesting for their illustration of which organics could be produced naturally on the early Earth, the delivery of extraterrestrial organics is actually observed and is real. The total amount of organics delivered to Earth externally is estimated to be $10^{16}-10^{18}$ kg [546], which is much larger than the total amount of of organic carbon in the biosphere (2×10^{15} kg, Table 7.2). The amount of organic carbon stored in the forms of coal and oil is more difficult to estimate. Extrapolating from existing reserves, the potential total reserve can be as high as 4×10^{15} kg (Table 7.2). Assuming that they are fossil fuels of biological origin, the total amount of carbon generated from plant photosynthesis can be estimated from the total amount of oxygen in the atmosphere, yielding a total reserve of 5×10^{17} kg [547]. The amount of externally delivered organic carbon is therefore comparable to the total reserve of fossil fuels. However, if we include kerogen into account of organic matter that has been generated by biological means, then the total becomes 1.5×10^{19} kg. This amount is higher than the estimated total amount of externally delivered organics.

In addition to external delivery after the formation of the Earth, it is possible that the planetesimals that collided to create the Earth still contain remnants of the complex organics that were part of the primordial solar nebula. If this is the case, then the amount of primordial organics would far exceed the amount later delivered through external bombardments and could easily account for all the organic reserves today [548].

12.3
The Future

While there has been tremendous progress in this field on the observational and experimental fronts, I do not want to minimize the challenges ahead. As of 2010, we still do not have a full understanding of the chemical structures of organic matter in space. We know that they are in the forms of molecules and solids. However, we do not know what the maximum complexity of an individual gas-phase interstellar molecule can be, nor do we have a chemical structure for the organic solids. Progress must be made in the experimental front where new forms of organic substance are created artificially and their properties measured in the laboratory. The pioneering efforts of Mayo Greenberg and Carl Sagan showed us that the conditions in interstellar space are very different from those in our terrestrial surroundings, and we cannot assume that all organic (and inorganic) matter in the Universe has a counterpart on Earth. We should keep in mind that the element helium, the second most abundant element in the Universe, was discovered in the Sun before it was found on Earth. Will the identification of the DIB someday lead to the discovery of new molecular species, or will the identification of the 21 μm feature lead to the discovery of a new organic solid? What will be the practical implications of these discoveries?

As discussed above, much of the progress in our understanding of organic matter in space has been driven by advancing technology. Millimeter-wave spectroscopy made the detection of rotational transitions of molecules possible, and infrared spectroscopy revealed the vibrational modes of molecules and solids.

Molecules and solids are detected either through spontaneous emission from a higher to a lower state, or in absorption against a continuum background by raising the molecule or solid to an excited state. For organic matter in a background of X-ray radiation, the high-energy photons can excite the electrons to a higher state, leaving an imprint on the X-ray spectrum that reflects the chemical composition of the molecule or solid. Past X-ray all-sky surveys have identified thousands of bright X-ray sources which can serve as a background to study interstellar materials along the lines of sight. Most of the K and L transitions of the abundant elements between carbon and iron lie in the X-ray part of the spectrum, and the High-Energy Transmission Grating-Advanced CCD Imaging Spectrometer (HETG-ACIS) of the Chandra Observatory has detected the K_α transitions of different ionization stages of oxygen [549]. Future X-ray telescopes with good throughput and spectral resolution ($R > 1000$) can probe the solid-state component of the diffuse ISM towards line of sights of bright X-ray sources [550].

Robotic space probes to asteroids, comets, planets, and planetary satellites open the possibility of on-site sample analysis or sample return, with the prospect of making the study of organic matter in space an experimental science. Obviously, there is no better alternative than to perform direct physical explorations and analysis of organic matter in Solar System objects within reach. The chemical structures of the organic matter determined from Solar System objects will shed light on the chemistry of organic matter in interstellar space and in distant galaxies.

The question of the origin of life, important as it is, may not be understood until we have a better grasp of the origin and distribution of organic matter in space. What form does it take? How ubiquitous is it? How much and how complex is the primordial organic matter in the early Earth? Our ability to answer these questions with new observational and laboratory studies will help us eventually address the question of the origin of life.

Appendix A
Glossary

Scientific nomenclature is created in order to provide the precise definitions of concepts. Some scientific terms have their origin in everyday usage (e.g., force and energy), but have specific meanings which may differ from their meaning in everyday usage. The definitions may also evolve with time as we gain a clearer understanding of the phenomenon. For example, the term "planets" was used to refer to the five "wandering" celestial objects (Mercury, Venus, Mars, Jupiter and Saturn) that have different sky trajectories from stars. After the invention of the telescope, Uranus, Neptune, and Pluto joined the ranks of planets as major bodies that revolve around the Sun. However, as more minor bodies in the outer Solar System were discovered, it was realized that we needed a separate category to label these objects and that Pluto is better fitted into this group. Thus, the definition of planets was more precisely defined in 2006 by the International Astronomical Union.

Also, different scientific disciplines have a different set of specialized terms ("jargons") which are not comprehensible to outsiders. Since this book has a diverse readership, a list of specialized terms are listed here with a short definition. This glossary list is intended as an introductory guide for easy reference. Readers who wish to have a more in-depth understanding of these concepts should refer to the respective specialized literature.

aerosol solid particles or liquid droplets in the atmosphere. This term is commonly used in the planetary science community to refer to droplets of hydrocarbon molecules and polymers of C_2H_2, HCN, as well as more complex organic solids in planetary atmospheres.

albedo ratio of reflected to incident radiation. Assuming sunlight is the source of incident radiation, an object with high albedo is bright and an object with low albedo is dark. For planets, satellites and asteroids, the value of albedo of the object can be used to infer its surface constitution.

amide compounds containing the $-CONR_2$ functional group.

aphelion the farthest point to the Sun in the elliptical orbit of a Sun-orbiting object.

asymptotic giant branch (AGB) stars the late stage of stellar evolution when the star is burning H and He in shells above a C–O electron-degenerate core. AGB stars are very luminous and have a minimum luminosity of about 3000 $L_\odot$.

Organic Matter in the Universe, First Edition. Sun Kwok.
© 2012 WILEY-VCH Verlag GmbH & Co. KGaA. Published 2012 by WILEY-VCH Verlag GmbH & Co. KGaA.

biosphere parts of the Earth where life exists. These include the surface of the Earth, the oceans, and the lower atmosphere.

carbon stars evolved stars whose surface abundance of C exceeds that of O. This occurs at the very late stages of stellar evolution as the element carbon is synthesized in the interior and brought up to the surface by convection.

chiral molecules when a carbon atom is attached to four different groups in a tetrahedral form, there are two geometric configurations which are mirror images of each other. One cannot be rotated to resemble the other.

chondrites meteorites that have not been modified by melting. Of particular interest are the carbonaceous chondrites, which represent about 5% of all chondrites.

chondrules millimeter-size round droplets that condensed in the solar nebula and then accreted into chondrites. Chondrules are not found in terrestrial rocks.

circumstellar envelope material consisting of gas and solid-state particles surrounding stars. These envelopes are often formed as the result of stellar ejection. Examples include supernovae, stellar winds from AGB stars and massive stars (OB stars and Wolf-Rayet stars), planetary nebulae, and novae.

color temperature temperature of a blackbody that will best fit the observed spectral distribution of radiative output of an astronomical object.

column density amount of matter along a line of sight in units of cm^{-3} cm. Since depth is difficult to measure in astronomy, column density is often all one can measure.

condon triplet of nucleotides that codes a specific amino acid.

conformation rotation about a single bond leading to different spatial arrangements of a molecule.

conformers equilibrium structure at specific conformations.

continuum radiation light that covers a large spread of wavelengths in a continuous manner. One example is Planck (thermal) radiation that theoretically covers the entire electromagnetic spectrum. This is in contrast to line (quantum) radiation that radiates over a small wavelength range.

cumulenes sp-bonded cumulenic chains $(C{=}C{=}C)_n$.

dredge up the lifting of newly synthesized elements in the core of evolved stars to the surface through the process of convection.

dust in astronomical literature, the term "dust" is used to refer to inorganic and organic solid-state particles. The "dust" in our daily environment contains a large fraction of biological materials, which are not assumed to be present in interstellar or stellar "dust".

eon time scale used in geology and paleontology. For example, the Hadean eon refers to the period 3.8 to 4.6 billion years ago (Ga), and the Archean eon refers to the period of 2.5–3.8 Ga.

extrasolar planets an extrasolar planet (or sometimes referred to as exoplanet) is a planet found outside the Solar System.

forbidden lines atomic lines that arise from metastable states that would, under terrestrial conditions, be collisionally de-excited before undergoing spontaneous emission. In a low-density environment, the collisional rates are low, allowing the atoms to decay radiatively even the process is very slow. In astronomical notation, a forbidden line is designated by [], for example, [O$_{III}$] stands for a forbidden line emitted by doubly ionized oxygen (O^{2+}).

functional groups group of atoms within a larger molecule makes the molecule behave a certain way in a chemical reaction.

genes segment of a DNA molecule that contains the information needed to synthesize a protein or a RNA.

Hadean the Hadean (Greek for underworld) refers to the geological period between 4.6 and 3.8 Gyr ago when the Earth was still in a hot and violent state.

haze common term used to refer to solid particles in the atmosphere that reduce visibility through scattering. Haze seen on Earth (in particular in cities) is often the result of human activities.

heavy elements term (also the term "metals") used in astronomy to refer to chemical elements heavier than helium.

H$_{II}$ regions gaseous nebulae photoionized by one or more massive central stars. H$_{II}$ regions generally show strong emission-line spectrum, in particular from recombination lines of H.

homologue chemistry term used to describe a compound belonging to a series of compounds differing from each other by a repeating unit, for example, a methylene group, a peptide residue, and so on.

isomers molecules that consist of the same number of atoms of each element and have the same molecular weight, but with different geometric structures.

Kuiper Belt disc-shape region in the outer Solar System from the orbit of Neptune to about 55 AU from the Sun. Over a thousand objects in the Kuiper Belt have been found, of which Pluto is the most well-known member.

lipids water insoluble hydrocarbons.

luminous infrared galaxies galaxies that emit a great amount (>90%) of light in the infrared.

macromolecules polymers with molecular weights above ~5000.

magnitude an inverse logarithmic scale of measuring stellar brightness. A magnitude 5 difference corresponds to a factor of 100. A star of magnitude 10 is 100 times brighter than a star of magnitude 15.

main sequence stars that burn hydrogen in the core follow a luminosity-temperature relationship which is only a function of stellar masses. On a log L_*–log T_* plot, stars of different masses form a band on the plot which is referred to as the main sequence. Astronomers also use the term "main sequence" to refer to the evolutionary stage when the stars are burning H in the core.

mantle the mantle of the Earth is a solid layer about 3000 km thick between the core and the crust of the Earth.

membrane the structure made up of lipid and protein molecules that defines the periphery of the cell and is selectively permeable.

meteors meteors refer to the optical phenomenon of flashes of light in the sky due to solid particles passing through the atmosphere and being vaporized in the process.

meteoroids meteoroids are the general term that refer to solid particles in interplanetary space.

meteorites meteorites are remnants of solid objects that pass through the atmosphere, strike the surface of the Earth, and survive the impact.

minerals a naturally occurring element or chemical compound, usually an inorganic substance with crystalline structure.

nanoparticles particles of sizes in the range of 1–100 nm, intermediate between molecules and bulk materials. They are different from bulk materials because their properties are dependent on their sizes. Due to the relatively large fraction of atoms on the surface, the effects of quantum confinement is much more significant than in bulk materials.

nuclei acids the nucleic acids DNA and RNA are polymers of nucleotides that store and transmit genetic information.

nucleobases heterocyclic aromatic organic compounds. The structure of the DNA and RNA molecules are built on pairs of nucleobases, which are cytosine (C), guanine (G), adenine (A), and thymine (T) for DNA. In RNA, uracil (U) is used in place of thymine.

oligomer consists of a few monomers instead of unlimited numbers as in polymer.

Oort Cloud a spherical region of radius up to 50 000 AU from the Sun. The Oort Cloud is believed to be a reservoir of comets.

optical depth a measure of the degree of opaqueness of a stellar atmosphere or an interstellar cloud. Its value depends on the density, absorption coefficient (often frequency dependent), as well as the physical distance along the line of sight.

pathways sequences of consecutive chemical reactions.

perihelion the nearest point to the Sun in an elliptical orbit of a Sun-orbiting object.

photoionization the ejection of one or more electrons from an atom as the result of the absorption of high-energy photons. For example, the H atom can be photoionized by Lyman continuum ($\lambda < 912$ Å photons), which have enough energy to remove an electron at the ground state (1s).

photolysis treatment with light.

photosphere the light surface below which the star is opaque. Technically, it is defined as the surface of optical depth 2/3.

planetary nebulae gaseous nebulae ejected and photoionized by a star evolving between the AGB and white dwarf phases of stellar evolution.

polysaccharides polymers of simple sugars.

polyynes sp-bonded polyyne chains $(-C{\equiv}C)_n$.

proteins long polymers of amino acids.

protoplanetary nebulae objects evolving between the asymptotic giant branch and planetary nebulae phases. The central stars of protoplanetary nebulae are not hot enough to photoionize the circumstellar gas and the visible light of the nebulae are due to scattered star light.

pyrolysis treatment with heat.

radiation transfer   the change in radiation as the result of passing through matter. The equation of transfer governs the behavior of light having passed through matter of certain density and absorption properties and is one of the most fundamental equations in astrophysics.

red giants   the evolutionary stage after the main sequence. When H in a stellar core is exhausted by nuclear burning, a star begins to burn H in an envelope surrounding the He core. The stellar envelope expands and the star becomes more luminous.

redshift z is defined by $\Delta\lambda = \lambda_0 z$ where $\Delta\lambda$ is the change in the wavelength of a spectral line of wavelength λ_0 as the result of the expansion of the Universe. Since there is a direct relationship between the receding speed of a galaxy and its distance, astronomers also use the redshift of a galaxy to suggest how far away it is.

reflection nebulae   gaseous nebulae that do not emit light on its own but reflect light from stars embedded inside the nebulae.

satellites   natural bodies that revolve around planets.

spectral types   Stars are classified according to the strengths of absorption lines in their photospheric spectra. The spectral types (or spectra classes) are arranged in order of temperature, and are labeled as O, B, A, F, G, K, and M. The term "early" refers to hotter members of the group, and "late" refers to the cooler members.

stereoisomers   a subclass of isomers where the atoms in the molecule have similar bonding patterns but different spatial arrangements.

tar   a black liquid of high viscosity derived from coal.

trans-Neptunian objects   refer to objects whose orbits around the Sun are larger than that of Neptune. These include objects in the Kuiper Belt and the Oort Cloud.

white dwarfs   the last stage of evolution for a star with an initial mass less than $8\,M_\odot$. The luminosity of a white dwarf is derived from gravitational contraction and not by nuclear fusion.

Appendix B
Astronomical Infrared and Submillimeter Spectroscopic Observational Facilities

The current understanding of organic matter in the Universe is based on technological advances in infrared and mm/submm astronomy. These advances are made possible with the combination of building of high-frequency receivers capable of high-spectral resolution observations as well as the capability of putting telescopes in space and therefore avoiding the obstruction of the atmosphere. Vibrational absorption of molecules such as H_2O (with absorption bands at 0.94, 1.13, 1.37, 1.87, 2.7, 3.2, 6.3, and $\lambda > 16\,\mu m$), O_2 (with bands at 0.63, 0.69, 0.76, 1.06, 1.27, 1.58 μm), CO_2 (2.0, 4.3, 15 μm), N_2O (4.5, 17 μm), CH_4 (3.3, 7.7 μm), O_3 (0.6 μm), and so on remove much of the celestial infrared radiations and prevent them from reaching the Earth. However, there exist a number of atmospheric windows where the opacity is relatively low through which observations can be carried out from the ground. In the near infrared, observations are carried out through the J (1.1–1.3 μm), H (1.5–1.8 μm), K (2–2.4 μm), L (3.2–3.8 μm), M (4.6–5 μm), and N (8–13 μm), Q (16–24 μm) band windows. In the submm region, pressure broadened rotational transitions of H_2O and O_2 are primarily responsible for the terrestrial atmospheric opacity. On a high and dry site, observations in the submm wavelengths can be made through a number of windows (Figure B.1). For most parts of the electromagnetic spectrum between 25 and 600 μm, observations have to be made from airborne, balloon, or space-based platforms. The development of observing facilities from these platforms have been one of the main reasons for our improved knowledge of organic matter in the Universe. In this appendix, we list a number of ground-based and space-based observatories that have played a major role in the development of infrared and mm/submm astronomy.

B.1
Single-Dish Millimeter and Submillimeter Wave Telescopes

Radio telescopes equipped with high-frequency receivers and spectrometers were the major impetus for the development of molecular astronomy. Increasing size of the dishes coupled with decreasing noise temperatures of the receivers led to better sensitivity needed to detect weak signals. Better surface accuracy of the dishes and the developments of solid-state junctions operating at higher frequencies

Organic Matter in the Universe, First Edition. Sun Kwok.
© 2012 WILEY-VCH Verlag GmbH & Co. KGaA. Published 2012 by WILEY-VCH Verlag GmbH & Co. KGaA.

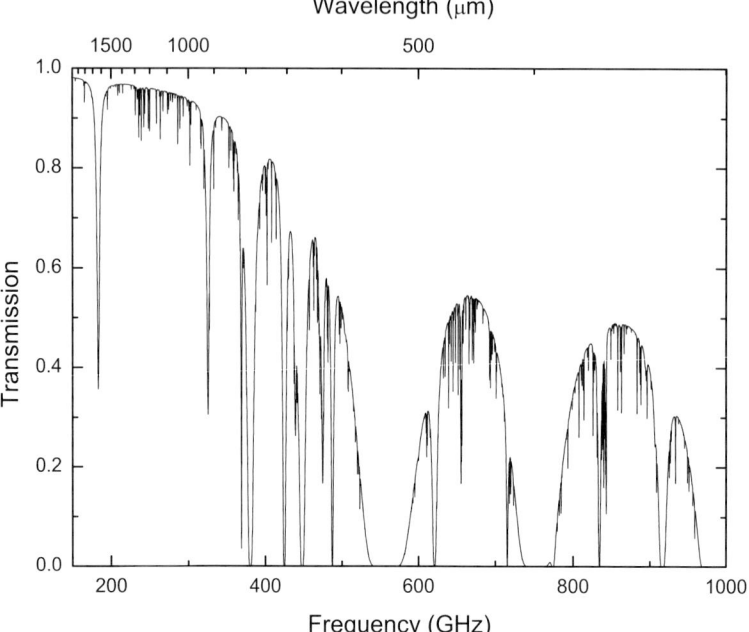

Figure B.1 Model of zenith atmospheric transmission on the summit of Mauna Kea (elevation 4200 m) for 0.5 mm of precipitable water vapor. A number of atmospheric win-dows with relatively high transmission can allow ground-based telescopes partial access to the submillimeter-wave spectrum.

allow observational coverage from mm to submm wavelengths. The need for superior atmospheric conditions at high frequencies has also moved the observing sites to higher and dryer mountain sites.

B.1.1
Arizona Radio Observatory

The Arizona Radio Observatory operates the Kitt Peak 12-m Telescope on Kitt Peak (1914 m elevation), Arizona, and the 10-m Submillimeter Telescope on Mt. Graham (3191 m), Arizona. The 12-m telescope began as a national facility of the National Radio Astronomy Observatory, but is now operated by the University of Arizona. Currently, the 12-m telescope is equipped with receivers covering the entire 68–180 GHz range.

B.1.2
Nobeyama 45-m Telescope

The Nobeyama Radio Observatory is located at an altitude of 1300 m in central Japan. The 45-m telescope began its operation in 1981 and is capable of observations at mm wavelengths [555].

B.1.3
Caltech Submillimeter Observatory

The Caltech Submillimeter Observatory (CSO) is the first submillimeter telescope located at Mauna Kea (4200 m). It has a 10.4-m telescope and is equipped with a variety of submillimeter-wave receivers operating between λ 1 to 0.3 mm. CSO began operations in 1988.

B.1.4
James–Clerk–Maxwell Telescope

The James–Clerk–Maxwell Telescope (JCMT) is a 15-m submillimeter-wave telescope located at Mauna Kea. It has receivers operating at 1.2, 0.8, and 0.5 µm [556]. JCMT began operations in 1987.

B.1.5
IRAM 30-m Telescope

The Institut de Radioastronomie Millimétrique (IRAM) 30-m telescope is located on Pico Veleta in the Spanish Sierra Nevada at an altitude of 2850 m. It is equipped with receivers and bolometers operating at 3, 2, 1, and 0.8 mm. The IRAM 30-m telescope began operations in 1984.

B.1.6
Green Bank Telescope

The Robert C. Byrd 100-m Green Bank Telescope (GBT) is located in Green Bank, West Virginia, USA and is currently (2010) the largest fully steerable radio telescope in the world. The GBT has many receivers covering the frequency range from 0.29 to 90 GHz and is particularly suitable for observations of heavy molecules with weak and low-frequency rotational lines [557]. The GBT began operations in 2002.

B.1.7
Atacama Submillimeter Telescope Experiment

The Atacama Submillimeter Telescope Experiment (ASTE) is a 10-m telescope located at the Atacama desert of altitude 4860 m. Its main instrument is a 350 GHz receiver with an array camera operating at 270, 350, and 650 GHz under development [558]. ASTE began scientific observations in 2004.

B.1.8
Atacama Pathfinder Experiment

The Atacama Pathfinder Experiment (APEX) is a 12-m telescope located at Llano de Chajnantor (5105 m altitude) in Northern Chile. It has a set of SIS heterodyne receivers covering 211–270 GHz, 275–370 GHz, and 375–500 GHz bands, and a heterodyne HEB receiver operating at 1.25–1.38 THz, and two bolometer arrays at λ 0.9 and 0.35 mm [559].

B.2
Ground-Based Infrared Telescopes

A ground-based telescope located at a high, dry site can observe through the atmospheric windows around 1–3, 10, and 20 µm. Within these windows, one can observe the 3.3 µm aromatic C–H stretch, the 3.4 µm aliphatic C–H stretch, and the 11.3 µm out-of-plane C–H bending modes. The stretching modes of many simple molecules (e.g., the first overtone of CO at 2.3 µm, the ν_3 C–H stretch of methane) also lie in the near infrared region accessible by ground-based spectroscopy.

B.2.1
United Kingdom Infrared Telescope

The United Kingdom Infrared Telescope (UKIRT) is a 3.8-m dedicated infrared telescope on the summit of Mauna Kea at an altitude of 4194 m. The Cooled Grating Spectrometer 4 (CGS4) has been the main spectrometer for much of the history of the telescope. The CGS4 is based on a 256×256 InSb array and is capable of four different gratings in the 1–5 µm range, achieving spectral resolution from 500–2200 to 37 000. In mid-infrared wavelengths, MICHELLE is based on a Si:As 320×240 pixel array and offers spectroscopic capability with $R \sim 200-30\,000$ in the 8–25 µm range. UKIRT began operations in October 1979.

B.2.2
The NASA Infrared Telescope Facility

The NASA Infrared Telescope Facility (IRTF) is a 3.0-m infrared telescope located at the summit of Mauna Kea. It is primarily used for Solar System observations, although part of the observing time has been used for astronomy.

B.2.3
Keck Telescope

The NIRSPEC on the Keck 10-m telescope is a near-infrared single-slit echelle and grating spectrometer [560]. It has a 1024×1024 InSb detector array and can carry out spectroscopy at resolving powers of $R \sim 1500-3000$ or $R \sim 15\,000-75\,000$ over the 1–5 µm region.

B.2.4
Gemini Telescopes

The twin 8-m Gemini telescopes (one in Mauna Kea and one in Chile) were de-
signed as infrared-optimized telescopes. Although in recent years, they have increasingly been used as optical telescopes [561]. The infrared spectrometers that
have been used on Gemini include PHOENIX, TEXAS, and MICHELLE. Phoenix
is a near infrared (1–5 μm) echelle spectrometer capable of very high spectral resolutions ($R \sim 50\,000-80\,000$).

B.3
Space Infrared and Submillimeter Telescopes

The ultimate way to avoid the interference of the atmosphere is to get above it
entirely. Remote-controlled telescopes based on artificial satellite platforms allow
complete wavelength coverage not possible from a ground-based telescope. However, the need for cooling of the detectors by liquid helium limits the lifetime of
operation of the telescopes as the satellite can only carry a limited supply of cryo-
genics. The weight and size limits of the launch vehicle also restricts the size of the
telescopes to be carried to space. Once launched, there is no possibility of repair or
replacement of malfunctioned parts, except in cases where the satellites are in low
orbit (e.g., the Hubble Space Telescope), reachable by the space shuttle. The angular resolution of space-based telescopes would be limited by diffraction, rather than
by the atmosphere.

B.3.1
Infrared Astronomical Satellite

The Infrared Astronomical Satellite (IRAS) is best known as the first infrared satellite that surveyed 96% of the sky in four (12.5, 25, 60 and 100 μm) photometric
bands [562]. What is less well known is that it also carried a slitless spectrometer called the Low Resolution Spectrometer (LRS) which also carried out an all-sky
survey in spectral mode (Section 3.18.2). The IRAS was launched on January 25,
1983.

B.3.2
Infrared Space Observatory

The Infrared Space Observatory is a 0.6-m infrared telescope with both imaging
and spectroscopic capabilities [563]. Its main spectroscopic instruments are the
Short Wavelength Spectrometer (SWS) [564] and the Long Wavelength Spectrometer (LWS) [565].

The SWS has two grating spectrometers, one covering the wavelength range of
2.4–12 μm and the other 12–45 μm. The resolution achieved in the grating mode

is $\sim$1000–2000. With the insertion of Fabry–Pérot filters, the resolution can be 20 times higher. Using different combinations of the grating order, aperture and detector array, a number of astronomical observing templates (AOT) can be used. The most commonly used AOT is S01 which scans the full SWS wavelength range. The LWS operates between 43–197 μm in either grating ($R \sim 150-200$) or Fabry–Pérot ($R \sim 6800-9700$) modes. ISO was launched on November 17, 1995.

B.3.3
Infrared Telescope in Space

The Infrared Telescope in Space (IRTS) is a 0.15-m infrared telescope that surveyed $\sim$7% of the sky [566]. The IRTS carried two infrared spectrometers: a Near-Infrared Spectrometer (NIRS) covering the wavelength range 1.4–25 μm and 2.8–3.9 μm with a resolution of $\sim$0.13 μm and a Mid-InfraRed Spectrometer (MIRS) operating between 4.5–11.7 μm with a resolution of $\sim$0.3 μm. The IRTS was launched on March 18, 1995, and operated until the exhaustion of the helium coolant on April 24, 1995.

B.3.4
AKARI

AKARI is a 0.7-m telescope that performed an all-sky survey in the mid-infrared [567]. The two instruments on board, the InfraRed Camera (IRC, covering 2.5–26 μm) and the Far-Infrared surveyor (FIS, covering 50–180 μm), have low-resolution spectroscopic capability [568]. The IRC has a spectroscopic mode which can provide slitless spectroscopy with resolution $R \sim 20-120$ over a field of $10' \times 10'$. Spectroscopic observations were made under pointed altitude control mode during short intermittent periods interrupting the broad-band all-sky survey [569]. AKARI was launched on February 21, 2006.

B.3.5
Submillimeter Wave Astronomy Satellite

The Submillimeter Wave Astronomy Satellite (SWAS) is a 0.6-m submm telescope equipped with two heterodyne receivers to cover the O_2 (487 GHz), C_I (492 GHz), H_2O (557 GHz), ^{13}CO (551 GHz) and $H_2^{18}O$ (548 GHz) lines [570]. SWAS was the first submm telescope in space. SWAS was launched in December of 1998.

B.3.6
Odin

Odin is a 1.1-m submillimeter telescope equipped with four tunable submm receivers (541–558 GHz, 486–504 GHz, 547–564 GHz, 563–581 GHz) and one fixed-frequency (119 GHz) receiver [571]. It covers a number of submm transitions of

CI, CO, ^{13}CO, C^{18}O, CS, H$_2$O, H$_2$^{18}O, H$_2$^{17}O, NH$_3$, ^{15}NH$_3$, O$_2$, etc. and is capable of performing spectroscopic scans. Odin was launched on February 20, 2001.

B.3.7
Spitzer Space Telescope

The Spitzer Space Telescope has a 0.85-m telescope with three cryogenically-cooled instruments [572]. Its spectroscopic instrument is the Infrared Spectrograph (IRS) [573]. The IRS has four separate modules: a low resolution, short-wavelength (SL) mode covering 5.3–14 μm; a high resolution, short-wavelength (SH) mode covering 10–19.5 μm; a low-resolution, long-wavelength (LL) mode for observations at 14–40 μm; and a high-resolution, long-wavelength (LH) mode for 19–37 μm. The SL and SH modules are supported by two 128 × 128 Si:As detectors, whereas the LL and LH modules are supported by two 128 × 128 Si:Sb detectors. Spitzer IRS is much more sensitive than ISO SWS/LWS, but its highest spectral resolution ($R \sim 600$) does not match what can be achieved by ISO. The Spitzer Space Telescope was launched on August 25, 2003.

B.3.8
Herschel Space Telescope

The Herschel Space Telescope is a 3.5-m telescope with two instruments for imaging and low-resolution spectroscopy [574]. Its spectroscopic instruments include a Heterodyne instrument for the far infrared (HIFI) [575] and Spectral and Photometric Imaging Receiver (SPIRE) [576]. HIFI covers the wavelength ranges of 240–625 μm with SiS mixers and 157–210 μm with hot electron bolometer mixers. SPIRE has an imaging Fourier Transform Spectrometer (FTS) which covers the spectral range 194–671 μm (447–1550 GHz). Herschel was launched on May 14, 2009.

B.4
Airborne Telescopes

Although not flying as high as an orbiting satellite, telescopes based on airplanes have the advantage of being able to be maintained and repaired after each flight. This allows new instrumentation, taking advantage of the novel technology to be used.

B.4.1
Kuiper Airborne Observatory

The Kuiper Airborne Observatory (KAO) was the first flying observatory based on a modified C-141 military transport plane. It carried a 0.9 m telescope and was capable of operating at 14 km altitude and therefore avoiding most of the water vapor in the Earth's atmosphere. Because of its ability to upgrade and replace instruments, KAO has operated a number of facility and PI spectrometers. These include prism, grating, heterodyne, Fabry–Pérot, Michaelson, and echelle spectrometers.

The first exploration of the infrared spectrum was performed with spectrophometry using narrow filters, achieving a resolution of $R \sim 50$. Such a low-resolution instrument was enough to discover the aromatic infrared bands. An example of a mid-spectral resolution instrument was the High Efficiency Infrared Faint Object Grating Spectrograph (HIFOGS) which is capable of spectral resolution of 100–1000 between 3 and 30 µm [577]. The high spectral resolution capability was provided by the Ames Cryogenic Grating Spectrometer (CGS) [578]. The CGS covers the wavelength range of 16–206 µm with resolution between 1500–5000. The KAO was based at the Ames Research Center at Moffett field, California, and was in operation between 1974 and 1995.

B.4.2
SOFIA

The Stratrospheric Observatory for Infrared Astronomy (SOFIA) is a Boeing 747SP aircraft modified to house a 2.5-m infrared telescope. The aircraft can fly to an altitude of 12 500 m and can achieve diffraction-limited imaging at wavelengths longer than 15 µm [579].

First generation spectroscopic instruments include an echelon spectrometer Echelon–Cross–Echelle Spectrograph (EXES, 5–28 µm, $R \sim 10^5$, 10^4, or 3000), a heterodyne spectrometer Caltech Submillimeter Interstellar Medium Investigations Receiver (CASMIR, 250–600 µm), an imaging grating spectrometer Field Imaging Far-Infrared Line Spectrometer (FIFI-LS, 42–210 µm, $R \sim 1700$), a heterodyne spectrometer German Receiver for Astronomy at Terahertz Frequencies (GREAT, 60–200 µm, $R \sim 10^6$ or 10^8), and an imaging Fabry–Pérot bolometer array spectrometer Submillimeter and Far InfraRed Experiment (SAFIRE, 145–655 µm, $R \sim 1000-1900$). SOFIA will begin full operations in 2014.

B.5
Millimeter and Submillimeter Arrays

An array of linked-telescopes can employ the technique of aperture synthesis based on the principle of interferometry to create images with angular resolutions corresponding to the spatial separation of the telescopes. Equipped with spectrometers, such arrays can obtain a series of images at each spectral channels, therefore providing kinematic in addition to spatial information. Millimeter and submm array observations centered around specific molecular line frequencies can give us the spatial distribution of the molecular species, therefore avoiding the problem of line confusion suffered by single-dish telescopes.

B.5.1
Nobeyama Millimeter-Wave Array

The Nobeyama Array is part of the Nobeyama Radio Observatory and consists of six 10-m telescopes with a baseline of 600 m [580]. The Nobeyama Millimeter Array began operations in 1994.

B.5.2
The Plateau de Bure Interferometer

The Plateau de Bure Interferometer (PdBI) is located on the Plateau de Bure at 2550 m altitude in the French Alps. It consists of six 15 m antennas and is capable of observations up to λ 1 mm. The maximum baseline between two antennas is about 760 m, corresponding to an angular resolution of $\sim$0.5 at λ 1.3 mm [581].

B.5.3
Submillimeter Array

The Submillimeter Array (SMA) consists of eight 6-m telescopes located at 4080 m above sea level on the summit of Mauna Kea. It has a number of antenna stations capable of providing variable antenna configurations with maximum baselines of $\sim$25–500 m, corresponding to angular resolutions of 5 to 0.25 at 350 GHz. It has a number of receivers covering the 230, 345, and 650 GHz bands. The SMA began scientific observations in 2004 [582].

B.5.4
Combined Array for Research in Millimeter-Wave Astronomy

Combined Array for Research in Millimeter-wave Astronomy (CARMA) consists of six 10.4 m, nine 6.1 m, and eight 3.5 m antennas equipped with receivers covering the 7 mm, 3 mm, and 1.3 mm bands [583]. CARMA is located on the Cedar Flat site at an altitude of 2200 m. CARMA began scientific operations in January of 2007.

B.5.5
Atacama Large Millimeter Array

The Atacama Large Millimeter Array (ALMA) is located on the Chajnantor plain of the Chilean Andes at an altitude of 5000 m. The full array, when completed, will consist of fifty 12 m antennas plus a compact array consisting of twelve 7 m and four 12 m antennas, providing baselines ranging from 150 m to 15 km. A total of 10 band receivers will cover the wavelength range of 0.3 to 3.6 mm (84 to 950 GHz). The 12 m primary beam will provide a field of view of $\sim 21''$ (300 GHz/ν) and the angular resolution achievable will be $0.02''$ (300 GHz/ν) (10 km/maximum baseline) [584]. Scientific operation is expected to commence in 2012.

Appendix C
Unit Conversions

Various communities have different ways of measuring the same physical quantities. For example, optical astronomers refer to the wavelength of a spectral line, whereas radio astronomers tend to use frequencies. In optical astronomy, wavelengths are expressed in units of nanometers (nm, or 10^{-9} m) or Angstroms ($\text{Å} = 10^{-10}$ m), whereas in infrared astronomy, one uses micrometers (μm, or 10^{-6} m). Frequencies are measured in Hz (cycles per second, in units of s^{-1}), kHz (10^3 Hz), MHz (10^6 Hz), GHz (10^9 Hz), and THz (10^{12} Hz). The conversion between frequency (ν) and wavelength (λ) is given by

$$\nu\lambda = c , \tag{C1}$$

where c is the speed of light. The ground-state rotational transition of CO has a frequency of 115 GHz, which corresponds to a wavelength of

$$\lambda = \frac{3 \times 10^{10} \text{ cm s}^{-1}}{115 \times 10^9 \text{ Hz}} \tag{C2}$$

$$= 2.6 \text{ mm} . \tag{C3}$$

In the chemistry community, energy of a spectral line is expressed in $hc(\lambda)^{-1}$ where h is the Plack constant and $(\lambda)^{-1}$ is called the wave number and is in units of inverse centimeters (cm^{-1}). A line of wave number 500 corresponds to the wavelength of $1/(500 \text{ cm}^{-1}) = 0.002 \text{ cm} = 20 \ \mu$m. The energy of the photon is

$$E = hc(1/\lambda) \tag{C4}$$

$$= \left(6.626 \times 10^{-27} \text{ erg s}\right) \left(3 \times 10^{10} \text{ cm s}^{-1}\right) \left(500 \text{ cm}^{-1}\right) \tag{C5}$$

$$= 1.99 \times 10^{-16} \text{ erg cm} \left(500 \text{ cm}^{-1}\right) \tag{C6}$$

$$= 9.93 \times 10^{-14} \text{ erg} . \tag{C7}$$

In atomic and molecular physics, energy is often expressed in units of electron volts (eV $= 1.60 \times 10^{-12}$ erg). The energy required to free an electron from the ground state of H is 13.6 eV and most electronic states of atoms and ions have

Organic Matter in the Universe, First Edition. Sun Kwok.
© 2012 WILEY-VCH Verlag GmbH & Co. KGaA. Published 2012 by WILEY-VCH Verlag GmbH & Co. KGaA.

energies on the order of an eV. In X-ray and γ-ray astronomy, energies of photons are also expressed in units of eV, keV, or MeV. A keV photon has a wavelength of

$$\lambda = \frac{hc}{1\,\text{keV}} \tag{C8}$$

$$= 1.24\,\text{nm} . \tag{C9}$$

Molecular astronomers often use temperatures in units of kelvin as a measure of energy separations between rotational states. The relationship between energy and temperature is given by $E = kT$ where k is the Boltzmann constant. If an excited state has energy (in units of K) that is much higher than the kinetic temperature, then this state is unlikely to be excited by collisions. In an interstellar cloud where the kinetic temperature is 100 K, only rotational states with energies $\sim 10^{-2}$ eV will be excited collisionally.

In chemical reactions, the unit of kilocalorie (kcal) is used for energy released or required for reactions. One calorie (1 cal $= 4.2$ J) is the amount of energy needed to raise the temperature of 1 g of water by 1 °C.

For the distance to Solar System objects, the astronomical unit (AU) is used, which is the mean distance between the Earth and the Sun. One AU is 1.50×10^{13} cm. For stars and galaxies, astronomers usually use parsec (1 pc $= 3.086 \times 10^{18}$ cm) as a distance unit. Stars are typically tens or hundreds of parsecs away and the distance from the Sun to the center of the Milky Way Galaxy is 8.5 kpc. Distances to external galaxies are measured in units of Mpc. Because of the expansion of the Universe, distant galaxies are also receding faster, and the receding speed (v) is measured by redshifts to spectral lines according to Doppler's law. The redshift (z), defined by $\Delta\lambda = \lambda_0 z$ where $\Delta\lambda$ is the change in the wavelength of a spectral line of wavelength λ_0, is also used as an indication of distance. A galaxy at $z = 0.1$ is receding at 10% of the speed of light. Assuming the Hubble constant $H_0 = v/D$ has a value of 71 km s^{-1} Mpc^{-1}, the distance to the galaxy is 422 Mpc. A precise conversion from z to distance and look-back time is dependent on the cosmological model. Because light travels at finite speed, a more distant galaxy needs more time to reach us. For a standard model of a Universe 13.67 Gyr old, a $z = 0.1$ galaxy is seen by us at 12.38 Gyr after the Big Bang. Light from a galaxy of redshift 3 was emitted when the Universe was 2.2 Gyr old and took 11.5 Gyr to reach us.

The total energy radiated by a star or galaxy is measured in units of solar luminosity ($L_\odot$), which has a value of 3.83×10^{33} erg s^{-1} or 3.83×10^{26} Watt. The rate of energy received at a distance (e.g., on Earth) is expressed as flux (F), which has units of erg cm^{-2} s^{-1}. The flux per frequency (or wavelength) is F_ν (or F_λ), which has units of erg cm^{-2} s^{-1} Hz^{-1} (Watt m^{-2} Hz^{-1}) or erg cm^{-2} s^{-1} cm^{-1} (Watt m^{-2} µm^{-1}). The two quantities are related by

$$\nu F_\nu = \lambda F_\lambda . \tag{C10}$$

A plot of λF_λ versus λ provides information on how much energy is radiated by an astronomical object at different parts of the electromagnetic spectrum (e.g., as plotted in Figures 5.1, 6.1, 6.4).

Solar flux as received on Earth is

$$F = \frac{L_\odot}{4\pi AU^2} \tag{C11}$$

$$= 1.36 \times 10^6 \, \mathrm{erg \, cm^{-2} \, s^{-1}} \tag{C12}$$

$$= 1366 \, \mathrm{Watt \, m^{-2}} \, . \tag{C13}$$

The basic unit of time is second. In the astronomical community, the unit year ($= 3.156 \times 10^7$ s) is often used. In the geological community, the units Myr (10^6 yr) and Gyr (10^9 yr) are used, and Ma and Ga are used to represent mega and giga years before present.

The basic units for mass are gram or kilogram. The unit of metric ton ($= 10^3$ kg) is also used in scientific literature, but it is not to be confused with the units of tons used in British (2240 pd) or American (2000 pd) daily usage. In astronomy, the unit of solar mass ($M_\odot$, 1 $M_\odot$ — 1.99×10^{33} g) is often used to express the mass of stars, galaxies and other astronomical objects. In biochemistry, the term dalton (Da), defined as 1/12th of the mass of the atom ^{12}C, is used to express the mass of a molecule when its molecular mass is large.

The basic unit for force is newton (kg m s^{-2}). On the surface of the Earth, the gravitational acceleration is 9.8 m s^{-2} and a 1 kg object is subjected to a force of 9.8 N. The atmospheric pressure unit is bar, which is defined as 10^5 N m^{-2} = 10^6 dyn cm^{-2}, and is approximately equal to one standard atmospheric pressure at sea level.

The geological impact community and explosive chemists often express the amount of energy delivered in an explosion in units of tons of TNT, which is defined as 10^9 cal, and corresponds approximately to 4.2×10^9 J.

References

1 Oró, J. (1960) Synthesis of adenine from ammonium cyanide. *Biochemical and Biophysical Research Communications*, **2**, 407–412.

2 Sanchez, R.A., Ferris, J.P., and Orgel, L.E. (1967) Studies in prebiotic synthesis. II. synthesis of purine precursors and amino acids from aqueous hydrogen cyanide. *Journal of Molecular Biology*, **30**, 223–253.

3 Ferris, J.P., Sanchez, R.A., and Orgel, L.E. (1968) Studies in prebiotic synthesis: III. synthesis of pyrimidines from cyanoacetylene and cyanate. *Journal of Molecular Biology*, **33**, 693–704.

4 Burbidge, E.M., Burbidge, G.R., Fowler, W.A., and Hoyle, F. (1957) Synthesis of the elements in stars. *Reviews of Modern Physics*, **29**, 547–650.

5 Wallerstein, G., Iben, I., Parker, P., Boesgaard, A.M., Hale, G.M., Champagne, A.E., Barnes, C.A., Käppeler, F., Smith, V.V., Hoffman, R.D., Timmes, F.X., Sneden, C., Boyd, R.N., Meyer, B.S., and Lambert, D.L. (1997) Synthesis of the elements in stars: forty years of progress. *Reviews of Modern Physics*, **69**, 995–1084.

6 Lattanzio, J. (2003) *Nucleosynthesis in AGB Stars: The Role of Dredge-Up and Hot Bottom Burning (Invited Review)*. Proceedings of the International Astronomical Union Symposium 209: Planetary Nebulae: Their Evolution and Role in the Universe, pp. 73–81.

7 Iye, M., Ota, K., Kashikawa, N., Furusawa, H., Hashimoto, T., Hattori, T., Matsuda, Y., Morokuma, T., Ouchi, M., and Shimasaku, K. (2006) A galaxy at a redshift $z = 6.96$. *Nature*, **443**, 186–188.

8 Ferkinhoff, C., Hailey-Dunsheath, S., Nikola, T., Parshley, S.C., Stacey, G.J., Benford, D.J., and Staguhn, J.G. (2010) First detection of the [OIII] 88 µm line at high redshifts: characterizing the starburst and narrow-line regions in extreme luminosity systems. *The Astrophysical Journal*, **714**, L147–L151.

9 Cox, P., Omont, A., Djorgovski, S.G., Bertoldi, F., Pety, J., Carilli, C.L., Isaak, K.G., Beelen, A., McMahon, R.G., and Castro, S. (2002) CO and dust in PSS 2322+1944 at a redshift of 4.12. *Astronomy and Astrophysics*, **387**, 406–411.

10 Klamer, I.J., Ekers, R.D., Sadler, E.M., Weiss, A., Hunstead, R.W., and Breuck, C.D. (2005) CO (1-0) and CO (5-4) observations of the most distant known radio galaxy at $z = 5.2$. *The Astrophysical Journal*, **621**, L1–L4.

11 Walter, F., Bertoldi, F., Carilli, C., Cox, P., Lo, K.Y., Neri, R., Fan, X., Omont, A., Strauss, M.A., and Menten, K.M. (2003) Molecular gas in the host galaxy of a quasar at redshift $z = 6.42$. *Nature*, **424**, 406–408.

12 Leipski, C., Meisenheimer, K., Klaas, U., Walter, F., Nielbock, M., Krause, O., Dannerbauer, H., Bertoldi, F., Besel, M.-A., de Rosa, G., Fan, X., Haas, M., Hutsemekers, D., Jean, C., Lemke, D., Rix, H.-W., and Stickel, M. (2010) Herschel-PACS far-infrared photometry of two $z > 4$ quasars. *Astronomy and Astrophysics*, **518**, L34.

Organic Matter in the Universe, First Edition. Sun Kwok.
© 2012 WILEY-VCH Verlag GmbH & Co. KGaA. Published 2012 by WILEY-VCH Verlag GmbH & Co. KGaA.

13 Fischer, J., Luhman, M.L., Satyapal, S., Greenhouse, M.A., Stacey, G.J., Bradford, C.M., Lord, S.D., Brauher, J.R., Unger, S.J., Clegg, P.E., Smith, H.A., Melnick, G., Colbert, J.W., Malkan, M.A., Spinoglio, L., Cox, P., Harvey, V., Suter, J.P., and Strelnitski, V. (1999) ISO far-IR spectroscopy of IR-bright galaxies and ULIRGs. *Astrophysics and Space Science*, **266**, 91–98.

14 Aller, L.H. (1963) *Astrophysics. The Atmospheres of the Sun and Stars*, Ronald Press.

15 Dunham, T.J., Jr. (1937) Interstellar neutral potassium and neutral calcium. *Publications of the Astronomical Society of the Pacific*, **49**, 26–28.

16 Swings, P. and Rosenfeld, L. (1937) Considerations regarding interstellar molecules. *The Astrophysical Journal*, **86**, 483–486.

17 McKellar, A. (1940) Evidence for the molecular origin of some hitherto unidentified interstellar lines. *Publications of the Astronomical Society of the Pacific*, **52**, 187–192.

18 Douglas, A.E. and Herzberg, G. (1941) Notes on CH$^+$ in interstellar space and in the laboratory. *The Astrophysical Journal*, **94**, 381.

19 Federman, S.R., Danks, A.C., and Lambert, D.L. (1984) The CN radical in diffuse interstellar clouds. *The Astrophysical Journal*, **287**, 219–227.

20 Weinreb, S., Barrett, A.H., Meeks, M.L., and Henry, J.C. (1963) Radio observations of OH in the interstellar medium. *Nature*, **200**, 829–831.

21 Cheung, A.C., Rank, D.M., Townes, C.H., Thornton, D.D., and Welch, W.J. (1968) Detection of NH_3 molecules in the interstellar medium by their microwave emission. *Physical Review Letters*, **21**, 1701–1705.

22 Cheung, A.C., Rank, D.M., Thornton, D.D., and Welch, W.J. (1969) Detection of water in interstellar regions by its microwave radiation. *Nature*, **221**, 626–628.

23 Snyder, L.E., Buhl, D., Zuckerman, B., and Palmer, P. (1969) Microwave detection of interstellar formaldehyde. *Physical Review Letters*, **22**, 679–681.

24 Olofsson, H. (1996) *The Neutral Envelopes around AGB and Post-AGB Objects*. Proceedings of International Astronomical Union Symposium 178: Molecules in Astrophysics: Probes & Processes, pp. 457–468.

25 Russell, R.W., Soifer, B.T., and Willner, S.P. (1977) The 4 to 8 micron spectrum of NGC 7027. *The Astrophysical Journal*, **217**, L149–L153.

26 Russell, R.W., Soifer, B.T., and Willner, S.P. (1978) The infrared spectra of CRL 618 and HD 44179 (CRL 915). *The Astrophysical Journal*, **220**, 568–572.

27 Duley, W.W. and Williams, D.A. (1981) The infrared spectrum of interstellar dust: surface functional groups on carbon. *Monthly Notices of the Royal Astronomical Society*, **196**, 269–274.

28 Henning, T. and Salama, F. (1998) Carbon in the universe. *Science*, **282**, 2204–2210.

29 Ehrenfreund, P. and Charnley, S.B. (2000) Organic molecules in the interstellar medium, comets, and meteorites: a voyage from dark clouds to the early Earth. *Annual Review of Astronomy and Astrophysics*, **38**, 427–483.

30 Novoselov, K.S., Geim, A.K., Morozov, S.V., Jiang, D., Zhang, Y., Dubonos, S.V., Grigorieva, I.V., and Firsov, A.A. (2004) Electric field effect in atomically thin carbon films. *Science*, **306**, 666–669.

31 Geim, A.K. and Novoselov, K.S. (2007) The rise of graphene. *Nature Materials*, **6**, 183–191.

32 Kroto, H.W., Heath, J.R., O'Brien, S.C., Curl, R.F., and Smalley, R.E. (1985) C_{60}: buckminsterfullerene. *Nature*, **318**, 162–163.

33 Henrard, L., Lambin, P., and Lucas, A.A. (1997) Carbon onions as possible carriers of the 2175 Å interstellar absorption bump. *The Astrophysical Journal*, **487**, 719–727.

34 Rohlfing, E.A. (1990) High resolution time-of-flight mass spectrometry of carbon and carbonaceous clusters. *The Journal of Chemical Physics*, **93**, 7851–7862.

35 Paquette, L.A., Ternansky, R.J., Balogh, D.W., and Kentgen, G. (1983) Total synthesis of dodecahedrane. *Jour-*

nal of the American Chemical Society, **105**, 5446–5450.

36 Kroto, H.W. and Jura, M. (1992) Circumstellar and interstellar fullerenes and their analogues. *Astronomy and Astrophysics*, **263**, 275–280.

37 Buseck, P.R., Tsipursky, S.J., and Hettich, R. (1992) Fullerenes from the geological environment. *Science*, **257**, 215–217.

38 Becker, L., Poreda, R.J., Hunt, A.G., Bunch, T.E., and Rampino, M. (2001) Impact event at the Permian-Triassic boundary: evidence from extraterrestrial noble gases in fullerenes. *Science*, **291**, 1530–1533.

39 Buseck, P.R. (2002) Geological fullerenes: review and analysis. *Earth and Planetary Science Letters*, **203**, 781–792.

40 Iijima, S. (1991) Helical microtubules of graphitic carbon. *Nature*, **354**, 56–58.

41 Heimann, R.B., Evsyukov, S.E., and Kavan, L. (1999) *Carbyne and Carbynoid Structures: Physics and Chemistry of Materials with Low-Dimensional Structures*, Springer.

42 McCarthy, M.C., Travers, M.J., Kovács, A., Gottlieb, C.A., and Thaddeus, P. (1997) Eight new carbon chain molecules. *The Astrophysical Journal Supplement Series*, **113**, 105–120.

43 Van Orden, A. and Saykally, R.J. (1998) Small carbon clusters: spectroscopy, structure, and energetics. *Chemical Reviews*, **98**, 2313–2357.

44 McCarthy, M.C., Travers, M.J., Kovács, A., Chen, W., Novick, S.E., Gottlieb, C.A., and Thaddeus, P. (1997) Detection and characterization of the cumulene carbenes H_2C_5 and H_2C_6. *Science*, **275**, 518–520.

45 Cataldo, F. (2006) Polyynes and cyanopolyynes: their synthesis with the carbon arc gives the same abundances occurring in carbon-rich stars. *Origins of Life and Evolution of the Biosphere*, **36**, 467–475.

46 Van Krevelen, D.W. (1993) *Coal: Typology – Physics – Chemistry – Constitution*, 3rd edn, Elsevier.

47 Speight, J.G. (1994) Application of spectroscopic techniques to the structural

analysis of petroleum. *Applied Spectroscopy Reviews*, **29**, 269–307.

48 Cataldo, F. (2002) The impact of a fullerene-like concept in carbon black science. *Carbon*, **40**, 157–162.

49 Cataldo, F. (2002) An investigation on the optical properties of carbon black, fullerite, and other carbonaceous materials in relation to the spectrum of interstellar extinction of light. *Fullerenes, Nanotubes and Carbon Nanostructures*, **10**, 155–170.

50 Rouleau, F. and Martin, P.G. (1991) Shape and clustering effects on the optical properties of amorphous carbon. *The Astrophysical Journal*, **377**, 526–540.

51 Weber, A.L. and Miller, S.L. (1981) Reasons for the occurrence of the twenty coded protein amino acids. *Journal of Molecular Evolution*, **17**, 273–284.

52 Huggins, W. (1881) Preliminary note on the photographic spectrum of Comet b 1881. *Proceedings of the Royal Society of London*, **33**, 1–3.

53 Douglas, A.E. (1951) Laboratory studies of the λ 4050 group of cometary spectra. *The Astrophysical Journal*, **114**, 466–468.

54 Scoville, N.Z., Hall, D.N.B., Kleinmann, S.G., and Ridgway, S.T. (1979) Detection of CO band emission in the Becklin–Neugebauer object. *The Astrophysical Journal*, **232**, L121–L124.

55 Wilson, R.W., Jefferts, K.B., and Penzias, A.A. (1970) Carbon monoxide in the Orion Nebula. *The Astrophysical Journal*, **161**, L43–L44.

56 Bernath, P. (2005) Atmospheric chemistry experiment: spectroscopy from orbit. *Optics and Photonics News*, **16**, 24–27.

57 Lacy, J.H., Carr, J.S., Evans II, N.J., Baas, F., Achtermann, J.M., and Arens, J.F. (1991) Discovery of interstellar methane: observations of gaseous and solid CH_4 absorption toward young stars in molecular clouds. *The Astrophysical Journal*, **376**, 556–560.

58 Betz, A.L. (1981) Ethylene in IRC+10216. *The Astrophysical Journal*, **244**, L103–L105.

59 Lacy, J.H., Evans II, N.J., Achtermann, J.M., Bruce, D.E., Arens, J.F., and Carr, J.S. (1989) Discovery of interstellar acetylene. *The Astrophysical Journal*, **342**, L43–L46.

60 Ball, J.A., Gottlieb, C.A., Lilley, A.E., and Radford, H.E. (1970) Detection of methyl alcohol in Sagittarius. *The Astrophysical Journal*, **162**, L203–L210.

61 Turner, B.E. and Apponi, A.J. (2001) Microwave detection of interstellar vinyl alcohol, $CH_2=CHOH$. *The Astrophysical Journal*, **561**, L207–L210.

62 Zuckerman, B., Turner, B.E., Johnson, D.R., Clark, F.O., Lovas, F.J., Fourikis, N., Palmer, P., Morris, M., Lilley, A.E., Ball, J.A., Gottlieb, C.A., Litvak, M.M., and Penfield, H. (1975) Detection of interstellar *trans*-ethyl alcohol. *The Astrophysical Journal*, **196**, L99–L102.

63 Hollis, J.M., Lovas, F.J., Jewell, P.R., and Coudert, L.H. (2002) Interstellar antifreeze: ethylene glycol. *The Astrophysical Journal*, **571**, L59–L62.

64 Cooper, G., Kimmich, N., Belisle, W., Sarinana, J., Brabham, K., and Garrel, L. (2001) Carbonaceous meteorites as a source of sugar-related organic compounds for the early Earth. *Nature*, **414**, 879–883.

65 Zuckerman, B., Ball, J.A., and Gottlieb, C.A. (1971) Microwave detection of interstellar formic acid. *The Astrophysical Journal*, **163**, L41–L45.

66 Winnewisser, G. and Churchwell, E. (1975) Detection of formic acid in Sagittarius B2 by its $2_{11}-2_{12}$ transition. *The Astrophysical Journal*, **200**, L33–L36.

67 Krisher, L.C. and Saegebarth, E. (1971) Microwave spectrum of acetic acid, CH_3COOH and CD_3COOH. *The Journal of Chemical Physics*, **54**, 4553–4558.

68 Mehringer, D.M., Snyder, L.E., Miao, Y., and Lovas, F.J. (1997) Detection and confirmation of interstellar acetic acid. *The Astrophysical Journal*, **480**, L71–L74.

69 Townes, C.H. and Cheung, A.C. (1969) A pumping mechanism for anomalous microwave absorption in formaldehyde in interstellar space. *The Astrophysical Journal*, **157**, L103–L108.

70 Thaddeus, P. (1972) Interstellar formaldehyde. I. the collisional pumping mechanism for anomalous 6-centimeter absorption. *The Astrophysical Journal*, **173**, 317–342.

71 Remijan, A.J., Hollis, J.M., Lovas, F.J., Stork, W.D., Jewell, P.R., and Meier, D.S. (2008) Detection of interstellar cyanoformaldehyde (CNCHO). *The Astrophysical Journal*, **675**, L85–L88.

72 Bell, M.B., Matthews, H.E., and Feldman, P.A. (1983) Observations of microwave transitions of A-state acetaldehyde in Sgr B2. *Astronomy and Astrophysics*, **127**, 420–422.

73 Matthews, H.E., Friberg, P., and Irvine, W.M. (1985) The detection of acetaldehyde in cold dust clouds. *The Astrophysical Journal*, **290**, 609–614.

74 Irvine, W.M., Brown, R.D., Cragg, D.M., Friberg, P., Godfrey, P.D., Kaifu, N., Matthews, H.E., Ohishi, M., Suzuki, H., and Takeo, H. (1988) A new interstellar polyatomic molecule: detection of propynal in the cold cloud TMC-1. *The Astrophysical Journal*, **335**, L89–L93.

75 Hollis, J.M., Jewell, P.R., Lovas, F.J., Remijan, A., and Møllendal, H. (2004) Green Bank Telescope detection of new interstellar aldehydes: propenal and propanal. *The Astrophysical Journal*, **610**, L21–L24.

76 Turner, B.E. (1977) Microwave detection of interstellar ketene. *The Astrophysical Journal*, **213**, L75–L79.

77 Ruiterkamp, R., Charnley, S.B., Butner, H.M., Huang, H.-C., Rodgers, S.D., Kuan, Y.-J., and Ehrenfreund, P. (2007) Organic astrochemistry: observations of interstellar ketene. *Astrophysics and Space Science*, **310**, 181–188.

78 Combes, F., Gerin, M., Wootten, A., Wlodarczak, G., Clausset, F., and Encrenaz, P.J. (1987) Acetone in interstellar space. *Astronomy and Astrophysics*, **180**, L13–L16.

79 Friedel, D.N., Snyder, L.E., Remijan, A.J., and Turner, B.E. (2005) Detection of interstellar acetone toward the Orion-KL hot core. *The Astrophysical Journal*, **632**, L95–L98.

80 Snyder, L.E., Buhl, D., Schwartz, P.R., Clark, F.O., Johnson, D.R., Lovas, F.J., and Giguere, P.T. (1974) Radio detection of interstellar dimethyl ether. *The Astrophysical Journal*, **191**, L79–L82.

81 Fuchs, G.W., Fuchs, U., Giesen, T.F., and Wyrowski, F. (2005) Trans-ethyl methyl ether in space. *Astronomy and Astrophysics*, **444**, 521–530.

82 Brown, R.D., Crofts, J.G., Gardner, F.F., Godfrey, P.D., Robinson, B.J., and Whiteoak, J.B. (1975) Discovery of interstellar methyl formate. *The Astrophysical Journal*, **197**, L29–L31.

83 Senent, M.L., Villa, M., Meléndez, F.J., and Domínguez-Gómez, R. (2005) Ab initio study of the rotational-torsional spectrum of methyl formate. *The Astrophysical Journal*, **627**, 567–576.

84 Keene, J., Blake, G.A., and Phillips, T.G. (1983) First detection of the ground-state $J_K = 1_0 \rightarrow 0_0$ submillimeter transition of interstellar ammonia. *The Astrophysical Journal*, **271**, L27–L30.

85 Snyder, L.E. and Buhl, D. (1971) Observations of radio emission from interstellar hydrogen cyanide. *The Astrophysical Journal*, **163**, L47–L52.

86 Thorwirth, S., Wyrowski, F., Schilke, P., Menten, K.M., Brünken, S., Müller, H.S.P., and Winnewisser, G. (2003) Detection of HCN direct *l*-type transitions probing hot molecular gas in the proto-planetary nebula CRL 618. *The Astrophysical Journal*, **586**, 338–343.

87 Godfrey, P.D., Brown, R.D., Robinson, B.J., and Sinclair, M.W. (1973) Discovery of interstellar methanimine (formaldimine). *Astrophysical Letters*, **13**, 119–121.

88 Dickens, J.E., Irvine, W.M., DeVries, C.H., and Ohishi, M. (1997) Hydrogenation of interstellar molecules: a survey for methylenimine (CH_2NH). *The Astrophysical Journal*, **479**, 307–312.

89 Tenenbaum, E.D., Dodd, J.L., Milam, S.N., Woolf, N.J., and Ziurys, L.M. (2010) Comparative spectra of oxygen-rich versus carbon-rich circumstellar shells: VY Canis Majoris and IRC+10 216 at 215–285 GHz. *The Astrophysical Journal*, **720**, L102–L107.

90 Kaifu, N., Morimoto, M., Nagane, K., Akabane, K., Iguchi, T., and Takagi, K. (1974) Detection of interstellar methylamine. *The Astrophysical Journal*, **191**, L135–L137.

91 Turner, B.E., Kislyakov, A.G., Liszt, H.S., and Kaifu, N. (1975) Microwave detection of interstellar cyanamide. *The Astrophysical Journal*, **201**, L149–L152.

92 Rubin, R.H., Swenson Jr., G.W., Benson, R.C., Tigelaar, H.L., and Flygare, W.H. (1971) Microwave detection of interstellar formamide. *The Astrophysical Journal*, **169**, L39–L44.

93 Gottlieb, C.A., Palmer, P., Rickard, L.J., and Zuckerman, B. (1973) Studies of interstellar formamide. *The Astrophysical Journal*, **182**, 699–710.

94 Hollis, J.M., Lovas, F.J., Remijan, A.J., Jewell, P.R., Ilyushin, V.V., and Kleiner, I. (2006) Detection of acetamide (CH_3CONH_2): the largest interstellar molecule with a peptide bond. *The Astrophysical Journal*, **643**, L25–L28.

95 Lovas, F.J., Hollis, J.M., Remijan, A.J., and Jewell, P.R. (2006) Detection of ketenimine (CH_2CNH) in Sagittarius B2(N) hot cores. *The Astrophysical Journal*, **645**, L137–L140.

96 Belloche, A., Menten, K.M., Comito, C., Müller, H.S.P., Schilke, P., Ott, J., Thorwirth, S., and Hieret, C. (2008) Detection of amino acetonitrile in Sgr B2(N). *Astronomy and Astrophysics*, **482**, 179–196.

97 Turner, B.E. and Zuckerman, B. (1974) Microwave detection of interstellar CH. *The Astrophysical Journal*, **187**, L59–L62.

98 Ziurys, L.M. and Turner, B.E. (1985) Detection of interstellar rotationally excited CH. *The Astrophysical Journal*, **292**, L25–L29.

99 Stacey, G.J., Lugten, J.B., and Genzel, R. (1987) Detection of interstellar CH in the far-infrared. *The Astrophysical Journal*, **313**, 859–866.

100 Waelkens, C., VanWinckel, H., Trams, N.R., and Waters, L.B.F.M. (1992) High-resolution spectroscopy of the central star of the Red-Rectangle nebula. *Astronomy and Astrophysics*, **256**, L15–L18.

101 Hall, D.I., Miles, J.R., Sarre, P.J., and Fossey, S.J. (1992) CH^+ in the Red Rectangle. *Nature*, **358**, 629–630.

102 Bakker, E.J., van Dishoeck, E.F., Waters, L.B.F.M., and Schoenmaker, T. (1997) Circumstellar C_2, CN, and CH^+ in the optical spectra of post-AGB stars.

Astronomy and Astrophysics, **323**, 469–487.

103 Liu, X.-W., Barlow, M.J., Nguyen-Q-Rieu, T.-B., Cox, P., Péquignot, D., Clegg, P.E., Swinyard, B.M., Griffin, M.J., Baluteau, J.P., Lim, T., Skinner, C.J., Smith, H.A., Ade, P.A.R., Furniss, I., Towlson, W.A., Unger, S.J., King, K.J., Davis, G.R., Cohen, M., Emery, R.J., Fischer, J., Glencross, W.M., Caux, E., Greenhouse, M.A., Gry, C., Joubert, M., Lorenzetti, D., Nisini, B., Omont, A., Orfei, R., Saraceno, P., Serra, G., Walker, H.J., Armand, C., Burgdorf, M., di Giorgio, A., Molinari, S., Price, M., Texier, D., Sidher, S., and Trams, N. (1996) The ISO LWS grating spectrum of NGC 7027. *Astronomy and Astrophysics*, **315**, L257–L260.

104 Cernicharo, J., Liu, X.-W., González-Alfonso, E., Cox, P., Barlow, M.J., Lim, T., and Swinyard, B.M. (1997) Discovery of far-infrared pure rotational transitions of CH$^+$ in NGC 7027. *The Astrophysical Journal*, **483**, L65–L68.

105 Wesson, R., Cernicharo, J., Barlow, M.J., Matsuura, M., Decin, L., Groenewegen, M.A.T., Polehampton, E.T., Agundez, M., Cohen, M., Daniel, F., Exter, K.M., Gear, W.K., Gomez, H.L., Hargrave, P.C., Imhof, P., Ivison, R.J., Leeks, S.J., Lim, T.L., Olofsson, G., Savini, G., Sibthorpe, B., Swinyard, B.M., Ueta, T., Witherick, D.K., and Yates, J.A. (2010) Herschel-SPIRE FTS spectroscopy of the carbon-rich objects AFGL 2688, AFGL 618 and NGC 7027. *Astronomy and Astrophysics*, **518**, L144.

106 Herzberg, G. and Shoosmith, J. (1959) Spectrum and structure of the free methylene radical. *Nature*, **183**, 1801–1802.

107 Michael, E.A., Lewen, F., Winnewisser, G., Ozeki, H., Habara, H., and Herbst, E. (2003) Laboratory spectrum of the $1_{11}–2_{02}$ rotational transition of CH$_2$. *The Astrophysical Journal*, **596**, 1356–1362.

108 Sears, T.J., Bunker, P.R., McKellar, A.R.W., Evenson, K.M., Jennings, D.A., and Brown, J.M. (1982) The rotational spectrum and hyperfine structure of the methylene radical CH$_2$

studied by far-infrared laser magnetic resonance spectroscopy. *The Journal of Chemical Physics*, **77**, 5348–5362.

109 Ozeki, H. and Saito, S. (1995) Laboratory submillimeter-wave spectroscopy of the CH$_2$(^{3}B$_1$) radical. *The Astrophysical Journal*, **451**, L97–L99.

110 Brünken, S., Michael, E.A., Lewen, F., Giesen, Th., Ozeki, H., Winnewisser, G., Jensen, P., and Herbst, E. (2004) High-resolution terahertz spectrum of CH$_2$ – low J rotational transitions near 2 THz. *Canadian Journal of Chemistry*, **82**, 676–683.

111 Hollis, J.M., Jewell, P.R., and Lovas, F.J. (1995) Confirmation of interstellar methylene. *The Astrophysical Journal*, **438**, 259–264.

112 Polehampton, E.T., Menten, K.M., Brünken, S., Winnewisser, G., and Baluteau, J.-P. (2005) Far-infrared detection of methylene. *Astronomy and Astrophysics*, **431**, 203–213.

113 Lyu, C.-H., Smith, A.M., and Bruhweiler, F.C. (2001) A search for interstellar CH$_2$ toward HD 154368 and ζ Ophiuchi. *The Astrophysical Journal*, **560**, 865–870.

114 Feuchtgruber, H., Helmich, F.P., van Dishoeck, E.F., and Wright, C.M. (2000) Detection of interstellar CH$_3$. *The Astrophysical Journal*, **535**, L111–L114.

115 Wootten, A. and Turner, B.E. (2008) *A Search for Interstellar CH$_2$D$^+$*. Proceedings of the International Astronomical Union Symposium 251: Organic Matter in Space, pp. 33–34.

116 Bézard, B., Romani, P.N., Feuchtgruber, H., and Encrenaz, T. (1999) Detection of the methyl radical on Neptune. *The Astrophysical Journal*, **515**, 868–872.

117 Thaddeus, P., Gottlieb, C.A., Hjalmarson, Å., Johansson, L.E.B., Irvine, W.M., Friberg, P., and Linke, R.A. (1985) Astronomical identification of the C$_3$H radical. *The Astrophysical Journal*, **294**, L49–L53.

118 Guélin, M., Green, S., and Thaddeus, P. (1978) Detection of the C$_4$H radical toward IRC+10216. *The Astrophysical Journal*, **224**, L27–L30.

119 Guélin, M., Cernicharo, J., Travers, M.J., McCarthy, M.C., Gottlieb, C.A., Thaddeus, P., Ohishi, M., Saito, S., and Ya-mamoto, S. (1997) Detection of a new linear carbon chain radical: C_7H. *Astronomy and Astrophysics*, **317**, L1–L4.

120 Cernicharo, J. and Guélin, M. (1996) Discovery of the C_8H radical. *Astronomy and Astrophysics*, **309**, L27–L30.

121 Kawaguchi, K., Kaifu, N., Ohishi, M., Ishikawa, S., Hirahara, Y., Yamamoto, S., Saito, S., Takano, S., Murakami, A., Vr-tilek, J.M., Gottlieb, C.A., Thaddeus, P., and Irvine, W.M. (1991) Observations of cumulene carbenes, H_2CCCC and H_2CCC, in TMC-1. *Publications of the Astronomical Society of Japan*, **43**, 607–619.

122 Bell, M.B., Feldman, P.A., Travers, M.J., McCarthy, M.C., Gottlieb, C.A., and Thaddeus, P. (1997) Detection of $HC_{11}N$ in the cold dust cloud TMC-1. *The Astrophysical Journal*, **483**, L61–L64.

123 Kawaguchi, K., Ohishi, M., Ishikawa, S.-I., and Kaifu, N. (1992) Detection of isocyanoacetylene HCCNC in TMC-1. *The Astrophysical Journal*, **386**, L51–L53.

124 Kawaguchi, K., Takano, S., Ohishi, M., Ishikawa, S.-I., Miyazawa, K., Kaifu, N., Yamashita, K., Yamamoto, S., Saito, S., Ohshima, Y., and Endo, Y. (1992) Detection of HNCCC in TMC-1. *The Astrophysical Journal*, **396**, L49–L51.

125 Dalgarno, A. and McCray, R.A. (1973) The formation of interstellar molecules from negative ions. *The Astrophysical Journal*, **181**, 95–100.

126 Herbst, E. (1981) Can negative molecular ions be detected in dense interstellar clouds. *Nature*, **289**, 656–657.

127 Millar, T.J., Herbst, E., and Bettens, R.P.A. (2000) Large molecules in the envelope surrounding IRC+10216. *Monthly Notices of the Royal Astronomical Society*, **316**, 195–203.

128 Kawaguchi, K., Kasai, Y., Ishikawa, S.-I., and Kaifu, N. (1995) A spectral-line survey observation of IRC+10216 between 28 and 50 GHz. *Publications of the Astronomical Society of Japan*, **47**, 853–876.

129 McCarthy, M.C., Gottlieb, C.A., Gupta, H., and Thaddeus, P. (2006) Laboratory and astronomical identification of the negative molecular ion C_6H^-. *The Astrophysical Journal*, **652**, L141–L144.

130 Cernicharo, J., Guélin, M., Agúndez, M., Kawaguchi, K., McCarthy, M., and Thaddeus, P. (2007) Astronomical detection of C_4H^-, the second interstellar anion. *Astronomy and Astrophysics*, **467**, L37–L40.

131 Brünken, S., Gupta, H., Gottlieb, C.A., McCarthy, M.C., and Thaddeus, P. (2007) Detection of the carbon chain negative ion C_8H^- in TMC-1. *The Astrophysical Journal*, **664**, L43–L46.

132 Hinkle, K.W., Keady, J.J., and Bernath, P.F. (1988) Detection of C_3 in the circumstellar shell of IRC+10216. *Science*, **241**, 1319–1322.

133 Bernath, P.F., Hinkle, K.H., and Keady, J.J. (1989) Detection of C_5 in the circumstellar shell of IRC+10216. *Science*, **244**, 562–564.

134 Giesen, T.F., Van Orden, A.O., Cruzan, J.D., Provencal, R.A., Saykally, R.J., Gendriesch, R., Lewen, F., and Winnewisser, G. (2001) Interstellar detection of CCC and high-precision laboratory measurements near 2 THz. *The Astrophysical Journal*, **551**, L181–L184.

135 Cernicharo, J., Goicoechea, J.R., and Caux, E. (2000) Far-infrared detection of C_3 in Sagittarius B2 and IRC+10216. *The Astrophysical Journal*, **534**, L199–L202.

136 Goicoechea, J.R., Cernicharo, J., Masso, H., and Senent, M.L. (2004) A new unidentified far-infrared band in NGC 7027. *The Astrophysical Journal*, **609**, 225–230.

137 Maier, J.P., Lakin, N.M., Walker, G.A.H., and Bohlender, D.A. (2001) Detection of C_3 in diffuse interstellar clouds. *The Astrophysical Journal*, **553**, 267–273.

138 Maier, J.P., Walker, G.A.H., and Bohlender, D.A. (2002) Limits to interstellar C_4 and C_5 toward ζ Ophiuchi. *The Astrophysical Journal*, **566**, 332–335.

139 Maier, J.P., Walker, G.A.H., and Bohlender, D.A. (2004) On the possible role of carbon chains as carriers of diffuse interstellar bands. *The Astrophysical Journal*, **602**, 286–290.

140 Irvine, W.M., Höglund, B., Friberg, P., Askne, J., and Elldér, J. (1981) The increasing chemical complexity of the Tau-

rus dark clouds: detection of CH_3CCH and C_4H. *The Astrophysical Journal*, **248**, L113–L117.

141 Remijan, A.J., Hollis, J.M., Snyder, L.E., Jewell, P.R., and Lovas, F.J. (2006) Methyltriacetylene (CH_3C_6H) toward TMC-1: the largest detected symmetric top. *The Astrophysical Journal*, **643**, L37–L40.

142 Solomon, P.M., Jefferts, K.B., Penzias, A.A., and Wilson, R.W. (1971) Detection of millimeter emission lines from interstellar methyl cyanide. *The Astrophysical Journal*, **168**, L107–L110.

143 Matthews, H.E. and Sears, T.J. (1983) Detection of the $J = 1 \rightarrow 0$ transition of CH_3CN. *The Astrophysical Journal*, **267**, L53–L57.

144 Marten, A., Hidayat, T., Biraud, Y., and Moreno, R. (2002) New millimeter heterodyne observations of Titan: vertical distributions of nitriles HCN, HC_3N, CH_3CN, and the isotopic ratio $^{15}N/^{14}N$ in its atmosphere. *Icarus*, **158**, 532–544.

145 Broten, N.W., MacLeod, J.M., Avery, L.W., Irvine, W.M., Höglund, B., Friberg, P., and Hjalmarson, Å. (1984) The detection of interstellar methyl-cyanoacetylene. *The Astrophysical Journal*, **276**, L25–L29.

146 Snyder, L.E., Hollis, J.M., Jewell, P.R., Lovas, F.J., and Remijan, A. (2006) Confirmation of interstellar methylcyanodiacetylene (CH_3C_5N). *The Astrophysical Journal*, **647**, 412–417.

147 Lovas, F.J., Remijan, A.J., Hollis, J.M., Jewell, P.R., and Snyder, L.E. (2006) Hyperfine structure identification of interstellar cyanoallene toward TMC-1. *The Astrophysical Journal*, **637**, L37–L40.

148 Yamamoto, S., Saito, S., Ohishi, M., Suzuki, H., Ishikawa, S.-I., Kaifu, N., and Murakami, A. (1987) Laboratory and astronomical detection of the cyclic C_3H radical. *The Astrophysical Journal*, **322**, L55–L58.

149 Mangum, J.G. and Wootten, A. (1990) Observations of the cyclic C_3H radical in the interstellar medium. *Astronomy and Astrophysics*, **239**, 319–325.

150 Dateo, C.E. and Lee, T.J. (1997) An accurate ab initio quartic force field and vibrational frequencies for cyclopropenyli-

dene. *Spectrochimica Acta Part A: Molecular and Biomolecular Spectroscopy*, **53**, 1065–1077.

151 Bernstein, L.S. and Lynch, D.K. (2009) Small carbonaceous molecules, ethylene oxide (c-C_2H_4O) and cyclopropenylidene (c-C_3H_2): sources of the unidentified infrared bands? *The Astrophysical Journal*, **704**, 226–239.

152 Reisenauer, H.P., Maier, G., Riemann, A., and Hoffmann, R.W. (1984) Cyclopropenylidene. *Angewandte Chemie International Edition in English*, **23**, 641–641.

153 Lee, T.J., Bunge, A., and Schaefer III, H.F. (1985) Toward the laboratory identification of cyclopropenylidene. *Journal of the American Chemical Society*, **107**, 137–142.

154 Thaddeus, P., Vrtilek, J.M., and Gottlieb, C.A. (1985) Laboratory and astronomical identification of cyclopropenylidene, C_3H_2. *The Astrophysical Journal*, **299**, L63–L66.

155 Matthews, H.E. and Irvine, W.M. (1985) The hydrocarbon ring C_3H_2 is ubiquitous in the Galaxy. *The Astrophysical Journal*, **298**, L61–L65.

156 Hollis, J.M., Remijan, A.J., Jewell, P.R., and Lovas, F.J. (2006) Cyclopropenone (c-H_2C_3O): a new interstellar ring molecule. *The Astrophysical Journal*, **642**, 933–939.

157 Dickens, J.E., Irvine, W.M., Ohishi, M., Ikeda, M., Ishikawa, S., Nummelin, A., and Hjalmarson, Å. (1997) Detection of interstellar ethylene oxide (c-C_2H_4O). *The Astrophysical Journal*, **489**, 753–757.

158 Cunningham, M.R., Jones, P.A., Godfrey, P.D., Cragg, D.M., Bains, I., Burton, M.G., Calisse, P., Crighton, N.H.M., Curran, S.J., Davis, T.M., Dempsey, J.T., Fulton, B., Hidas, M.G., Hill, T., Kedziora-Chudczer, L., Minier, V., Pracy, M.B., Purcell, C., Shobbrook, J., and Travouillon, T. (2007) A search for propylene oxide and glycine in Sagittarius B2 (LMH) and Orion. *Monthly Notices of the Royal Astronomical Society*, **376**, 1201–1210.

159 Lebouteiller, V., Kuassivi, and Ferlet, R. (2005) Phosphorus in the diffuse inter-

stellar medium. *Astronomy and Astrophysics*, **443**, 509–517.

160 Dufton, P.L., Keenan, F.P., and Hibbert, A. (1986) The abundance of phosphorus in the interstellar medium. *Astronomy and Astrophysics*, **164**, 179–183.

161 Irwin, P.G.J., Parrish, P., Fouchet, T., Calcutt, S.B., Taylor, F.W., Simon-Miller, A.A., and Nixon, C.A. (2004) Retrievals of jovian tropospheric phosphine from Cassini/CIRS. *Icarus*, **172**, 37–49.

162 Turner, B.E. and Bally, J. (1987) Detection of interstellar PN: the first identified phosphorus compound in the interstellar medium. *The Astrophysical Journal*, **321**, L75–L79.

163 Ziurys, L.M. (1987) Detection of interstellar PN: the first phosphorus-bearing species observed in molecular clouds. *The Astrophysical Journal*, **321**, L81–L85.

164 Guélin, M., Cernicharo, J., Paubert, G., and Turner, B.E. (1990) Free CP in IRC+10216. *Astronomy and Astrophysics*, **230**, L9–L11.

165 Agúndez, M., Cernicharo, J., and Guélin, M. (2007) Discovery of phosphaethyne (HCP) in space: phosphorus chemistry in circumstellar envelopes. *The Astrophysical Journal*, **662**, L91–L94.

166 Tenenbaum, E.D., Woolf, N.J., and Ziurys, L.M. (2007) Identification of phosphorus monoxide $(X^2\Pi_r)$ in VY Canis Majoris: detection of the first P–O bond in space. *The Astrophysical Journal*, **666**, L29–L32.

167 Milam, S.N., Halfen, D.T., Tenenbaum, E.D., Apponi, A.J., Woolf, N.J., and Ziurys, L.M. (2008) Constraining phosphorus chemistry in carbon- and oxygen-rich circumstellar envelopes: observations of PN, HCP, and CP. *The Astrophysical Journal*, **684**, 618–625.

168 Salama, F. (1999) *Polycyclic Aromatic Hydrocarbons in the Interstellar Medium: A Review*. Proceedings of Les Houches Workshop, 2–6 February 1998, "Solid Interstellar Matter: The ISO Revolution", p. 65.

169 Moutou, C., Léger, A., and D'Hendecourt, L. (1996) Far-infrared emission of PAH molecules (14–40 µm):

a preparation for ISO spectroscopy. *Astronomy and Astrophysics*, **310**, 297–308.

170 Allamandola, L.J., Sandford, S.A., and Wopenka, B. (1987) Interstellar polycyclic aromatic hydrocarbons and carbon in interplanetary dust particles and meteorites. *Science*, **237**, 56–59.

171 Allamandola, L.J., Tielens, A.G.G.M., and Barker, J.R. (1989) Interstellar polycyclic aromatic hydrocarbons: the infrared emission bands, the excitation/emission mechanism, and the astrophysical implications. *The Astrophysical Journal Supplement Series*, **71**, 733–775.

172 Thorwirth, S., Theulé, P., Gottlieb, C.A., McCarthy, M.C., and Thaddeus, P. (2007) Rotational spectra of small PAHs: acenaphthene, acenaphthylene, azulene, and fluorene. *The Astrophysical Journal*, **662**, 1309–1314.

173 Cernicharo, J., Decin, L., Barlow, M.J., Agúndez, M., Royer, P., Vandenbussche, B., Wesson, R., Polehampton, E.T., de Beck, E., Blommaert, J.A.D.L., Daniel, F., De Meester, W., Exter, K.M., Feuchtgruber, H., Gear, W.K., Goicoechea, J.R., Gomez, H.L., Groenewegen, M.A.T., Hargrave, P.C., Huygen, R., Imhof, P., Ivison, R.J., Jean, C., Kerschbaum, F., Leeks, S.J., Lim, T.L., Matsuura, M., Olofsson, G., Posch, T., Regibo, S., Savini, G., Sibthorpe, B., Swinyard, B.M., Vandenbussche, B., and Waelkens, C. (2010) Detection of anhydrous hydrochloric acid, HCl, in IRC+10216 with the Herschel SPIRE and PACS spectrometers. detection of HCl in IRC+10216. *Astronomy and Astrophysics*, **518**, L136

174 Ziurys, L.M., Savage, C., Highberger, J.L., Apponi, A.J., Guélin, M., and Cernicharo, J. (2002) More metal cyanide species: detection of AlNC $(X^1\Sigma^+)$ toward IRC+10216. *The Astrophysical Journal*, **564**, L45–L48.

175 Yamazaki, E., Okabayashi, T., and Tanimoto, M. (2001) Laboratory rotational spectrum of nickel monochloride in the ground electronic $^2\Pi_{3/2}$ state. *The Astrophysical Journal*, **551**, L199–L201.

176 Halfen, D.T., Apponi, A.J., and Ziurys, L.M. (2002) Laboratory detection and pure rotational spectrum of the CaC radical ($X^3\Sigma^-$). *The Astrophysical Journal*, **577**, L67–L70.

177 Lovas, F.J., Kawashima, Y., Grabow, J.-U., Suenram, R.D., Fraser, G.T., and Hirota, E. (1995) Microwave spectra, hyperfine structure, and electric dipole moments for conformers I and II of glycine. *The Astrophysical Journal*, **455**, L201–L204.

178 Snyder, L.E. (1997) The search for interstellar glycine. *Origins of Life and Evolution of the Biosphere*, **27**, 115–133.

179 Kuan, Y.-J., Charnley, S.B., Huang, H.-C., Tseng, W.-L., and Kisiel, Z. (2003) Interstellar glycine. *The Astrophysical Journal*, **593**, 848–867.

180 Snyder, L.E., Lovas, F.J., Hollis, J.M., Friedel, D.N., Jewell, P.R., Remijan, A., Ilyushin, V.V., Alekseev, E.A., and Dyubko, S.F. (2005) A rigorous attempt to verify interstellar glycine. *The Astrophysical Journal*, **619**, 914–930.

181 Hollis, J.M., Lovas, F.J., and Jewell, P.R. (2000) Interstellar glycolaldehyde: the first sugar. *The Astrophysical Journal*, **540**, L107–L110.

182 Hollis, J.M., Jewell, P.R., Lovas, F.J., and Remijan, A. (2004) Green Bank Telescope observations of interstellar glycolaldehyde: low-temperature sugar. *The Astrophysical Journal*, **613**, L45–L48.

183 Beltrán, M.T., Codella, C., Viti, S., Neri, R., and Cesaroni, R. (2009) First detection of glycolaldehyde outside the galactic center. *The Astrophysical Journal*, **690**, L93–L96.

184 Widicus Weaver, S.L. and Blake, G.A. (2005) 1,3-Dihydroxyacetone in Sagittarius B2(N-LMH): the first interstellar ketose. *The Astrophysical Journal*, **624**, L33–L36.

185 Apponi, A.J., Halfen, D.T., Ziurys, L.M., Hollis, J.M., Remijan, A.J., and Lovas, F.J. (2006) Investigating the limits of chemical complexity in Sagittarius B2(N): a rigorous attempt to confirm 1,3-dihydroxyacetone. *The Astrophysical Journal*, **643**, L29–L32.

186 Kutner, M.L., Machnik, D.E., Tucker, K.D., and Dickman, R.L. (1980) Search for interstellar pyrrole and furan. *The Astrophysical Journal*, **242**, 541–544.

187 Irvine, W.M., Elldér, J., Hjalmarson, Å., Kollberg, E., Rydbeck, O.E.H., Sørensen, G.O., Bak, B., and Svanholt, H. (1981) Searches for interstellar imidazole and cyanoform. *Astronomy and Astrophysics*, **97**, 192–194.

188 Simon, M.N. and Simon, M. (1973) A search for interstellar acrylonitrile, pyrimidine, and pyridine. *The Astrophysical Journal*, **184**, 757–761.

189 Kuan, Y.-J., Yan, C.-H., Charnley, S.B., Kisiel, Z., Ehrenfreund, P., and Huang, H.-C. (2003) A search for interstellar pyrimidine. *Monthly Notices of the Royal Astronomical Society*, **345**, 650–656.

190 Guillois, O., Ledoux, G., and Reynaud, C. (1999) Diamond infrared emission bands in circumstellar media. *The Astrophysical Journal*, **521**, L133–L136.

191 Chang, H.-C., Lin, J.-C., Wu, J.-Y., and Chen, K.-H. (1995) Infrared spectroscopy and vibrational relaxation of CH_x and CD_x stretches on synthetic diamond nanocrystal surfaces. *The Journal of Physical Chemistry*, **99**, 11081–11088.

192 Chen, C.-F., Wu, C.-C., Cheng, C.-L., Sheu, S.-Y., and Chang, H.-C. (2002) The size of interstellar nanodiamonds revealed by infrared spectra of CH on synthetic diamond nanocrystal surfaces. *The Journal of Chemical Physics*, **116**, 1211–1214.

193 Ierofeieff, M.V. and Latchinoff, P.A. (1888) Météorite diamantére tombé le 10/22 Septembre 1886 en Russie, à Nowo Urei. *Government de Penze. Comptes Rendus Hebdomadaires des Seances de l'Academie des Science Paris*, **106**, 1679–1682.

194 Lewis, R.S., Ming, T., Wacker, J.F., Anders, E., and Steel, E. (1987) Interstellar diamonds in meteorites. *Nature*, **326**, 160–162.

195 Frum, C.I., Engelman Jr., R., Hedderich, H.G., Bernath, P.F., Lamb, L.D., and Huffman, D.R. (1991) The infrared emission spectrum of gas-phase C_{60} (buckmisterfullerene). *Chemical Physics Letters*, **176**, 504–508.

196 Nemes, L., Ram, R.S., Bernath, P.F., Tinker, F.A., Zumwalt, M.C., Lamb, L.D.,

and Huffman, D.R. (1994) Gas-phase infrared emission spectra of C_{60} and C_{70}. Temperature dependent studies. *Chemical Physics Letters*, **218**, 295–303.

197 Krätschmer, W., Lamb, L.D., Fostiropoulos, K., and Huffman, D.R. (1990) Solid C_{60}: a new form of carbon. *Nature*, **347**, 354–358.

198 Fulara, J., Jakobi, M., and Maier, J.P. (1993) Electronic and infrared spectra of C_{60}^{+} and C_{60}^{-} in neon and argon matrices. *Chemical Physics Letters*, **211**, 227–234.

199 Cami, J., Bernard-Salas, J., Peeters, E., and Malek, S.E. (2010) Detection of C_{60} and C_{70} in a young planetary nebula. *Science*, **329**, 1180–1182.

200 Somerville, W.B. and Bellis, J.G. (1989) An astronomical search for the molecule C_{60}. *Monthly Notices of the Royal Astronomical Society*, **240**, 41P–46P.

201 Snow, T.P. and Seab, C.G. (1989) A search for interstellar and circumstellar C_{60}. *Astronomy and Astrophysics*, **213**, 291–294.

202 Clayton, G.C., Kelly, D.M., Lacy, J.H., Little-Marenin, I.R., Feldman, P.A., and Bernath, P.F. (1995) A mid-infrared search for C_{60} in R Coronae Borealis stars and IRC+10216. *The Astronomical Journal*, **109**, 2096–2103.

203 Foing, B.H. and Ehrenfreund, P. (1994) Detection of two interstellar absorption bands coincident with spectral features of C_{60}^{+}. *Nature*, **369**, 296–298.

204 García-Hernández, D.A., Manchado, A., García-Lario, P., Stanghellini, L., Villaver, E., Shaw, R.A., Szczerba, R., and Perea-Calderón, J.V. (2010) Formation of fullerenes in H-containing planetary nebulae. *The Astrophysical Journal*, **724**, L39–L43.

205 Zhang, Y. and Kwok, S. (2011) Detection of C_{60} in the proto-planetary nebula iras 01005+7910. *The Astrophysical Journal*, **730**, 126.

206 Sellgren, K., Werner, M.W., Ingalls, J.G., Smith, J.D.T., Carleton, T.M., and Joblin, C. (2010) C_{60} in reflection nebulae. *The Astrophysical Journal*, **722**, L54–L57.

207 Ehrenfreund, P. and Foing, B.H. (2010) Fullerenes and cosmic carbon. *Science*, **329** (5996), 1159–1160.

208 Johansson, L.E.B., Andersson, C., Ellder, J., Friberg, P., Hjalmarson, Å., Hoglund, B., Irvine, W.M., Olofsson, H., and Rydbeck, G. (1984) Spectral scan of Orion A and IRC+10216 from 72 to 91 GHz. *Astronomy and Astrophysics*, **130**, 227–256.

209 Sutton, E.C., Blake, G.A., Masson, C.R., and Phillips, T.G. (1985) Molecular line survey of Orion A from 215 to 247 GHz. *The Astrophysical Journal Supplement Series*, **58**, 341–378.

210 Jewell, P.R., Hollis, J.M., Lovas, F.J., and Snyder, L.E. (1989) Millimeter- and submillimeter-wave surveys of Orion A emission lines in the ranges 200.7–202.3, 203.7–205.3, and 330–360 GHz. *The Astrophysical Journal Supplement Series*, **70**, 833–864.

211 White, G.J., Araki, M., Greaves, J.S., Ohishi, M., and Higginbottom, N.S. (2003) A spectral survey of the Orion Nebula from 455–507 GHz. *Astronomy and Astrophysics*, **407**, 589–607.

212 Schilke, P., Benford, D.J., Hunter, T.R., Lis, D.C., and Phillips, T.G. (2001) A line survey of Orion-KL from 607 to 725 GHz. *The Astrophysical Journal Supplement Series*, **132**, 281–364.

213 Comito, C., Schilke, P., Phillips, T.G., Lis, D.C., Motte, F., and Mehringer, D. (2005) A molecular line survey of Orion KL in the 350 micron band. *The Astrophysical Journal Supplement Series*, **156**, 127–167.

214 Olofsson, A.O.H., Persson, C.M., Koning, N., Bergman, P., Bernath, P.F., Black, J.H., Frisk, U., Geppert, W., Hasegawa, T.I., Hjalmarson, Å., Kwok, S., Larsson, B., Lecacheux, A., Nummelin, A., Olberg, M., Sandqvist, A., and Wirström, E.S. (2007) A spectral line survey of Orion KL in the bands 486–492 and 541–577 GHz with the Odintellite. *Astronomy and Astrophysics*, **476**, 791–806.

215 Koning, N., Kwok, S., Bernath, P., Hjalmarson, Å., and Olofsson, H. (2008) *Organic Molecules in the Spectral Line Survey of Orion KL with the Odin Satellite from 486–492 GHz and 541–577 GHz.* Proceedings of the International Astronomical Union Symposium 251: Organic Matter in Space, pp. 29–30.

216 Cummins, S.E., Linke, R.A., and Thaddeus, P. (1986) A survey of the millimeter-wave spectrum of Sagittarius B2. *The Astrophysical Journal Supplement Series*, **60**, 819–878.

217 Turner, B.E. (1991) A molecular line survey of Sagittarius B2 and Orion-KL from 70 to 115 GHz. II – analysis of the data. *The Astrophysical Journal Supplement Series*, **76**, 617–686.

218 Kaifu, N., Ohishi, M., Kawaguchi, K., Saito, S., Yamamoto, S., Miyaji, T., Miyazawa, K., Ishikawa, S., Noumaru, C., Harasawa, S., Okuda, M., and Suzuki, H. (2004) A 8.8–50 GHz complete spectral line survey toward TMC-1 I. survey data. *Publications of the Astronomical Society of Japan*, **56**, 69–173.

219 Friedel, D.N., Snyder, L.E., Turner, B.E., and Remijan, A. (2004) A spectral line survey of selected 3 millimeter bands toward Sagittarius B2(N-LMH) using the National Radio Astronomy Observatory 12 meter radio telescope and the Berkeley–Illinois–Maryland Association Array. I. the observational data. *The Astrophysical Journal*, **600**, 234–253.

220 Bell, M.B., Avery, L.W., and Watson, J.K.G. (1993) A spectral-line survey of W51 from 17.6 to 22.0 GHz. *The Astrophysical Journal Supplement Series*, **86**, 211–233.

221 Ceccarelli, C., Bacmann, A., Boogert, A., Caux, E., Dominik, C., Lefloch, B., Lis, D., Schilke, P., van der Tak, F., Caselli, P., Cernicharo, J., Codella, C., Comito, C., Fuente, A., Baudry, A., Bell, T., Benedettini, M., Bergin, E.A., Blake, G.A., Bottinelli, S., Cabrit, S., Castets, A., Coutens, A., Crimier, N., Demyk, K., Encrenaz, P., Falgarone, E., Gerin, M., Goldsmith, P.F., Helmich, F., Hennebelle, P., Henning, T., Herbst, E., Hily-Blant, P., Jacq, T., Kahane, C., Kama, M., Klotz, A., Langer, W., Lord, S., Lorenzani, A., Maret, S., Melnick, G., Neufeld, D., Nisini, B., Pacheco, S., Pagani, L., Parise, B., Pearson, J., Phillips, T., Salez, M., Saraceno, P., Schuster, K., Tielens, X., van der Wiel, M.H.D., Vastel, C., Viti, S., Wakelam, V., Walters, A., Wyrowski, F., Yorke, H., Liseau, R., Olberg, M., Szczerba, R., Benz, A.O., and Melchior, M. (2010) Herschel spectral surveys of star-forming regions. Overview of the 555–636 GHz range. *Astronomy and Astrophysics*, **521**, L22–1–8.

222 Johansson, L.E.B., Andersson, C., Elder, J., Friberg, P., Hjalmarson, Å., Hoglund, B., Olofsson, H., Rydbeck, G., and Irvine, W.M. (1985) The spectra of Orion A and IRC+10216 between 72.2 and 91.1 GHz. *Astronomy and Astrophysics Supplement Series*, **60**, 135–168.

223 Avery, L.W., Amano, T., Bell, M.B., Feldman, P.A., Johns, J.W.C., MacLeod, J.M., Matthews, H.E., Morton, D.C., Watson, J.K.G., Turner, B.E., Hayashi, S.S., Watt, G.D., and Webster, A.S. (1992) A spectral line survey of IRC+10216 at millimeter and submillimeter wavelengths. *The Astrophysical Journal Supplement Series*, **83**, 363–385.

224 Groesbeck, T.D., Phillips, T.G., and Blake, G.A. (1994) The molecular emission-line spectrum of IRC+10216 between 330 and 358 GHz. *The Astrophysical Journal Supplement Series*, **94**, 147–162.

225 Cernicharo, J., Guélin, M., and Kahane, C. (2000) A λ 2 mm molecular line survey of the C-star envelope IRC+10216. *Astronomy and Astrophysics Supplement Series*, **142**, 181–215.

226 He, J.H., Kwok, S., Müller, H.S.P., Zhang, Y., Hasegawa, T., Peng, T.C., and Huang, Y.C. (2008) A spectral line survey in the 2 and 1.3 mm windows toward the carbon-rich envelope of IRC+10216. *The Astrophysical Journal Supplement Series*, **177**, 275–325.

227 Ziurys, L.M., Milam, S.N., Apponi, A.J., and Woolf, N.J. (2007) Chemical complexity in the winds of the oxygen-rich supergiant star VY Canis Majoris. *Nature*, **447**, 1094–1097.

228 Zhang, Y., Kwok, S., and Dinh-V-Trung (2008) A spectral line survey of NGC 7027 at millimeter wavelengths. *The Astrophysical Journal*, **678**, 328–346.

229 Zhang, Y., Kwok, S., and Dinh-V-Trung (2009) A molecular line survey of the highly evolved carbon star CIT 6. *The Astrophysical Journal*, **691**, 1660–1677.

230 Zhang, Y., Kwok, S., and Nakashima, J.-I. (2009) A molecular line survey of the extreme carbon star CRL 3068 at millimeter wavelengths. *The Astrophysical Journal*, **700**, 1262–1281.

231 Zhang, Y., Kwok, S., and Dinh-V-Trung (2008) *Molecular Lines in the Envelopes of Evolved Stars*. Proceedings of the International Astronomical Union Symposium 251: Organic Matter in Space, pp. 169–170.

232 Kwok, S., Volk, K., and Bidelman, W.P. (1997) Classification and identification of *IRAS* sources with low-resolution spectra. *The Astrophysical Journal Supplement Series*, **112**, 557–584.

233 Yamamura, I., Onaka, T., Tanabe, T., Roellig, T.L., and Yuen, L. (1996) Mid-infrared spectral observations of point sources by IRTS. *Publications of the Astronomical Society of Japan*, **48**, L65–L69.

234 Thaddeus, P. (2006) The prebiotic molecules observed in the interstellar gas. *Philosophical Transactions of The Royal Society B: Biological Sciences*, **361**, 1681–1687.

235 Herbst, E. and van Dishoeck, E.F. (2009) Complex organic interstellar molecules. *Annual Review of Astronomy and Astrophysics*, **47**, 427–480.

236 Kalenskii, S.V., Slysh, V.I., Goldsmith, P.F., and Johansson, L.E.B. (2004) A 4–6 GHz spectral scan and 8–10 GHz observations of the dark cloud TMC-1. *The Astrophysical Journal*, **610**, 329–338.

237 Ohishi, M. (2008) *Molecular Spectral Line Surveys and the Organic Molecules in the Interstellar Molecular Clouds*. Proceedings of the International Astronomical Union Symposium 251: Organic Matter in Space, pp. 17–25.

238 Polehampton, E.T., Baluteau, J.-P., Swinyard, B.M., Goicoechea, J.R., Brown, J.M., White, G.J., Cernicharo, J., and Grundy, T.W. (2007) The *ISO* LWS high-resolution spectral survey towards Sagittarius B2. *Monthly Notices of the Royal Astronomical Society*, **377**, 1122–1150.

239 Kwok, S. (2007) *Physics and chemistry of the interstellar medium*, University Science Books, p. 282.

240 Genzel, R. and Stutzki, J. (1989) The Orion Molecular Cloud and star-forming region. *Annual Review of Astronomy and Astrophysics*, **27**, 41–85.

241 Aitken, D.K., Roche, P.F., Spenser, P.M., and Jones, B. (1979) Infrared spatial and spectral studies of an ionization front region in the Orion Nebula. *Astronomy and Astrophysics*, **76**, 60–64.

242 van den Bergh, S. (1966) A study of reflection nebulae. *The Astronomical Journal*, **71**, 990–998.

243 Sellgren, K., Werner, M.W., and Dinerstein, H.L. (1983) Extended near-infrared emission from visual reflection nebulae. *The Astrophysical Journal*, **271**, L13–L17.

244 Sellgren, K., Allamandola, L.J., Bregman, J.D., Werner, M.W., and Wooden, D.H. (1985) Emission features in the 4–13 micron spectra of the reflection nebulae NGC 7023 and NGC 2023. *The Astrophysical Journal*, **299**, 416–423.

245 Sellgren, K., Luan, L., and Werner, M.W. (1990) The excitation of 12 micron emission from very small particles. *The Astrophysical Journal*, **359**, 384–391.

246 Uchida, K.I., Sellgren, K., Werner, M.W., and Houdashelt, M.L. (2000) *Infrared Space Observatory* mid-infrared spectra of reflection nebulae. *The Astrophysical Journal*, **530**, 817–833.

247 Soifer, B.T., Russell, R.W., and Merrill, K.M. (1976) 2–4 micron spectrophotometric observations of the galactic center. *The Astrophysical Journal*, **207**, L83–L85.

248 Willner, S.P., Russell, R.W., Puetter, R.C., Soifer, B.T., and Harvey, P.M. (1979) The 4 to 8 micron spectrum of the galactic center. *The Astrophysical Journal*, **229**, L65–L68.

249 Wickramasinghe, D.T. and Allen, D.A. (1980) The 3.4-μm interstellar absorption feature. *Nature*, **287**, 518–519.

250 Adamson, A.J., Whittet, D.C.B., and Duley, W.W. (1990) The 3.4-μm interstellar absorption feature in Cyg OB2 no. 12. *Monthly Notices of the Royal Astronomical Society*, **243**, 400–404.

251 Sandford, S.A., Allamandola, L.J., Tielens, A.G.G.M., Sellgren, K., Tapia, M., and Pendleton, Y. (1991) The interstellar C–H stretching band near 3.4 microns:

constraints on the composition of organic material in the diffuse interstellar medium. *The Astrophysical Journal*, **371**, 607–620.

252 Pendleton, Y.J., Sandford, S.A., Allamandola, L.J., Tielens, A.G.G.M., and Sellgren, K. (1994) Near-infrared absorption spectroscopy of interstellar hydrocarbon grains. *The Astrophysical Journal*, **437**, 683–696.

253 Whittet, D.C.B., Boogert, A.C.A., Gerakines, P.A., Schutte, W., Tielens, A.G.G.M., de Graauw, T., Prusti, T., van Dishoeck, E.F., Wesselius, P.R., and Wright, C.M. (1997) Infrared spectroscopy of dust in the diffuse interstellar medium toward Cygnus OB2 no. 12. *The Astrophysical Journal*, **490**, 729–734.

254 Duley, W.W. and Williams, D.A. (1983) A 3.4 μm absorption band in amorphous carbon: implications for interstellar dust. *Monthly Notices of the Royal Astronomical Society*, **205**, 67P–70P.

255 Duley, W.W., Scott, A.D., Seahra, S., and Dadswell, G. (1998) Integrated absorbances in the 3.4 μm CH_n band in hydrogenated amorphous carbon. *The Astrophysical Journal*, **503**, L183–L185.

256 Chiar, J.E., Tielens, A.G.G.M., Whittet, D.C.B., Schutte, W.A., Boogert, A.C.A., Lutz, D., van Dishoeck, E.F., and Bernstein, M.P. (2000) The composition and distribution of dust along the line of sight toward the galactic center. *The Astrophysical Journal*, **537**, 749–762.

257 Mennella, V. (2006) Activation energy of C–H bond formation in carbon grains irradiated with hydrogen atoms. *The Astrophysical Journal*, **647**, L49–L52.

258 Boulanger, F., Baud, B., and van Albada, G.D. (1985) Warm dust in the neutral interstellar medium. *Astronomy and Astrophysics*, **144**, L9–L12.

259 Giard, M., Serra, G., Caux, E., Pajot, F., and Lamarre, J.M. (1988) First detection of the aromatic 3.3-micron feature in the diffuse emission of the galactic disk. *Astronomy and Astrophysics*, **201**, L1–L4.

260 Tanaka, M., Matsumoto, T., Murakami, H., Kawada, M., Noda, M., and Matsuura, S. (1996) IRTS observation of the unidentified 3.3-micron band in the diffuse galactic emission. *Publications of the Astronomical Society of Japan*, **48**, L53–L57.

261 Onaka, T., Yamamura, I., Tanabe, T., Roellig, T.L., and Yuen, L. (1996) Detection of the mid-infrared unidentified bands in the diffuse galactic emission by IRTS. *Publications of the Astronomical Society of Japan*, **48**, L59–L63.

262 Onaka, T. (2000) Interstellar dust: what do space observations tell us? *Advances in Space Research*, **25**, 2167–2176.

263 Sellgren, K. (2001) Aromatic hydrocarbons, diamonds, and fullerenes in interstellar space: puzzles to be solved by laboratory and theoretical astrochemistry. *Spectrochimica Acta Part A: Molecular and Biomolecular Spectroscopy*, **57**, 627–642.

264 Witt, A.N., Mandel, S., Sell, P.H., Dixon, T., and Vijh, U.P. (2008) Extended red emission in high galactic latitude interstellar clouds. *The Astrophysical Journal*, **679**, 497–511.

265 Kwok, S. (2007) *Physics and Chemistry of the Interstellar Medium*, University Science Books, p. 521.

266 Groves, B., Dopita, M.A., Sutherland, R.S., Kewley, L.J., Fischera, J., Leitherer, C., Brandl, B., and van Breugel, W. (2008) Modeling the pan-spectral energy distribution of starburst galaxies. IV. The controlling parameters of the starburst SED. *The Astrophysical Journal Supplement Series*, **176**, 438–456.

267 Brandl, B.R., Bernard-Salas, J., Spoon, H.W.W., Devost, D., Sloan, G.C., Guilles, S., Wu, Y., Houck, J.R., Weedman, D.W., Armus, L., Appleton, P.N., Soifer, B.T., Charmandaris, V., Hao, L., Marshall, J.A., Higdon, S.J., and Herter, T.L. (2006) The mid-infrared properties of starburst galaxies from *Spitzer*-IRS spectroscopy. *The Astrophysical Journal*, **653**, 1129–1144.

268 Genzel, R. and Cesarsky, C.J. (2000) Extragalactic results from the Infrared Space Observatory. *Annual Review of Astronomy and Astrophysics*, **38**, 761–814.

269 Sturm, E., Lutz, D., Tran, D., Feuchtgruber, H., Genzel, R., Kunze, D., Moorwood, A.F.M., and Thornley, M.D. (2000) ISO-SWS spectra of galaxies: continuum

and features. *Astronomy and Astrophysics*, **358**, 481–493.

270 Bernard-Salas, J., Peeters, E., Sloan, G.C., Gutenkunst, S., Matsuura, M., Tielens, A.G.G.M., Zijlstra, A.A., and Houck, J.R. (2009) Unusual dust emission from planetary nebulae in the Magellanic Clouds. *The Astrophysical Journal*, **699**, 1541–1552.

271 Wright, G.S., Bridger, A., Geballe, T.R., and Pendleton, Y. (1996) *Studies of NIR Dust Absorption Features in the Nuclei of Active and IRAS Galaxies*. Proceedings of the International Conference on Cold Dust and Galaxy Morphology, January 22–26, 1996, "New Extragalactic Perspectives in the New South Africa", pp. 143–150.

272 Imanishi, M. (2000) The 3.4 μm absorption feature towards three obscured active galactic nuclei. *Monthly Notices of the Royal Astronomical Society*, **319**, 331–336.

273 Imanishi, M. and Dudley, C.C. (2000) Energy diagnoses of nine infrared luminous galaxies based on 3–4 micron spectra. *The Astrophysical Journal*, **545**, 701–711.

274 Bellamy, L.J. (1975) *The Infrared Spectra of Complex Molecules*, Chapman and Hall.

275 Dartois, E., Geballe, T.R., Pino, T., Cao, A.-T., Jones, A., Deboffle, D., Guerrini, V., Bréchignac, P., and d'Hendecourt, L. (2007) IRAS 08572+3915: constraining the aromatic versus aliphatic content of interstellar HACs. *Astronomy and Astrophysics*, **463**, 635–640.

276 Spoon, H.W.W., Armus, L., Cami, J., Tielens, A.G.G.M., Chiar, J.E., Peeters, E., Keane, J.V., Charmandaris, V., Appleton, P.N., Teplitz, H.I., and Burgdorf, M.J. (2004) Fire and ice: *Spitzer* Infrared Spectrograph (IRS) mid-infrared spectroscopy of IRAS F00183-7111. *The Astrophysical Journal Supplement Series*, **154**, 184–187.

277 J.-M. Perrin, Darbon, S., and Sivan, J.-P. (1995) Observation of extended red emission (ERE) in the halo of M82. *Astronomy and Astrophysics*, **304**, L21–L24.

278 Kwok, S. (1975) Radiation pressure on grains as a mechanism for mass loss

in red giants. *The Astrophysical Journal*, **198**, 583–591.

279 Volk, K., Xiong, G.-Z., and Kwok, S. (2000) *Infrared Space Observatory* spectroscopy of extreme carbon stars. *The Astrophysical Journal*, **530**, 408–417.

280 Kwok, S. (2000) *The origin and evolution of planetary nebulae*, Cambridge University Press.

281 Kwok, S. (1993) Proto-planetary nebulae. *Annual Review of Astronomy and Astrophysics*, **31**, 63–92.

282 VanWinckel, H. (2003) Post-AGB stars. *Annual Review of Astronomy and Astrophysics*, **41**, 391–427.

283 Kwok, S. (1982) From red giants to planetary nebulae. *The Astrophysical Journal*, **258**, 280–288.

284 Kwok, S. (1990) An infrared sequence in the late stages of stellar evolution. *Monthly Notices of the Royal Astronomical Society*, **244**, 179–183.

285 Geballe, T.R., Tielens, A.G.G.M., Kwok, S., and Hrivnak, B.J. (1992) Unusual 3 micron emission features in three proto-planetary nebulae. *The Astrophysical Journal*, **387**, L89–L91.

286 Hrivnak, B.J., Geballe, T.R., and Kwok, S. (2007) A study of the 3.3 and 3.4 μm emission features in proto-planetary nebulae. *The Astrophysical Journal*, **662**, 1059–1066.

287 Kwok, S., Volk, K., and Hrivnak, B.J. (1999) Chemical evolution of carbonaceous materials in the last stages of stellar evolution. *Astronomy and Astrophysics*, **350**, L35–L38.

288 Hudgins, D.M. and Allamandola, L.J. (1999) Interstellar PAH emission in the 11–14 micron region: new insights from laboratory data and a tracer of ionized PAHs. *The Astrophysical Journal*, **516**, L41–L44.

289 Kwok, S., Volk, K., and Bernath, P. (2001) On the origin of infrared plateau features in proto-planetary nebulae. *The Astrophysical Journal*, **554**, L87–L90.

290 Cernicharo, J., Heras, A.M., Tielens, A.G.G.M., Pardo, J.R., Herpin, F., Guélin, M., and Waters, L.B.F.M. (2001) *Infrared Space Observatory*'s discovery of C_4H_2, C_6H_2, and benzene in CRL 618.

The Astrophysical Journal, **546**, L123–L126.

291 Kwok, S. (2004) The synthesis of organic and inorganic compounds in evolved stars. *Nature*, **430**, 985–991.

292 Cody, G.D., Ade, H., Alexander, C.M.O'D., Araki, T., Butterworth, A., Fleckenstein, H., Flynn, G., Gilles, M.K., Jacobsen, C., Kilcoyne, A.L.D., Messenger, K., Sandford, S.A., Tyliszczak, T., Westphal, A.J., Wirick, S., and Yabuta, H. (2008) Quantitative organic and light-element analysis of comet 81P/Wild 2 particles using C-, N-, and O-μ-XANES. *Meteoritics and Planetary Science*, **43**, 353–365.

293 Falkowski, P., Scholes, R.J., Boyle, E., Canadell, J., Canfield, D., Elser, J., Gruber, N., Hibbard, K., Hogberg, P., Linder, S., Mackenzie, F.T., Moore, B., Pedersen, T., Rosenthal, Y., Seitzinger, S., Smetacek, V., and Steffen, W. (2000) The global carbon cycle: a test of our knowledge of Earth as a system. *Science*, **290**, 291–296.

294 Sagan, C., Thompson, W.R., Carlson, R., Gurnett, D., and Hord, C. (1993) A search for life on Earth from the Galileo spacecraft. *Nature*, **365**, 715–721.

295 Kuiper, G.P. (1944) Titan: a satellite with an atmosphere. *The Astrophysical Journal*, **100**, 378–383.

296 Ridgway, S.T., Larson, H.P., and Fink, U. (1976) *The Infrared Spectrum of Jupiter*. Proceedings of the Colloquium, Tucson, Arizona, May 19–21, 1975, "Jupiter: Studies of the interior, atmosphere, magnetosphere, and satellites", pp. 384–417.

297 Coustenis, A. and Taylor, F. (1999) *Titan: The Earth-Like Moon*, World Scientific.

298 Coustenis, A., Salama, A., Schulz, B., Ott, S., Lellouch, E., Encrenaz, T.H., Gautier, D., and Feuchtgruber, H. (2003) Titan's atmosphere from ISO mid-infrared spectroscopy. *Icarus*, **161**, 383–403.

299 Coustenis, A., Achterberg, R.K., Conrath, B.J., Jennings, D.E., Marten, A., Gautier, D., Nixon, C.A., Flasar, F.M., Teanby, N.A., Bézard, B., Samuelson, R.E., Carlson, R.C., Lellouch, E., Bjoraker, G.L., Romani, P.N., Taylor, F.W., Irwin, P.G.J., Fouchet, T., Hubert, A., Orton, G.S., Kunde, V.G., Vinatier, S., Mondellini, J., Abbas, M.M., and Courtin, R. (2007) The composition of Titan's stratosphere from Cassini/CIRS mid-infrared spectra. *Icarus*, **189**, 35–62.

300 Waite Jr., J.H. (2007) The process of tholin formation in Titan's upper atmosphere. *Science*, **316**, 870–875.

301 Grundy, W.M., Buie, M.W., Stansberry, J.A., Spencer, J.R., and Schmitt, B. (1999) Near-infrared spectra of icy outer Solar System surfaces: remote determination of H_2O ice temperatures. *Icarus*, **142** (2), 536–549.

302 de Bergh, C., Schmitt, B., Moroz, L.V., Quirico, E., and Cruikshank, D.P. (2008) Laboratory data on ices, refractory carbonaceous materials, and minerals relevant to transneptunian objects and Centaurs, in *The Solar System Beyond Neptune*, The University of Arizona Press, eds M. Antonietta Barucci *et al.*, pp. 19–35.

303 Cruikshank, D.P., Wegryn, E., DalleOre, C.M., Brown, R.H., Bibring, J.-P., Buratti, B.J., Clark, R.N., McCord, T.B., Nicholson, P.D., Pendleton, Y.J., Owen, T.C., Filacchione, G., Coradini, A., Cerroni, P., Capaccioni, F., Jaumann, R., Nelson, R.M., Baines, K.H., Sotin, C., Bellucci, G., Combes, M., Langevin, Y., Sicardy, B., Matson, D.L., Formisano, V., Drossart, P., and Mennella, V. (2008) Hydrocarbons on Saturn's satellites Iapetus and Phoebe. *Icarus*, **193**, 334–343.

304 Buratti, B.J., Cruikshank, D.P., Brown, R.H., Clark, R.N., Bauer, J.M., Jaumann, R., McCord, T.B., Simonelli, D.P., Hibbitts, C.A., Hansen, G.B., Owen, T.C., Baines, K.H., Bellucci, G., Bibring, J.-P., Capaccioni, F., Cerroni, P., Coradini, A., Drossart, P., Formisano, V., Langevin, Y., Matson, D.L., Mennella, V., Nelson, R.M., Nicholson, P.D., Sicardy, B., Sotin, C., Roush, T.L., Soderlund, K., and Muradyan, A. (2005) Cassini Visual and Infrared Mapping Spectrometer observations of Iapetus: detection of CO_2. *The Astrophysical Journal*, **622**, L149–L152.

305 Nguyen, M.J., Raulin, F., Coll, P., Derenne, S., Szopa, C., Cernogora, G., Israël, G., and Bernard, J.M. (2007) Carbon isotopic enrichment in Titan's tholins? implications for Titan's aerosols. *Planetary and Space Science*, **55**, 2010–2014.

306 Brown, R.H., Soderblom, L.A., Soderblom, J.M., Clark, R.N., Jaumann, R., Barnes, J.W., Sotin, C., Buratti, B., Baines, K.H., and Nicholson, P.D. (2008) The identification of liquid ethane in Titan's Ontario Lacus. *Nature*, **454**, 607–610.

307 Lorenz, R.D., Mitchell, K.L., Kirk, R.L., Hayes, A.G., Aharonson, O., Zebker, H.A., Paillou, P., Radebaugh, J., Lunine, J.I., Janssen, M.A., Wall, S.D., Lopes, R.M., Stiles, B., Ostro, S., Mitri, G., and Stofan, E.R. (2008) Titan's inventory of organic surface materials. *Geophysical Research Letters*, **35**, L02201–L02206.

308 Kvenvolden, K., Lawless, J., Pering, K., Peterson, E., Flores, J., and Ponnamperuma, C. (1970) Evidence for extraterrestrial amino-acids and hydrocarbons in the Murchison meteorite. *Nature*, **228**, 923–926.

309 Cronin, J.R., Pizzarello, S., and Frye, J.S. (1987) ^{13}C NMR spectroscopy of the insoluble carbon of carbonaceous chondrites. *Geochimica et Cosmochimica Acta*, **51**, 299–303.

310 Shimoyama, A., Naraoka, H., Komiya, M., and Harada, K. (1989) Analyses of carboxylic acids and hydrocarbons in Antarctic carbonaceous chondrites, Yamato-74662 and Yamato-793321. *Geochemical Journal*, **23**, 181–193.

311 Cronin, J.R., Cooper, G.W., and Pizzarello, S. (1995) Characteristics and formation of amino acids and hydroxy acids of the Murchison meteorite. *Advances in Space Research*, **15**(3), 91–97.

312 Gilmour, I. and Pillinger, C.T. (1994) Isotopic compositions of individual polycyclic aromatic hydrocarbons from the Murchison meteorite. *Monthly Notices of the Royal Astronomical Society*, **269**, 235–240.

313 Stoks, P.G. and Schwartz, A.W. (1981) Nitrogen-heterocyclic compounds in meteorites: significance and mechanisms of formation. *Geochimica et Cosmochimica Acta*, **45**, 563–569.

314 Cronin, J.R. and Pizzarello, S. (1990) Aliphatic hydrocarbons of the Murchison meteorite. *Geochimica et Cosmochimica Acta*, **54**, 2859–2868.

315 Pizzarello, S., Feng, X., Epstein, S., and Cronin, J.R. (1994) Isotopic analyses of nitrogenous compounds from the murchison meteorite: ammonia, amines, amino acids, and polar hydrocarbons. *Geochimica et Cosmochimica Acta*, **58**, 5579–5587.

316 Schmitt-Kopplin, P., Gabelica, Z., Gougeon, R.D., Fekete, A., Kanawati, B., Harir, M., Gebefuegi, I., Eckel, G., and Hertkorn, N. (2010) High molecular diversity of extraterrestrial organic matter in Murchison meteorite revealed 40 years after its fall. *Proceedings of the National Academy of Science*, **107**, 2763–2768.

317 Botta, O. and Bada, J.L. (2002) Extraterrestrial organic compounds in meteorites. *Surveys in Geophysics*, **23**, 411–467.

318 Cody, G.D., Alexander, C.M.O'D., Kilcoyne, A.L.D., and Yabuta, H. (2008) *Unraveling the Chemical History of the Solar System as Recorded in Extraterrestrial Organic Matter*. Proceedings of the International Astronomical Union Symposium 251: Organic Matter in Space, pp. 277–284.

319 Cody, G.D. and Alexander, C.M.O'D. (2005) NMR studies of chemical structural variation of insoluble organic matter from different carbonaceous chondrite groups. *Geochimica et Cosmochimica Acta*, **69**, 1085–1097.

320 Kerridge, J.F. (1999) Formation and processing of organics in the early Solar System. *Space Science Reviews*, **90**, 275–288.

321 Ehrenfreund, P., Robert, F., d'Hendencourt, L., and Behar, F. (1991) Comparison of interstellar and meteoritic organic matter at 3.4 μm. *Astronomy and Astrophysics*, **252**, 712–717.

322 Pendleton, Y.J. and Allamandola, L.J. (2002) The organic refractory material in the diffuse interstellar medium: mid-infrared spectroscopic constraints. *The Astrophysical Journal Supplement Series*, **138**, 75–98.

323 Becker, L., Poreda, R.J., and Bunch, T.E. (2000) Fullerenes: an extraterrestrial carbon carrier phase for noble gases. *Proceedings of the National Academy of Science*, **97**, 2979–2983.

324 Martins, Z., Botta, O., Fogel, M.L., Sephton, M.A., Glavin, D.P., Watson, J.S., Dworkin, J.P., Schwartz, A.W., and Ehrenfreund, P. (2008) Extraterrestrial nucleobases in the Murchison meteorite. *Earth and Planetary Science Letters*, **270**, 130–136.

325 Nakamura-Messenger, K., Messenger, S., Keller, L.P., Clemett, S.J., and Zolensky, M.E. (2006) Organic globules in the Tagish Lake meteorite: remnants of the protosolar disk. *Science*, **314**, 1439–1442.

326 Love, S.G. and Brownlee, D.E. (1993) A direct measurement of the terrestrial mass accretion rate of cosmic dust. *Science*, **262**, 550–553.

327 Sandford, S.A. and Walker, R.M. (1985) Laboratory infrared transmission spectra of individual interplanetary dust particles from 2.5 to 25 microns. *The Astrophysical Journal*, **291**, 838–851.

328 Schramm, L.S., Brownlee, D.E., and Wheelock, M.M. (1989) Major element composition of stratospheric micrometeorites. *Meteoritics*, **24**, 99–112.

329 Flynn, G.J., Keller, L.P., Feser, M., Wirick, S., and Jacobsen, C. (2003) The origin of organic matter in the Solar System: evidence from the interplanetary dust particles. *Geochimica et Cosmochimica Acta*, **67**, 4791–4806.

330 Flynn, G.J., Keller, L.P., Wirick, S., and Jacobsen, C. (2008) *Organic Matter in Interplanetary Dust Particles*. Proceedings of the International Astronomical Union Symposium 251: Organic Matter in Space, pp. 267–276.

331 Messenger, S. (2000) Identification of molecular-cloud material in interplanetary dust particles. *Nature*, **404**, 968–971.

332 Mumma, M.J., Weaver, H.A., Larson, H.P., Davis, D.S., and Williams, M. (1986) Detection of water vapor in Halley's comet. *Science*, **232**, 1523–1528.

333 Schloerb, F.P., Kinzel, W.M., Swade, D.A., and Irvine, W.M. (1986) HCN production from comet Halley. *The Astrophysical Journal*, **310**, L55–L60.

334 Bockelée-Morvan, D., Lis, D.C., Wink, J.E., Despois, D., Crovisier, J., Bachiller, R., Benford, D.J., Biver, N., Colom, P., Davies, J.K., Gérard, E., Germain, B., Houde, M., Mehringer, D., Moreno, R., Paubert, G., Phillips, T.G., and Rauer, H. (2000) New molecules found in comet C/1995 O1 (Hale-Bopp). Investigating the link between cometary and interstellar material. *Astronomy and Astrophysics*, **353**, 1101–1114.

335 Crovisier, J., Bockelée-Morvan, D., Colom, P., Biver, N., Despois, D., Lis, D.C., and the Team for target-of-opportunity radio observations of comets (2004) The composition of ices in comet C/1995 O1 (Hale-Bopp) from radio spectroscopy. Further results and upper limits on undetected species. *Astronomy and Astrophysics*, **418**, 1141–1157.

336 Crovisier, J., Bockelée-Morvan, D., Biver, N., Colom, P., Despois, D., and Lis, D.C. (2004) Ethylene glycol in comet C/1995 O1 (Hale-Bopp). *Astronomy and Astrophysics*, **418**, L35–L38.

337 Mumma, M.J., DiSanti, M.A., DelloRusso, N., Magee-Sauer, K., Gibb, E., and Novak, R. (2003) Remote infrared observations of parent volatiles in comets: a window on the early Solar System. *Advances in Space Research*, **31**, 2563–2575.

338 Mumma, M.J., McLean, I.S., DiSanti, M.A., Larkin, J.E., DelloRusso, N., Magee-Sauer, K., Becklin, E.E., Bida, T., Chaffee, F., Conrad, A.R., Figer, D.F., Gilbert, A.M., Graham, J.R., Levenson, N.A., Novak, R.E., Reuter, D.C., Teplitz, H.I., Wilcox, M.K., and Xu, L.-H. (2001) A survey of organic volatile species in comet C/1999 H1 (Lee) using NIRSPEC at the Keck Observatory. *The Astrophysical Journal*, **546**, 1183–1193.

339 Kissel, J., Sagdeev, R.Z., Bertaux, J.L., Angarov, V.N., Audouze, J., Blamont, J.E., Büchler, K., Evlanov, E.N., Fechtig, H., Fomenkova, M.N., von Hoerner, H., Inogamov, N.A., Khromov, V.N., Knabe, W., Krueger, F.R., Langevin, Y., Leonas, V.B., Levasseur-Regourd, A.C., Managadze, G.G., Podkolzin, S.N., Shapiro, V.D., Tabaldyev, S.R., and Zubkov, B.V. (1986) Composition of comet Halley dust particles from Vega observations. *Nature*, **321**, 280–282.

340 Kissel, J., Brownlee, D.E., Büchler, K., Clark, B.C., Fechtig, H., Grün, E., Hornung, K., Igenbergs, E.B., Jessberger, E.K., Krueger, F.R., Kuczera, H., McDonnell, J.A.M., Morfill, G.M., Rahe, J., Schwehm, G.H., Sekanina, Z., Utterback, N.G., Völk, H.J., and Zook, H.A. (1986) Composition of comet Halley dust particles from Giotto observations. *Nature*, **321**, 336–337.

341 Sandford, S.A., Aléon, J., Alexander, C.M.O'D., Araki, T., Bajt, S., Baratta, G.A., Borg, J., Bradley, J.P., Brownlee, D.E., Brucato, J.R., Burchell, M.J., Busemann, H., Butterworth, A., Clemett, S.J., Cody, G., Colangeli, L., Cooper, G., d'Hendecourt, L., Djouadi, Z., Dworkin, J.P., Ferrini, G., Fleckenstein, H., Flynn, G.J., Franchi, I.A., Fries, M., Gilles, M.K., Glavin, D.P., Gounelle, M., Grossemy, F., Jacobsen, C., Keller, L.P., Kilcoyne, A.L.D., Leitner, J., Matrajt, G., Meibom, A., Mennella, V., Mostefaoui, S., Nittler, L.R., Palumbo, M.E., Papanastassiou, D.A., Robert, F., Rotundi, A., Snead, C.J., Spencer, M.K., Stadermann, F.J., Steele, A., Stephan, T., Tsou, P., Tyliszczak, T., Westphal, A.J., Wirick, S., Wopenka, B., Yabuta, H., Zare, R.N., and Zolensky, M.E. (2006) Organics captured from comet 81P/Wild 2 by the Stardust spacecraft. *Science*, **314**, 1720–1724.

342 Elsila, J.E., Glavin, D.P., and Dworkin, J.P. (2009) Cometary glycine detected in samples returned by Stardust. *Meteoritics and Planetary Science*, **44**, 1323–1330.

343 Clemett, S.J., Sandford, S.A., Nakamura-Messenger, K., Hörz, F., and McKay, D.S. (2010) Complex aromatic hydrocarbons in Stardust samples collected from comet 81P/Wild 2. *Meteoritics and Planetary Science*, **45**, 701–722.

344 Keller, L.P., Bajt, S., Baratta, G.A., Borg, J., Bradley, J.P., Brownlee, D.E., Busemann, H., Brucato, J.R., Burchell, M., Colangeli, L., D'Hendecourt, L., Djouadi, Z., Ferrini, G., Flynn, G., Franchi, I.A., Fries, M., Grady, M.M., Graham, G.A., Grossemy, F., Kearsley, A., Matrajt, G., Nakamura-Messenger, K., Mennella, V., Nittler, L., Palumbo, M.E., Stadermann, F.J., Tsou, P., Rotundi, A., Sandford, S.A., Snead, C., Steele, A., Wooden, D., and Zolensky, M. (2006) Infrared spectroscopy of comet 81P/Wild 2 samples returned by Stardust. *Science*, **314**, 1728–1731.

345 Hsieh, H.H. and Jewitt, D. (2006) A population of comets in the main asteroid belt. *Science*, **312**, 561–563.

346 Gradie, J. and Veverka, J. (1980) The composition of the Trojan asteroids. *Nature*, **283**, 840–842.

347 Roush, T.L. and Cruikshank, D.P. (2004) Observations and laboratory data of planetary organics, in *Astrobiology: Future Perspectives*, Vol. 305, p. 149.

348 Jewitt, D. and Luu, J. (1993) Discovery of the candidate Kuiper Belt object 1992 QB1. *Nature*, **362**, 730–732.

349 Cruikshank, D.P., Roush, T.L., Bartholomew, M.J., Geballe, T.R., Pendleton, Y.J., White, S.M., Bell, J.F., Davies, J.K., Owen, T.C., de Bergh, C., Tholen, D.J., Bernstein, M.P., Brown, R.H., Tryka, K.A., and Dalle-Ore, C.M. (1998) The composition of Centaur 5145 Pholus. *Icarus*, **135**, 389–407.

350 Kwok, S. (2007) *Physics and Chemistry of the Interstellar Medium*, University Science Books, pp. 319–320.

351 Jewitt, D.C. and Luu, J.X. (2001) Colors and spectra of Kuiper Belt objects. *The Astronomical Journal*, **122**, 2099–2114.

352 Merlin, F., Alvarez-Candal, A., Delsanti, A., Fornasier, S., Barucci, M.A., DeMeo, F.E., de Bergh, C., Doressoundiram, A., Quirico, E., and Schmitt, B. (2009) Stratification of methane ice on Eris' surface. *The Astronomical Journal*, **137**, 315–328.

353 Brown, M.E., Barkume, K.M., Blake, G.A., Schaller, E.L., Rabinowitz, D.L., Roe, H.G., and Trujillo, C.A. (2007) Methane and ethane on the bright Kuiper Belt object 2005 FY9. *The Astronomical Journal*, **133**, 284–289.

354 Barucci, M.A., Brown, M.E., Emery, J.P., and Merlin, F. (2008) Composition and surface properties of trans-Neptunian objects and Centaurs, in *The Solar System Beyond Neptune* (eds M. Antonietta Barucci *et al.*), The University of Arizona Press, pp. 143–160.

355 Swain, M.R., Vasisht, G., and Tinetti, G. (2008) The presence of methane in the atmosphere of an extrasolar planet. *Nature*, **452**, 329–331.

356 Sellgren, K. (1984) The near-infrared continuum emission of visual reflection nebulae. *The Astrophysical Journal*, **277**, 623–633.

357 Tokunaga, A.T. (1997) *A Summary of the "UIR" Bands.* Proceedings of ISAS International Symposium: Diffuse Infrared Radiation and the IRTS, pp. 149–160.

358 Geballe, T.R. (1997) *Spectroscopy of the Unidentified Infrared Emission Bands.* Proceedings of the Symposium held as part of the 108th Annual Meeting of the ASP, June 24–26, 1996, "From Stardust to Planetesimals", pp. 119–128.

359 Peeters, E., Hony, S., VanKerckhoven, C., Tielens, A.G.G.M., Allamandola, L.J., Hudgins, D.M., and Bauschlicher, C.W. (2002) The rich 6 to 9 μm spectrum of interstellar PAHs. *Astronomy and Astrophysics*, **390**, 1089–1113.

360 Beintema, D.A., van den Ancker, M.E., Molster, F.J., Waters, L.B.F.M., Tielens, A.G.G.M., Waelkens, C., de Jong, T., de Graauw, T., Justtanont, K., Yamamura, I., Heras, A., Lahuis, F., and Salama, A. (1996) The rich spectrum of circumstellar PAHs. *Astronomy and Astrophysics*, **315**, L369–L372.

361 Sellgren, K., Uchida, K.I., and Werner, M.W. (2007) The 15–20 μm *Spitzer* spectra of interstellar emission features in NGC 7023. *The Astrophysical Journal*, **659**, 1338–1351.

362 Smith, J.D.T., Draine, B.T., Dale, D.A., Moustakas, J., and Kennicutt Jr., R.C. (2007) The mid-infrared spectrum of star-forming galaxies: global properties of polycyclic aromatic hydrocarbon emission. *The Astrophysical Journal*, **656**, 770–791.

363 VanKerckhoven, C., Hony, S., Peeters, E., Tielens, A.G.G.M., Allamandola, L.J., Hudgins, D.M., Cox, P., Roelfsema, P.R., Voors, R.H.M., Waelkens, C., Waters, L.B.F.M., and Wesselius, P.R. (2000) The C–C–C bending modes of PAHs: a new emission plateau from 15 to 20 μm. *Astronomy and Astrophysics*, **357**, 1013–1019.

364 Zhang, Y., Kwok, S., and Hrivnak, B.J. (2010) A Spitzer/infrared spectrograph spectral study of a sample of galactic carbon-rich proto-planetary nebulae. *The Astrophysical Journal*, **725**, 990–1001.

365 Sloan, G.C., Jura, M., Duley, W.W., Kraemer, K.E., Bernard-Salas, J., Forrest, W.J., Sargent, B., Li, A., Barry, D.J., Bohac, C.J., Watson, D.M., and Houck, J.R. (2007) The unusual hydrocarbon emission from the early carbon star HD 100764: the connection between aromatics and aliphatics. *The Astrophysical Journal*, **664**, 1144–1153.

366 Heger, M.L. (1922) Further study of the sodium lines in class B stars. *Lick Observatory Bulletin*, **337**, 141–145.

367 Merrill, P.W. (1934) Unidentified interstellar lines. *Publications of the Astronomical Society of the Pacific*, **46**, 206–207.

368 Ehrenfreund, P., Cami, J., Jiménez-Vicente, J., Foing, B.H., Kaper, L., van der Meer, A., Cox, N., d'Hendecourt, L., Maier, J.P., Salama, F., Sarre, P.J., Snow, T.P., and Sonnentrucker, P. (2002) Detection of diffuse interstellar bands in the Magellanic Clouds. *The Astrophysical Journal*, **576**, L117–L120.

369 Sollerman, J., Cox, N., Mattila, S., Ehrenfreund, P., Kaper, L., Leibundgut, B., and Lundqvist, P. (2005) Diffuse interstellar bands in NGC 1448. *Astronomy and Astrophysics*, **429**, 559–567.

370 Sarre, P.J. (2006) The diffuse interstellar bands: a major problem in astronom-

ical spectroscopy. *Journal of Molecular Spectroscopy*, **238**, 1–10.

371 Maier, J.P., Walker, G.A.H., Bohlender, D.A., Mazzotti, F.J., Raghunandan, R., Fulara, J., Garkusha, I., and Nagy, A. (2011) Identification of H_2CCC as a diffuse interstellar band carrier. *The Astrophysical Journal*, **726**, 41–50.

372 Snow, T.P. and McCall, B.J. (2006) Diffuse atomic and molecular clouds. *Annual Review of Astronomy and Astrophysics*, **44**, 367–414.

373 Stecher, T.P. (1965) Interstellar extinction in the ultraviolet. *The Astrophysical Journal*, **142**, 1683–1684.

374 Stecher, T.P. (1969) Interstellar extinction in the ultraviolet. II. *The Astrophysical Journal*, **157**, L125–L126.

375 Elíasdóttir, Á., Fynbo, J.P.U., Hjorth, J., Ledoux, C., Watson, D.J., Andersen, A.C., Malesani, D., Vreeswijk, P.M., Prochaska, J.X., Sollerman, J., and Jaunsen, A.O. (2009) Dust extinction in high-z galaxies with gamma-ray burst afterglow spectroscopy: the 2175 Å feature at $z = 2.45$. *The Astrophysical Journal*, **697**, 1725–1740.

376 Nandy, K. and Thompson, G.I. (1975) The correlation between the ultraviolet lambda 2200 feature and the diffuse lambda 4430 band. *Monthly Notices of the Royal Astronomical Society*, **173**, 237–243.

377 Witt, A.N., Bohlin, R.C., and Stecher, T.P. (1983) The diffuse interstellar feature at 4430 Å and interstellar extinction in the far-ultraviolet. *The Astrophysical Journal*, **267**, L47–L51.

378 Cardelli, J.A., Clayton, G.C., and Mathis, J.S. (1989) The relationship between infrared, optical, and ultraviolet extinction. *The Astrophysical Journal*, **345**, 245–256.

379 Joblin, C., Leger, A., and Martin, P. (1992) Contribution of polycyclic aromatic hydrocarbon molecules to the interstellar extinction curve. *The Astrophysical Journal*, **393**, L79–L82.

380 Duley, W.W. and Seahra, S. (1998) Graphite, polycyclic aromatic hydrocarbons, and the 2175 Å extinction feature. *The Astrophysical Journal*, **507**, 874–888.

381 Malloci, G., Mulas, G., Cecchi-Pestellini, C., and Joblin, C. (2008) De-

hydrogenated polycyclic aromatic hydrocarbons and UV bump. *Astronomy and Astrophysics*, **489**, 1183–1187.

382 Steglich, M., Jäger, C., Rouillé, G., Huisken, F., Mutschke, H., and Henning, T. (2010) Electronic spectroscopy of medium-sized polycyclic aromatic hydrocarbons: implications for the carriers of the 2175 Å UV bump. *The Astrophysical Journal*, **712**, L16–L20.

383 Mennella, V., Colangeli, L., Bussoletti, E., Palumbo, P., and Rotundi, A. (1998) A new approach to the puzzle of the ultraviolet interstellar extinction bump. *The Astrophysical Journal*, **507**, L177–L180.

384 de Heer, W.A. and Ugarte, D. (1993) Carbon onions produced by heat treatment of carbon soot and their relation to the 217.5 nm interstellar absorption feature. *Chemical Physics Letters*, **207**, 480–486.

385 Chhowalla, M., Wang, H., Sano, N., Teo, K.B.K., Lee, S.B., and Amaratunga, G.A.J. (2003) Carbon onions: carriers of the 217.5 nm interstellar absorption feature. *Physical Review Letters*, **90**, 155504-1–4.

386 Papoular, R.J. and Papoular, R. (2009) A polycrystalline graphite model for the 2175 Å interstellar extinction band. *Monthly Notices of the Royal Astronomical Society*, **394**, 2175–2181.

387 Cohen, M., Anderson, C.M., Cowley, A., Coyne, G.V., Fawley, W.M., Gull, T.R., Harlan, E.A., Herbig, G.H., Holden, F., Hudson, H.S., Jakoubek, R.O., Johnson, H.M., Merrill, K.M., Schiffer III, F.H., Soifer, B.T., and Zuckerman, B. (1975) The peculiar object HD 44179 ("The Red Rectangle"). *The Astrophysical Journal*, **196**, 179–189.

388 Schmidt, G.D., Cohen, M., and Margon, B. (1980) Discovery of optical molecular emission from the bipolar nebula surrounding HD 44179. *The Astrophysical Journal*, **239**, L133–L138.

389 Witt, A.N. and Schild, R.E. (1988) Hydrogenated amorphous carbon grains in reflection nebulae. *The Astrophysical Journal*, **325**, 837–845.

390 Witt, A.N. and Boroson, T.A. (1990) Spectroscopy of extended red emission in reflection nebulae. *The Astrophysical Journal*, **355**, 182–189.

391 Schmidt, G.D. and Witt, A.N. (1991) X marks the SPOT – distribution and excitation of unidentified molecules in the Red Rectangle. *The Astrophysical Journal*, **383**, 698–704.

392 Vijh, U.P., Witt, A.N., and Gordon, K.D. (2004) Discovery of blue luminescence in the Red Rectangle: possible fluorescence from neutral polycyclic aromatic hydrocarbon molecules? *The Astrophysical Journal*, **606**, L65–L68.

393 Smith, T.L. and Witt, A.N. (2002) The photophysics of the carrier of extended red emission. *The Astrophysical Journal*, **565**, 304–318.

394 Duley, W.W. (1985) Evidence for hydrogenated amorphous carbon in the Red Rectangle. *Monthly Notices of the Royal Astronomical Society*, **215**, 259–263.

395 Sakata, A., Wada, S., Narisawa, T., Asano, Y., Iijima, Y., Onaka, T., and Tokunaga, A.T. (1992) Quenched carbonaceous composite: fluorescence spectrum compared to the extended red emission observed in reflection nebulae. *The Astrophysical Journal*, **393**, L83–L86.

396 Webster, A. (1993) The extended red emission and the fluorescence of C_{60}. *Monthly Notices of the Royal Astronomical Society*, **264**, L1–L2.

397 Witt, A.N., Gordon, K.D., and Furton, D.G. (1998) Silicon nanoparticles: source of extended red emission? *The Astrophysical Journal*, **501**, L111–L115.

398 Ledoux, G., Ehbrecht, M., Guillois, O., Huisken, F., Kohn, B., Laguna, M.A., Nenner, I., Paillard, V., Papoular, R., Porterat, D., and Reynaud, C. (1998) Silicon as a candidate carrier for ERE. *Astronomy and Astrophysics*, **333**, L39–L42.

399 Chang, H.-C., Chen, K., and Kwok, S. (2006) Nanodiamond as a possible carrier of extended red emission. *The Astrophysical Journal*, **639**, L63–L66.

400 Duley, W.W. (2001) The 3.3 micron and extended red emissions in interstellar clouds: further evidence for carbon nanoparticles. *The Astrophysical Journal*, **553**, 575–580.

401 Hoyle, F. and Wickramasinghe, N.C. (1999) Biofluorescence and the extended red emission in astrophysical sources. *Astrophysics and Space Science*, **268**, 321–325.

402 VanWinckel, H., Cohen, M., and Gull, T.R. (2002) The ERE of the "Red Rectangle" revisited. *Astronomy and Astrophysics*, **390**, 147–154.

403 Duley, W.W. (1988) Sharp emission lines from diamond dust in the Red Rectangle? *Astrophysics and Space Science*, **150**, 387–390.

404 Kwok, S., Volk, K.M., and Hrivnak, B.J. (1989) A 21 micron emission feature in four proto-planetary nebulae. *The Astrophysical Journal*, **345**, L51–L54.

405 Volk, K., Kwok, S., and Hrivnak, B.J. (1999) High-resolution *Infrared Space Observatory* spectroscopy of the unidentified 21 micron feature. *The Astrophysical Journal*, **516**, L99–L102.

406 Hrivnak, B.J., Volk, K., Geballe, T.R., and Kwok, S. (2008) *Aromatic, Aliphatic, and the Unidentified 21 Micron Emission Features in Proto-Planetary Nebulae*. Proceedings of the International Astronomical Union Symposium 251: Organic Matter in Space, pp. 213–214.

407 Volk, K., Matsuura, M., Bernard-Salas, J., Sloan, G.C., Szczerba, R., Kemper, F., Woods, P.M., Hrivnak, B.J., Tielens, X., Meixner, M., Gordon, K., Indebetouw, R., van Loon, J., and Marengo, M. (2010) Discovery of 21 micron sources in the Magellanic Clouds. *Bulletin of the American Astronomical Society*, **215**, 486–486.

408 Hrivnak, B.J., Volk, K., and Kwok, S. (2009) A Spitzer study of 21 and 30 µm emission in several galactic carbon-rich protoplanetary nebulae. *The Astrophysical Journal*, **694**, 1147–1160.

409 Buss Jr., R.H. (1990) Hydrocarbon emission features in the infrared spectra of warm supergiants. *The Astrophysical Journal*, **365**, L23–L26.

410 Webster, A. (1995) The lowest of the strongly infrared-active vibrations of the fulleranes and an astronomical emission band at a wavelength of 21 µm. *Monthly Notices of the Royal Astronomical Society*, **277**, 1555–1566.

411 Hill, H.G.M., Jones, A.P., and d'Hendecourt, L.B. (1998) Diamonds in carbon-rich proto-planetary nebulae.

Astronomy and Astrophysics, **336**, L41–L44.

412 von Helden, G., Tielens, A.G.G.M., van Heijnsbergen, D., Duncan, M.A., Hony, S., Waters, L.B.F.M., and Meijer, G. (2000) Titanium carbide nanocrystals in circumstellar environments. *Science*, **288**, 313–316.

413 Papoular, R. (2000) The contribution of oxygen to the "30", "26" and "20" μm features. *Astronomy and Astrophysics*, **362**, L9–L12.

414 Posch, T., Mutschke, H., and Andersen, A. (2004) Reconsidering the origin of the 21 micron feature: oxides in carbon-rich protoplanetary nebulae? *The Astrophysical Journal*, **616**, 1167–1180.

415 Speck, A.K. and Hofmeister, A.M. (2004) Processing of presolar grains around post-asymptotic giant branch stars: silicon carbide as the carrier of the 21 micron feature. *The Astrophysical Journal*, **600**, 986–991.

416 Forrest, W.J., Houck, J.R., and McCarthy, J.F. (1981) A far-infrared emission feature in carbon-rich stars and planetary nebulae. *The Astrophysical Journal*, **248**, 195–200.

417 Hrivnak, B.J., Volk, K., and Kwok, S. (2000) 2-45 micron infrared spectroscopy of carbon-rich proto-planetary nebulae. *The Astrophysical Journal*, **535**, 275–292.

418 Öpik, E. (1931) On the physical interpretation of color-excess in early type stars. *Harvard College Observatory Circular*, **359**, 1–17.

419 Barnard, E.E. (1919) On the dark markings of the sky, with a catalogue of 182 such objects. *The Astrophysical Journal*, **49**, 1–24.

420 Mathis, J.S., Rumpl, W., and Nordsieck, K.H. (1977) The size distribution of interstellar grains. *The Astrophysical Journal*, **217**, 425–433.

421 Henning, T.K. (ed.) (2003) *Astromineralogy*, Springer-Verlag.

422 Douglas, A.E. (1977) Origin of diffuse interstellar lines. *Nature*, **269**, 130–132.

423 Jochnowitz, E.B. and Maier, J.P. (2008) *Electronic Spectra of Carbon Chains and Rings: Astrophysical Relevance?* Proceedings of the International Astronomical Union Symposium 251: Organic Matter in Space, pp. 395–402.

424 Puget, J.L. and Léger, A. (1989) A new component of the interstellar matter: small grains and large aromatic molecules. *Annual Review of Astronomy and Astrophysics*, **27**, 161–198.

425 Tielens, A.G.G.M. (2008) Interstellar polycyclic aromatic hydrocarbon molecules. *Annual Review of Astronomy and Astrophysics*, **46**, 289–337.

426 Lovas, F.J., McMahon, R.J., Grabow, J.-U., Schnell, M., Mack, J., Scott, L.T., and Kuczkowski, R.L. (2005) Interstellar chemistry: a strategy for detecting polycyclic aromatic hydrocarbons in space. *Journal of the American Chemical Society*, **127**, 4345–4349.

427 Pilleri, P., Herberth, D., Giesen, T.F., Gerin, M., Joblin, C., Mulas, G., Malloci, G., Grabow, J.-U., Brünken, S., Surin, L., Steinberg, B.D., Curtis, K.R., and Scott, L.T. (2009) Search for corannulene ($C_{20}H_{10}$) in the Red Rectangle. *Monthly Notices of the Royal Astronomical Society*, **397**, 1053–1060.

428 Jones, A.P., Duley, W.W., and Williams, D.A. (1990) The structure and evolution of hydrogenated amorphous carbon grains and mantles in the interstellar medium. *Quarterly Journal of the Royal Astronomical Society*, **31**, 567–582.

429 Scott, A. and Duley, W.W. (1996) The decomposition of hydrogenated amorphous carbon: a connection with polycyclic aromatic hydrocarbon molecules. *The Astrophysical Journal*, **472**, L123–L125.

430 Scott, A., Duley, W.W., and Pinho, G.P. (1997) Polycyclic aromatic hydrocarbons and fullerenes as decomposition products of hydrogenated amorphous carbon. *The Astrophysical Journal*, **489**, L193–L195.

431 Allamandola, L.J., Tielens, A.G.G.M., and Barker, J.R. (1985) Polycyclic aromatic hydrocarbons and the unidentified infrared emission bands: auto exhaust along the Milky Way! *The Astrophysical Journal*, **290**, L25–L28.

432 Colangeli, L., Mennella, V., Palumbo, P., Rotundi, A., and Bussoletti, E. (1995) Mass extinction coefficients of various submicron amorphous carbon grains: tabulated values from 40 nm to 2 mm. *Astronomy and Astrophysics Supplement Series*, **113**, 561–577.

433 Mennella, V., Brucato, J.R., Colangeli, L., and Palumbo, P. (2002) C–H bond formation in carbon grains by exposure to atomic hydrogen: the evolution of the carrier of the interstellar 3.4 micron band. *The Astrophysical Journal*, **569**, 531–540.

434 Herlin, N., Bohn, I., Reynaud, C., Cauchetier, M., Galvez, A., and Rouzaud, J.-N. (1998) Nanoparticles produced by laser pyrolysis of hydrocarbons: analogy with carbon cosmic dust. *Astronomy and Astrophysics*, **330**, 1127–1135.

435 Jäger, C., Mutschke, H., Henning, T., and Huisken, F. (2008) Spectral properties of gas-phase condensed fullerene-like carbon nanoparticles from far-ultraviolet to infrared wavelengths. *The Astrophysical Journal*, **689**, 249–259.

436 Hu, A. and Duley, W.W. (2007) Laboratory simulation of 11–15 μm spectra associated with polycyclic aromatic hydrocarbon molecules. *The Astrophysical Journal*, **660**, L137–L140.

437 Duley, W.W. and Hu, A. (2009) Polyynes and interstellar carbon nanoparticles. *The Astrophysical Journal*, **698**, 808–811.

438 Jäger, C., Huisken, F., Mutschke, H., Jansa, I.L., and Henning, T. (2009) Formation of polycyclic aromatic hydrocarbons and carbonaceous solids in gas-phase condensation experiments. *The Astrophysical Journal*, **696**, 706–712.

439 Sakata, A., Wada, S., Tanabé, T., and Onaka, T. (1984) Infrared spectrum of the laboratory-synthesized quenched carbonaceous composite (QCC): comparison with the infrared unidentified emission bands. *The Astrophysical Journal*, **287**, L51–L54.

440 Sakata, A., Wada, S., Onaka, T., and Tokunaga, A.T. (1987) Infrared spectrum of quenched carbonaceous composite (QCC). II. a new identification of the 7.7 and 8.6 micron unidentified infrared emission bands. *The Astrophysical Journal*, **320**, L63–L67.

441 Sakata, A., Wada, S., Okutsu, Y., Shintani, H., and Nakada, Y. (1983) Does a 2,200 Å hump observed in an artificial carbonaceous composite account for UV interstellar extinction? *Nature*, **301**, 493–494.

442 Wada, S., Mizutani, Y., Narisawa, T., and Tokunaga, A.T. (2009) On the carrier of the extended red emission and blue luminescence. *The Astrophysical Journal*, **690**, 111–119.

443 Durand, B. (1980) Sedimentary organic matter and kerogen. definition and quantitative importance of kerogen, in *Kerogen: Insoluble Organic Matter from Sedimentary Rocks* (ed. B. Durand), Editions technip, pp. 13–34.

444 Papoular, R. (2001) The use of kerogen data in understanding the properties and evolution of interstellar carbonaceous dust. *Astronomy and Astrophysics*, **378**, 597–607.

445 Papoular, R., Conard, J., Giuliano, M., Kister, J., and Mille, G. (1989) A coal model for the carriers of the unidentified IR bands. *Astronomy and Astrophysics*, **217**, 204–208.

446 Papoular, R., Conard, J., Guillois, O., Nenner, I., Reynaud, C., and Rouzaud, J.-N. (1996) A comparison of solid-state carbonaceous models of cosmic dust. *Astronomy and Astrophysics*, **315**, 222–236.

447 Guillois, O., Nenner, I., Papoular, R., and Reynaud, C. (1996) Coal models for the infrared emission spectra of proto–planetary nebulae. *The Astrophysical Journal*, **464**, 810–817.

448 Puget, J.L., Leger, A., and D'Hendecourt, L. (1995) Comment on the coal model of interstellar dust. *Astronomy and Astrophysics*, **293**, 559–561.

449 Papoular, R., Guillois, O., Nenner, I., and Reynaud, C. (1995) Reply to "Comment on the coal model of interstellar dust". *Astronomy and Astrophysics*, **293**, 562–564.

450 Dow, W.G. (1977) Kerogen studies and geological interpretations. *Journal of Geochemical Exploration*, **7**, 79–99.

451 Helgeson, H.C., Richard, L., McKenzie, W.F., Norton, D.L., and Schmitt, A. (2009) A chemical and thermodynamic model of oil generation in hydrocarbon source rocks. *Geochimica et Cosmochimica Acta*, **73**, 594–695.

452 Kenney, J.F., Kutcherov, V.A., Bendeliani, N.A., and Alekseev, V.A. (2002) The evolution of multicomponent systems at high pressures: VI. the thermodynamic stability of the hydrogen¡vcarbon system: the genesis of hydrocarbons and the origin of petroleum. *Proceedings of the National Academy of Sciences of the United States of America*, **99**, 10976–10981.

453 Porfir'ev, V.B. (1974) Inorganic origin of petroleum. *American Association of Petroleum Geologists Bulletin*, **58**, 3–33.

454 Gold, T. (2001) *The Deep Hot Biosphere: The Myth of Fossil Fuels*, Copernicus Books (imprint of Springer-Verlag, 1999).

455 Hoyle, F. (1955) *Frontiers of Astronomy*, Harper, p. 37.

456 Cataldo, F. and Keheyan, Y. (2003) Heavy petroleum fractions as possible analogues of carriers of the unidentified infrared bands. *International Journal of Astrobiology*, **2**, 41–50.

457 Cataldo, F., Keheyan, Y., and Heymann, D. (2004) Complex organic matter in space: about the chemical composition of carriers of the unidentified infrared bands (UIBs) and protoplanetary emission spectra recorded from certain astrophysical objects. *Origins of Life and Evolution of Biospheres*, **34**, 13–24.

458 Proust, M. (1808) XLIX. Materials for a history of the prussiates. *The Philosophical Magazine*, **32**, 336–357.

459 Oró, J. (2002) Historical understanding of life's beginnings, in *Life's Origin: The Beginnings of Biological Evolution* (ed. J.W. Schopf), University of California Press, NATO-ASI Series C, pp. 7–45.

460 Matthews, C.N. and Moser, R.E. (1967) Peptide synthesis from hydrogen cyanide and water. *Nature*, **215**, 1230–1234.

461 Oró, J. and Kimball, A.P. (1961) Synthesis of purines under possible primitive Earth conditions. I. Adenine from hydrogen cyanide. *Archives of Biochemistry and Biophysics*, **94**, 217–227.

462 Matthews, C., Nelson, J., Varma, P., and Minard, R. (1977) Deuterolysis of amino acid precursors: evidence for hydrogen cyanide polymers as protein ancestors. *Science*, **198**, 622–625.

463 Matthews, C.N. and Minard, R.D. (2006) Hydrogen cyanide polymers, comets and the origin of life. *Faraday Discussions*, **133**, 393–401.

464 Matthews, C.N. and Minard, R.D. (2008) *Hydrogen Cyanide Polymers Connect Cosmochemistry and Biochemistry*. Proceedings of the International Astronomical Union Symposium 251: Organic Matter in Space, pp. 453–457.

465 Sagan, C. and Khare, B.N. (1979) Tholins: organic chemistry of interstellar grains and gas. *Nature*, **277**, 102–107.

466 Khare, B.N., Sagan, C., Arakawa, E.T., Suits, F., Callcott, T.A., and Williams, M.W. (1984) Optical constants of organic tholins produced in a simulated Titanian atmosphere: from soft x-ray to microwave frequencies. *Icarus*, **60**, 127–137.

467 Coll, P., Coscia, D., Smith, N., Gazeau, M.C., Ramírez, S.I., Cernogora, G., Israël, G., and Raulin, F. (1999) Experimental laboratory simulation of Titan's atmosphere: aerosols and gas phase. *Planetary and Space Science*, **47**, 1331–1340.

468 Ferris, J., Tran, B., Joseph, J., Vuitton, V., Briggs, R., and Force, M. (2005) The role of photochemistry in Titan's atmospheric chemistry. *Advances in Space Research*, **36**, 251–257.

469 Quirico, E., Montagnac, G., Lees, V., McMillan, P.F., Szopa, C., Cernogora, G., Rouzaud, J.-N., Simon, P., Bernard, J.-M., Coll, P., Fray, N., Minard, R.D., Raulin, F., Reynard, B., and Schmitt, B. (2008) New experimental constraints on the composition and structure of tholins. *Icarus*, **198**, 218–231.

470 Cruikshank, D.P., Imanaka, H., and DalleOre, C.M. (2005) Tholins as coloring agents on outer Solar System bodies. *Advances in Space Research*, **36**, 178–183.

471 Bernard, J.-M., Quirico, E., Brissaud, O., Montagnac, G., Reynard, B.,

McMillan, P., Coll, P., Nguyen, M.-J., Raulin, F., and Schmitt, B. (2006) Reflectance spectra and chemical structure of Titan's tholins: application to the analysis of Cassini–Huygens observations. *Icarus*, **185**, 301–307.

472 Quirico, E., Szopa, C., Cernogora, G., Lees, V., Derenne, S., McMillan, P.F., Montagnac, G., Reynard, B., Rouzaud, J.-N., Fray, N., Coll, P., Raulin, F., Schmitt, B., and Minard, B. (2008) *Tholins and Their Relevance for Astrophysical Issues.* Proceedings of the International Astronomical Union Symposium 251: Organic Matter in Space, pp. 409–416.

473 Hoyle, F. and Wickramasinghe, N.C. (1977) Polysaccharides and the infrared spectrum of OH 26.5+0.6. *Monthly Notices of the Royal Astronomical Society*, **181**, 51P–55P.

474 Hoyle, F., Wickramasinghe, N.C., Al-Mufti, S., and Olavesen, A.H. (1982) Infrared spectroscopy of micro-organisms near 3,4 μm in relation to geology and astronomy. *Astrophysics and Space Science*, **81**, 489–492.

475 Wickramasinghe, N.C. and Hoyle, F. (1999) The astonishing redness of Kuiper-Belt objects. *Astrophysics and Space Science*, **268**, 369–372.

476 Holye, F. and Wickramasinghe, N.C. (1999) Identification of the λ 2200 Å interstellar absorption feature. *Astrophysics and Space Science*, **268**, 301–303.

477 Wickramasinghe, N.C., Hoyle, F., and Nandy, K. (1977) Organic molecules in interstellar dust: a possible spectral signature at λ 2200 Å? *Astrophysics and Space Science*, **47**, L9–L13.

478 Chalfie, M. and Kain, S.R. (eds) (2006) *Green Fluorescent Protein: Properties, Applications, and Protocols*, 2nd edn, Wiley-Interscience, Hoboken.

479 Matz, M.V., Lukyanov, K.A., and Lukyanov, S.A. (2002) Family of the green fluorescent protein: journey to the end of the rainbow. *BioEssays*, **24**, 953–959.

480 Duley, W.W. and Williams, D.A. (1984) *Interstellar chemistry*. Academic Press.

481 Winnewisser, G. and Herbst, E. (1993) Interstellar molecules. *Reports on Progress in Physics*, **56**, 1209–1273.

482 Miller, S.L. (1953) A production of amino acids under possible primitive Earth conditions. *Science*, **117**, 528–529.

483 Sagan, C. and Khare, B.N. (1971) Long-wavelength ultraviolet photoproduction of amino acids on the primitive Earth. *Science*, **173**, 417–420.

484 Harada, K. and Fox, S.W. (1964) Thermal synthesis of natural amino-acids from a postulated primitive terrestrial atmosphere. *Nature*, **201**, 335–336.

485 MayoGreenberg, J., Li, A., Mendoza-Gómez, C.X., Schutte, W.A., Gerakines, P.A., and de Groot, M. (1995) Approaching the interstellar grain organic refractory component. *The Astrophysical Journal*, **455**, L177–L180.

486 Gibb, E.L. and Whittet, D.C.B. (2002) The 6 micron feature in protostars: evidence for organic refractory material. *The Astrophysical Journal*, **566**, L113–L116.

487 Bernstein, M.P., Sandford, S.A., Allamandola, L.J., Chang, S., and Scharberg, M.A. (1995) Organic compounds produced by photolysis of realistic interstellar and cometary ice analogs containing methanol. *The Astrophysical Journal*, **454**, 327–344.

488 Bernstein, M.P., Dworkin, J.P., Sandford, S.A., Cooper, G.W., and Allamandola, L.J. (2002) Racemic amino acids from the ultraviolet photolysis of interstellar ice analogues. *Nature*, **416**, 401–403.

489 MuñozCaro, G.M., Meierhenrich, U.J., Schutte, W.A., Barbier, B., ArconesSegovia, A., Rosenbauer, H., Thiemann, W.H.-P., Brack, A., and Greenberg, J.M. (2002) Amino acids from ultraviolet irradiation of interstellar ice analogues. *Nature*, **416**, 403–406.

490 Kobayashi, K., Takano, Y., Masuda, H., Tonishi, H., Kaneko, T., Hashimoto, H., and Saito, T. (2004) Possible cometary organic compounds as sources of planetary biospheres. *Advances in Space Research*, **33**, 1277–1281.

491 Dworkin, J.P., Deamer, D.W., Sandford, S.A., and Allamandola, L.J. (2001) Spe-

cial feature: self-assembling amphiphilic molecules: synthesis in simulated interstellar/precometary ices. *Proceedings of the National Academy of Science*, **98**, 815–819.

492 Harris, S.J. and Weiner, A.M. (1985) Chemical kinetics of soot particle growth. *Annual Review of Physical Chemistry*, **36**, 31–52.

493 Frenklach, M. and Feigelson, E.D. (1989) Formation of polycyclic aromatic hydrocarbons in circumstellar envelopes. *The Astrophysical Journal*, **341**, 372–384.

494 Frenklach, M. (2002) Reaction mechanism of soot formation in flames. *Physical Chemistry Chemical Physics*, **4**, 2028–2037.

495 Fogel, M.E. and Leung, C.M. (1998) Modeling extinction and infrared emission from fractal dust grains: fractal dimension as a shape parameter. *The Astrophysical Journal*, **501**, 175–191.

496 Kennedy, I.M. (1997) Models of soot formation and oxidation. *Progress in Energy and Combustion Science*, **23**, 95–132.

497 Nakamura, K., Zolensky, M.E., Tomita, S., Nakashima, S., and Tomeoka, K. (2002) Hollow organic globules in the Tagish lake meteorite as possible products of primitive organic reactions. *International Journal of Astrobiology*, **1**, 179–189.

498 Saito, M. and Kimura, Y. (2009) Origin of organic globules in meteorites: laboratory simulation using aromatic hydrocarbons. *The Astrophysical Journal*, **703**, L147–L151.

499 Ney, E.P. and Hatfield, B.F. (1978) The isothermal dust condensation of Nova Vulpeculae 1976. *The Astrophysical Journal*, **219**, L111–L115.

500 Lynch, D.K., Woodward, C.E., Gehrz, R., Helton, L.A., Rudy, R.J., Russell, R.W., Pearson, R., Venturini, C.C., Mazuk, S., Rayner, J., Ness, J.U., Starrfield, S., Wagner, R.M., Osborne, J.P., Page, K., Puetter, R.C., Perry, R.B., Schwarz, G., Vanlandingham, K., Black, J., Bode, M., Evans, A., Geballe, T., Greenhouse, M., Hauschildt, P., Krautter, J., Liller, W., Lyke, J., Truran, J., Kerr, T., Eyres, S.P.S., and Shore, S.N. (2008) Nova V2362 Cygni (Nova Cygni 2006): Spitzer, Swift, and ground-based spectral evolution. *The Astronomical Journal*, **136**, 1815–1827.

501 Evans, A., Geballe, T.R., Rawlings, J.M.C., Eyres, S.P.S., and Davies, J.K. (1997) Infrared spectroscopy of Nova Cassiopeiae 1993. II – Evolution of the dust. *Monthly Notices of the Royal Astronomical Society*, **292**, 192–204.

502 Evans, A., Tyne, V.H., Smith, O., Geballe, T.R., Rawlings, J.M.C., and Eyres, S.P.S. (2005) Infrared spectroscopy of Nova Cassiopeiae 1993 – IV. A closer look at the dust. *Monthly Notices of the Royal Astronomical Society*, **360**, 1483–1492.

503 Prieto, J.L., Kistler, M.D., Thompson, T.A., Yüksel, H., Kochanek, C.S., Stanek, K.Z., Beacom, J.F., Martini, P., Pasquali, A., and Bechtold, J. (2008) Discovery of the dust-enshrouded progenitor of SN 2008S with Spitzer. *The Astrophysical Journal*, **681**, L9–L12.

504 Prieto, J.L., Sellgren, K., Thompson, T.A., and Kochanek, C.S. (2009) A Spitzer/IRS spectrum of the 2008 luminous transient in NGC 300: connection to proto-planetary nebulae. *The Astrophysical Journal*, **705**, 1425–1432.

505 Orgel, L.E. (1998) Polymerization on the rocks: theoretical introduction. *Origins of Life and Evolution of the Biosphere*, **28**, 227–234.

506 Ferris, J.P. and Hagan, W.J. (1984) HCN and chemical evolution: the possible role of cyano compounds in prebiotic synthesis. *Tetrahedron*, **40**, 1093–1120.

507 Oro, J. (1995) Chemical synthesis of lipids and the origin of life. *Journal of Biological Physics*, **20**, 135–147.

508 Butlerov, A.M. (1861) Einiges über die chemische Structur der Körper. *Zeitschrift für Chemie und Pharmacie*, **4**, 549–560.

509 Crick, F. (1981) *Life Itself: Its Origin and Nature*, Simon and Schuster, New York.

510 Oparin, A.I. (1938) *The Origin of Life*, Macmillan.

511 Haldane, J.B.S. (1929) The origin of life. *Rationalist Annual*, **148**, 3–10.

512 Miller, S.L. and Urey, H.C. (1959) Organic compound synthesis on the primitive Earth. *Science*, **130**, 245–251.

513 Arrhenius, S.A. and Borns, H. (1908) *Worlds in the Making; The Evolution of the Universe*, Harper & brothers.

514 Hoyle, F. and Wickramasinghe, N.C. (1977) Polysaccharides and infrared spectra of galactic sources. *Nature*, **268**, 610–612.

515 Hoyle, F. and Wickramasinghe, N.C. (1999) Comets – a vehicle for panspermia. *Astrophysics and Space Science*, **268**, 333–341.

516 Wickramasinghe, C. (2010) The astrobiological case for our cosmic ancestry. *International Journal of Astrobiology*, **9**, 119–129.

517 Cano, R.J. and Borucki, M.K. (1995) Revival and identification of bacterial spores in 25- to 40-million-year-old Dominican amber. *Science*, **268**, 1060–1064.

518 Vreeland, R.H., Rosenzweig, W.D., and Powers, D.W. (2000) Isolation of a 250 million-year-old halotolerant bacterium from a primary salt crystal. *Nature*, **407**, 897–900.

519 Grieve, R.A.F. (1998) Extraterrestrial impacts on Earth: the evidence and the consequences. *Geological Society, London, Special Publications*, **140**, 105–131.

520 Kring, D.A. and Cohen, B.A. (2002) Cataclysmic bombardment throughout the inner Solar System 3.9–4.0 Ga. *Journal of Geophysical Research*, **107**, 4–1–4–6.

521 Barringer, D.M. (1905) Coon Mountain and its crater. *Proceedings of the Academy of Natural Sciences of Philadelphia*, **57**, 861–886.

522 Grieve, R.A.F. (2001) Impact cratering on Earth. *Geological Survey of Canada Bulletin 548: A Synthesis of Geological Hazards in Canada*, pp. 207–224.

523 Therriault, A.M., Grieve, R.A.F., and Reimold, W.U. (1997) Original size of the Vredefort structure: implications for the geological evolution of the Witwatersrand Basin. *Meteoritics and Planetary Science*, **32**, 71–77.

524 Grieve, R.A.F., Stöffler, D., and Deutsch, A. (1991) The Sudbury structure: controversial or misunderstood? *Journal of Geophysical Research*, **96**, 22753–22764.

525 Alvarez, L.W., Alvarez, W., Asaro, F., and Michel, H.V. (1980) Extraterrestrial cause for the Cretaceous-Tertiary extinction. *Science*, **208**, 1095–1108.

526 Hildebrand, A.R., Penfield, G.T., Kring, D.A., Pilkington, M., Camargo, Z.A., Jacobsen, S.B., and Boynton, W.V. (1991) Chicxulub Crater: a possible Cretaceous/Tertiary boundary impact crater on the Yucatán Peninsula, Mexico. *Geology*, **19**, 867–871.

527 Schulte, P., Alegret, L., Arenillas, I., Arz, J.A., Barton, P.J., Bown, P.R., Bralower, T.J., Christeson, G.L., Claeys, P., Cockell, C.S., Collins, G.S., Deutsch, A., Goldin, T.J., Goto, K., Grajales-Nishimura, J.M., Grieve, R.A.F., Gulick, S.P.S., Johnson, K.R., Kiessling, W., Koeberl, C., Kring, D.A., MacLeod, K.G., Matsui, T., Melosh, J., Montanari, A., Morgan, J.V., Neal, C.R., Nichols, D.J., Norris, R.D., Pierazzo, E., Ravizza, G., Rebolledo-Vieyra, M., Reimold, W.U., Robin, E., Salge, T., Speijer, R.P., Sweet, A.R., Urrutia-Fucugauchi, J., Vajda, V., Whalen, M.T., and Willumsen, P.S. (2010) The Chicxulub asteroid impact and mass extinction at the Cretaceous-Paleogene Boundary. *Science*, **327**, 1214–1218.

528 Strom, R.G., Malhotra, R., Ito, T., Yoshida, F., and Kring, D.A. (2005) The origin of planetary impactors in the inner Solar System. *Science*, **309**, 1847–1850.

529 Cohen, B.A., Swindle, T.D., and Kring, D.A. (2000) Support for the lunar cataclysm hypothesis from lunar meteorite impact melt ages. *Science*, **290**, 1754–1756.

530 Kring, D.A. (2003) Environmental consequences of impact cratering events as a function of ambient conditions on Earth. *Astrobiology*, **3**, 133–152.

531 Love, S.G. and Brownlee, D.E. (1991) Heating and thermal transformation of micrometeoroids entering the Earth's atmosphere. *Icarus*, **89**, 26–43.

532 Delsemme, A.H. (1998) *Our Cosmic Origins: From the Big Bang to the Emergence of Life and Intelligence*, Cambridge University Press.

533 Oró, J. (1961) Comets and the formation of biochemical compounds on the primitive Earth. *Nature*, **190**, 389–390.

534 Anders, E. (1989) Pre-biotic organic matter from comets and asteroids. *Nature,* **342**, 255–257.

535 Deamer, D.W. (1997) The first living systems: a bioenergetic perspective. *Microbiology and Molecular Biology Reviews,* **61**, 239–261.

536 Bernatowicz, T., Fraundorf, G., Ming, T., Anders, E., Wopenka, B., Zinner, E., and Fraundorf, P. (1987) Evidence for interstellar SiC in the Murray carbonaceous meteorite. *Nature,* **330**, 728–730.

537 Nittler, L.R., Alexander, C.M.O'D., Gao, X., Walker, R.M., and Zinner, E. (1997) Stellar sapphires: the properties and origins of presolar Al_2O_3 in meteorites. *The Astrophysical Journal,* **483**, 475–495.

538 Nagashima, K., Krot, A.N., and Yurimoto, H. (2004) Stardust silicates from primitive meteorites. *Nature,* **428**, 921–924.

539 Zinner, E. (1998) Stellar nucleosynthesis and the isotopic composition of presolar grains from primitive meteorites. *Annual Review of Earth and Planetary Sciences,* **26**, 147–188.

540 Ehrenfreund, P., Irvine, W., Becker, L., Blank, J., Brucato, J.R., Colangeli, L., Derenne, S., Despois, D., Dutrey, A., Fraaije, H., Lazcano, A., Owen, T., Robert, F., and International Space Science Institute ISSI-Team (2002) Astrophysical and astrochemical insights into the origin of life. *Reports on Progress in Physics,* **65**, 1427–1487.

541 Alexander, C.M.O'D., Cody, G.D., Fogel, M., and Yabuta, H. (2008) *Organics in Meteorites – Solar or Interstellar?*. Proceedings of the International Astronomical Union Symposium 251: Organic Matter in Space, pp. 293–298.

542 Gillett, F.C., Forrest, W.J., and Merrill, K.M. (1973) 8–13-micron spectra of NGC 7027, BD +30° 3639, and NGC 6572. *The Astrophysical Journal,* **183**, 87–93.

543 Merrill, K.M., Soifer, B.T., and Russell, R.W. (1975) The 2–4 micron spectrum of NGC 7027. *The Astrophysical Journal,* **200**, L37–L39.

544 Knacke, R.F. (1977) Carbonaceous compounds in interstellar dust. *Nature,* **269**, 132–134.

545 Duley, W.W. and Williams, D.A. (1979) Are there organic grains in the interstellar medium. *Nature,* **277**, 40–41.

546 Chyba, C. and Sagan, C. (1992) Endogenous production, exogenous delivery and impact-shock synthesis of organic molecules: an inventory for the origins of life. *Nature,* **355**, 125–132.

547 Calculating Earth's total fossil fuel reserves. http://www.scribd.com/doc/27989321/Calculating-Earth-s-Total-Fossil-Fuel-Reserves (accessed 5 July 2011).

548 Kwok, S. (2002) Mining for cosmic coal. *Astronomy,* **30**, 46–50.

549 Yao, Y. and Wang, Q.D. (2006) X-ray absorption spectroscopy of the multiphase interstellar medium: oxygen and neon abundances. *The Astrophysical Journal,* **641**, 930–937.

550 Lee, J.C., Xiang, J., Ravel, B., Kortright, J., and Flanagan, K. (2009) Condensed matter astrophysics: a prescription for determining the species-specific composition and quantity of interstellar dust using X-rays. *The Astrophysical Journal,* **702**, 970–979.

551 Kwok, S., Su, K.Y.L., and Hrivnak, B.J. (1998) *Hubble Space Telescope* V-band imaging of the bipolar proto-planetary nebula IRAS 17150-3224. *The Astrophysical Journal,* **501**, L117–L121.

552 Su, K.Y.L., Volk, K., Kwok, S., and Hrivnak, B.J. (1998) *Hubble Space Telescope* imaging of IRAS 17441-2411: a case study of a bipolar nebula with a circumstellar disk. *The Astrophysical Journal,* **508**, 744–751.

553 Hrivnak, B.J., Kwok, S., and Su, K.Y.L. (1999) The discovery of two new bipolar proto-planetary nebulae: IRAS 16594-4656 and IRAS 17245-3951. *The Astrophysical Journal,* **524**, 849–856.

554 García-Lario, P., Manchado, A., Ulla, A., and Manteiga, M. (1999) *Infrared Space Observatory* observations of IRAS 16594-4656: a new proto-planetary nebula with a strong 21 micron dust feature. *The Astrophysical Journal,* **513**, 941–946.

555 Hirabayashi, H. (1985) *New 45 m Radio Telescope and Fourier-Transform Type Spectrometer at Nobeyama Radio Observatory.* Proceedings of the International Astronomical Union Symposium 112: The Search for Extraterrestrial Life: Recent Developments, pp. 425–433.

556 van der Veen, W. (1989) *The James Clerk Maxwell Telescope.* Proceedings of 22nd Eslab Symposium on Infrared Spectroscopy in Astronomy, pp. 567–569.

557 Prestage, R.M., Constantikes, K.T., Hunter, T.R., King, L.J., Lacasse, R.J., Lockman, F.J., and Norrod, R.D. (2009) The Green Bank Telescope. *Proceedings of the IEEE*, **97**, 1382–1390.

558 Ezawa, H., Kawabe, R., Kohno, K., and Yamamoto, S. (2004) The Atacama Submillimeter Telescope Experiment (ASTE). *Proceedings of the SPIE*, **5489**, 763–772.

559 Gusten, R., Booth, R.S., Cesarsky, C., Menten, K.M., Agurto, C., Anciaux, M., Azagra, F., Belitsky, V., Belloche, A., Bergman, P., DeBreuck, C., Comito, C., Dumke, M., Duran, C., Esch, W., Fluxa, J., Greve, A., Hafok, H., Haupl, W., Helldner, L., Henseler, A., Heyminck, S., Johansson, L.E., Kasemann, C., Klein, B., Korn, A., Kreysa, E., Kurz, R., Lapkin, I., Leurini, S., Lis, D., Lundgren, A., Mac-Auliffe, F., Martinez, M., Melnick, J., Morris, D., Muders, D., Nyman, L.A., Olberg, M., Olivares, R., Pantaleev, M., Patel, N., Pausch, K., Philipp, S.D., Philipps, S., Sridharan, T.K., Polehampton, E., Revereet, V., Risacher, C., Roa, M., Sauer, P., Schilke, P., Santana, J., Schneider, G., Sepulveda, J., Siringo, G., Spyromilio, J., Stenvers, K.H., van der Tak, F., Torres, D., Vanzi, L., Vassilev, V., Weiss, A., Willmeroth, K., Wunsch, A., and Wyrowski, F. (2006) APEX: the Atacama Pathfinder EXperiment. *Proceedings of the SPIE*, **6267**, 626714–1–26.

560 McLean, I.S., Becklin, E.E., Bendiksen, O., Brims, G., Canfield, J., Figer, D.F., Graham, J.R., Hare, J., Lacayanga, F., Larkin, J.E., Larson, S.B., Levenson, N.G., Magnone, N., Teplitz, H.I., and Wong, W. (1998) Design and development of NIRSPEC: a near-infrared echelle spectrograph for the Keck II telescope. *Proceedings of the SPIE*, **3354**, 566–578.

561 Gillett, F.C., Mountain, M., Kurz, R., Simons, D.A., Smith, M.G., and Boroson, T. (1996) The Gemini Telescopes Project. *Revista Mexicana de Astronomia y Astrofisica Conference Series*, **4**, 75–82.

562 IRAS Science Team (1983) IRAS, the Infrared Astronomical Satellite. *Nature*, **303**, 287–291.

563 Kessler, M.F., Steinz, J.A., Anderegg, M.E., Clavel, J., Drechsel, G., Estaria, P., Faelker, J., Riedinger, J.R., Robson, A., Taylor, B.G., and Ximénezde-Ferrán, S. (1996) The Infrared Space Observatory (ISO) mission. *Astronomy and Astrophysics*, **315**, L27–L31.

564 de Graauw, T., Haser, L.N., Beintema, D.A., Roelfsema, P.R., van Agthoven, H., Barl, L., Bauer, O.H., Bekenkamp, H.E.G., Boonstra, A.-J., Boxhoorn, D.R., Coté, J., de Groene, P., van Dijkhuizen, C., Drapatz, S., Evers, J., Feuchtgruber, H., Fericks, M., Genzel, R., Haerendel, G., Heras, A.M., van der Hucht, K.A., van der Hulst, T., Huygen, R., Jacobs, H., Jakob, G., Kamperman, T., Katterloher, R.O., Kester, D.J.M., Kunze, D., Kussendrager, D., Lahuis, F., Lamers, H.J.G.L.M., Leech, K., van der Lei, S., van der Linden, R., Luinge, W., Lutz, D., Melzner, F., Morris, P.W., van Nguyen, D., Ploeger, G., Price, S., Salama, A., Schaeidt, S.G., Sijm, N., Smoorenburg, C., Spakman, J., Spoon, H., Steinmayer, M., Stoecker, J., Valentijn, E.A., Vandenbussche, B., Visser, H., Waelkens, C., Waters, L.B.F.M., Wensink, J., Wesselius, P.R., Wiezorrek, E., Wieprecht, E., Wijnbergen, J.J., Wildeman, K.J., and Young, E. (1996) Observing with the ISO Short-Wavelength Spectrometer. *Astronomy and Astrophysics*, **315**, L49–L54.

565 Clegg, P.E., Ade, P.A.R., Armand, C., Baluteau, J.-P., Barlow, M.J., Buckley, M.A., Berges, J.-C., Burgdorf, M., Caux, E., Ceccarelli, C., Cerulli, R., Church, S.E., Cotin, F., Cox, P., Cruvellier, P., Culhane, J.L., Davis, G.R.,

DiGiorgio, A., Diplock, B.R., Drummond, D.L., Emery, R.J., Ewart, J.D., Fischer, J., Furniss, I., Glencross, W.M., Greenhouse, M.A., Griffin, M.J., Gry, C., Harwood, A.S., Hazell, A.S., Joubert, M., King, K.J., Lim, T., Liseau, R., Long, J.A., Lorenzetti, D., Molinari, S., Murray, A.G., Naylor, D.A., Nisini, B., Norman, K., Omont, A., Orfei, R., Patrick, T.J., Péquignot, D., Pouliquen, D., Price, M.C., Rogers, A.J., Robinson, F.D., Saisse, M., Saraceno, P., Serra, G., Sidher, S.D., Smith, A.F., Smith, H.A., Spinoglio, L., Swinyard, B.M., Texier, D., Towlson, W.A., Trams, N.R., Unger, S.J., and White, G.J. (1996) The ISO Long-Wavelength Spectrometer. *Astronomy and Astrophysics*, **315**, L38–L42.

566 Okuda, H. (1997) IRTS and IRIS. *The Far Infrared and Submillimetre Universe*, **401**, 207–212.

567 Murakami, H., Baba, H., Barthel, P., Clements, D.L., Cohen, M., Doi, Y., Enya, K., Figueredo, E., Fujishiro, N., Fujiwara, H., Fujiwara, M., Garcia-Lario, P., Goto, T., Hasegawa, S., Hibi, Y., Hirao, T., Hiromoto, N., Hong, S.S., Imai, K., Ishigaki, M., Ishiguro, M., Ishihara, D., Ita, Y., Jeong, W.-S., Jeong, K.S., Kaneda, H., Kataza, H., Kawada, M., Kawai, T., Kawamura, A., Kessler, M.F., Kester, D., Kii, T., Kim, D.C., Kim, W., Kobayashi, H., Koo, B.C., Kwon, S.M., Lee, H.M., Lorente, R., Makiuti, S., Matsuhara, H., Matsumoto, T., Matsuo, H., Matsuura, S., Müller, T.G., Murakami, N., Nagata, H., Nakagawa, T., Naoi, T., Narita, M., Noda, M., Oh, S.H., Ohnishi, A., Ohyama, Y., Okada, Y., Okuda, H., Oliver, S., Onaka, T., Ootsubo, T., Oyabu, S., Pak, S., Park, Y.-S., Pearson, C.P., Rowan-Robinson, M., Saito, T., Sakon, I., Salama, A., Sato, S., Savage, R.S., Serjeant, S., Shibai, H., Shirahata, M., Sohn, J., Suzuki, T., Takagi, T., Takahashi, H., Tanabé, T., Takeuchi, T.T., Takita, S., Thomson, M., Uemizu, K., Ueno, M., Usui, F., Verdugo, E., Wada, T., Wang, L., Watabe, T., Watarai, H., White, G.J., Yamamura, I., Yamauchi, C., and Yasuda, A. (2007) The Infrared Astronomical Mission AKARI.

Publications of the Astronomical Society of Japan, **59**, S369–S376.

568 Onaka, T. and Salama, A. (2009) AKARI: space infrared cooled telescope. *Experimental Astronomy*, **27**, 9–17.

569 Ohyama, Y., Onaka, T., Matsuhara, H., Wada, T., Kim, W., Fujishiro, N., Uemizu, K., Sakon, I., Cohen, M., Ishigaki, M., Ishihara, D., Ita, Y., Kataza, H., Matsumoto, T., Murakami, H., Oyabu, S., Tanabé, T., Takagi, T., Ueno, M., Usui, F., Watarai, H., Pearson, C.P., Takeyama, N., Yamamuro, T., and Ikeda, Y. (2007) Near-infrared and mid-infrared spectroscopy with the Infrared Camera (IRC) for AKARI. *Publications of the Astronomical Society of Japan*, **59**, S411–S422.

570 Melnick, G.J., Stauffer, J.R., Ashby, M.L.N., Bergin, E.A., Chin, G., Erickson, N.R., Goldsmith, P.F., Harwit, M., Howe, J.E., Kleiner, S.C., Koch, D.G., Neufeld, D.A., Patten, B.M., Plume, R., Schieder, R., Snell, R.L., Tolls, V., Wang, Z., Winnewisser, G., and Zhang, Y.F. (2000) The Submillimeter Wave Astronomy Satellite: science objectives and instrument description. *The Astrophysical Journal*, **539**, L77–L85.

571 Nordh, H.L., von Schéele, F., Frisk, U., Ahola, K., Booth, R.S., Encrenaz, P.J., Hjalmarson, Å., Kendall, D., Kyrölä, E., Kwok, S., Lecacheux, A., Leppelmeier, G., Llewellyn, E.J., Mattila, K., Mégie, G., Murtagh, D., Rougeron, M., and Witt, G. (2003) The Odin orbital observatory. *Astronomy and Astrophysics*, **402**, L21–L25.

572 Werner, M.W., Roellig, T.L., Low, F.J., Rieke, G.H., Rieke, M., Hoffmann, W.F., Young, E., Houck, J.R., Brandl, B., Fazio, G.G., Hora, J.L., Gehrz, R.D., Helou, G., Soifer, B.T., Stauffer, J., Keene, J., Eisenhardt, P., Gallagher, D., Gautier, T.N., Irace, W., Lawrence, C.R., Simmons, L., Cleve, J.E.V., Jura, M., Wright, E.L., and Cruikshank, D.P. (2004) The Spitzer space telescope mission. *The Astrophysical Journal Supplement Series*, **154**, 1–9.

573 Houck, J.R., Roellig, T.L., van Cleve, J., Forrest, W.J., Herter, T., Lawrence, C.R., Matthews, K., Reitsema, H.J.,

Soifer, B.T., Watson, D.M., Weedman, D., Huisjen, M., Troeltzsch, J., Barry, D.J., Bernard-Salas, J., Blacken, C.E., Brandl, B.R., Charmandaris, V., Devost, D., Gull, G.E., Hall, P., Henderson, C.P., Higdon, S.J.U., Pirger, B.E., Schoenwald, J., Sloan, G.C., Uchida, K.I., Appleton, P.N., Armus, L., Burgdorf, M.J., Fajardo-Acosta, S.B., Grillmair, C.J., Ingalls, J.G., Morris, P.W., and Teplitz, H.I. (2004) The Infrared Spectrograph (IRS) on the *Spitzer Space Telescope*. *The Astrophysical Journal Supplement Series*, **154**, 18–24.

574 Pilbratt, G.L., Riedinger, J.R., Passvogel, T., Crone, G., Doyle, D., Gageur, U., Heras, A.M., Jewell, C., Metcalfe, L., Ott, S., and Schmidt, M. (2010) Herschel Space Observatory. An ESA facility for far-infrared and submillimetre astronomy. *Astronomy and Astrophysics*, **518**, L1–1–6.

575 de Graauw, T., Helmich, F.P., Phillips, T.G., Stutzki, J., Caux, E., Whyborn, N.D., Dieleman, P., Roelfsema, P.R., Aarts, H., Assendorp, R., Bachiller, R., Baechtold, W., Barcia, A., Beintema, D.A., Belitsky, V., Benz, A.O., Bieber, R., Boogert, A., Borys, C., Bumble, B., Caïs, P., Caris, M., Cerulli-Irelli, P., Chattopadhyay, G., Cherednichenko, S., Ciechanowicz, M., Coeur-Joly, O., Comito, C., Cros, A., de Jonge, A., de Lange, G., Delforges, B., Delorme, Y., den Boggende, T., Desbat, J.-M., Diez-González, C., di Giorgio, A.M., Dubbeldam, L., Edwards, K., Eggens, M., Erickson, N., Evers, J., Fich, M., Finn, T., Franke, B., Gaier, T., Gal, C., Gao, J.R., Gallego, J.-D., Gauffre, S., Gill, J.J., Glenz, S., Golstein, H., Goulooze, H., Gunsing, T., Güsten, R., Hartogh, P., Hatch, W.A., Higgins, R., Honingh, E.C., Huisman, R., Jackson, B.D., Jacobs, H., Jacobs, K., Jarchow, C., Javadi, H., Jellema, W., Justen, M., Karpov, A., Kasemann, C., Kawamura, J., Keizer, G., Kester, D., Klapwijk, T.M., Klein, T., Kollberg, E., Kooi, J., Kooiman, P.-P., Kopf, B., Krause, M., Krieg, J.-M., Kramer, C., Kruizenga, B., Kuhn, T., Laauwen, W., Lai, R., Larsson, B., Leduc, H.G., Leinz, C., Lin, R.H.,

Liseau, R., Liu, G.S., Loose, A., López-Fernandez, I., Lord, S., Luinge, W., Marston, A., Martín-Pintado, J., Maestrini, A., Maiwald, F.W., McCoey, C., Mehdi, I., Megej, A., Melchior, M., Meinsma, L., Merkel, H., Michalska, M., Monstein, C., Moratschke, D., Morris, P., Muller, H., Murphy, J.A., Naber, A., Natale, E., Nowosielski, W., Nuzzolo, F., Olberg, M., Olbrich, M., Orfei, R., Orleanski, P., Ossenkopf, V., Peacock, T., Pearson, J.C., Peron, I., Phillip-May, S., Piazzo, L., Planesas, P., Rataj, M., Ravera, L., Risacher, C., Salez, M., Samoska, L.A., Saraceno, P., Schieder, R., Schlecht, E., Schlöder, F., Schmülling, F., Schultz, M., Schuster, K., Siebertz, O., Smit, H., Szczerba, R., Shipman, R., Steinmetz, E., Stern, J.A., Stokroos, M., Teipen, R., Teyssier, D., Tils, T., Trappe, N., van Baaren, C., van Leeuwen, B.-J., van de Stadt, H., Visser, H., Wildeman, K.J., Wafelbakker, C.K., Ward, J.S., Wesselius, P., Wild, W., Wulff, S., Wunsch, H.-J., Tielens, X., Zaal, P., Zirath, H., Zmuidzinas, J., and Zwart, F. (2010) The Herschel-Heterodyne Instrument for the Far-Infrared (HIFI). *Astronomy and Astrophysics*, **518**, L6–1–7.

576 Griffin, M.J., Abergel, A., Abreu, A., Ade, P.A.R., André, P., Augueres, J.-L., Babbedge, T., Bae, Y., Baillie, T., Baluteau, J.-P., Barlow, M.J., Bendo, G., Benielli, D., Bock, J.J., Bonhomme, P., Brisbin, D., Brockley-Blatt, C., Caldwell, M., Cara, C., Castro-Rodriguez, N., Cerulli, R., Chanial, P., Chen, S., Clark, E., Clements, D.L., Clerc, L., Coker, J., Communal, D., Conversi, L., Cox, P., Crumb, D., Cunningham, C., Daly, F., Davis, G.R., de Antoni, P., Delderfield, J., Devin, N., di Giorgio, A., Didschuns, I., Dohlen, K., Donati, M., Dowell, A., Dowell, C.D., Duband, L., Dumaye, L., Emery, R.J., Ferlet, M., Ferrand, D., Fontignie, J., Fox, M., Franceschini, A., Frerking, M., Fulton, T., Garcia, J., Gastaud, R., Gear, W.K., Glenn, J., Goizel, A., Griffin, D.K., Grundy, T., Guest, S., Guillemet, L., Hargrave, P.C., Harwit, M., Hastings, P., Hatziminaoglou, E., Herman, M., Hinde, B.,

Hristov, V., Huang, M., Imhof, P., Isaak, K.J., Israelsson, U., Ivison, R.J., Jennings, D., Kiernan, B., King, K.J., Lange, A.E., Latter, W., Laurent, G., Laurent, P., Leeks, S.J., Lellouch, E., Levenson, L., Li, B., Li, J., Lilienthal, J., Lim, T., Liu, S.J., Lu, N., Madden, S., Mainetti, G., Marliani, P., McKay, D., Mercier, K., Molinari, S., Morris, H., Moseley, H., Mulder, J., Mur, M., Naylor, D.A., Nguyen, H., O'Halloran, B., Oliver, S., Olofsson, G., Olofsson, H.-G., Orfei, R., Page, M.J., Pain, I., Panuzzo, P., Papageorgiou, A., Parks, G., Parr-Burman, P., Pearce, A., Pearson, C., Pérez-Fournon, I., Pinsard, F., Pisano, G., Podosek, J., Pohlen, M., Polehampton, E.T., Pouliquen, D., Rigopoulou, D., Rizzo, D., Roseboom, I.G., Roussel, H., Rowan-Robinson, M., Rownd, B., Saraceno, P., Sauvage, M., Savage, R., Savini, G., Sawyer, E., Scharmberg, C., Schmitt, D., Schneider, N., Schulz, B., Schwartz, A., Shafer, R., Shupe, D.L., Sibthorpe, B., Sidher, S., Smith, A., Smith, A.J., Smith, D., Spencer, L., Stobie, B., Sudiwala, R., Sukhatme, K., Surace, C., Stevens, J.A., Swinyard, B.M., Trichas, M., Tourette, T., Triou, H., Tseng, S., Tucker, C., Turner, A., Vaccari, M., Valtchanov, I., Vigroux, L., Virique, E., Voellmer, G., Walker, H., Ward, R., Waskett, T., Weilert, M., Wesson, R., White, G.J., Whitehouse, N., Wilson, C.D., Winter, B., Woodcraft, A.L., Wright, G.S., Xu, C.K., Zavagno, A., Zemcov, M., Zhang, L., and Zonca, E. (2010) The Herschel-SPIRE instrument and its in-flight performance. *Astronomy and Astrophysics*, **518**, L3–1–7.

577 Witteborn, F.C., Cohen, M., Bregman, J.D., Heere, K.R., Greene, T.P., and Wooden, D.H. (1995) *HIFOGS: Its Design, Operations and Calibration.* Proceedings of Airborne Astronomy Symposium on the Galactic Ecosystem: From Gas to Stars to Dust, pp. 573–578.

578 Erickson, E.F., Houck, J.R., Harwit, M.O., Rank, D.M., Haas, M.R., Hollenbach, D.J., Simpson, J.P., and Augason, G.C (1985) An FIR cooled grating spectrometer for the Kuiper Airborne Observatory. *Infrared Physics*, **25**, 513–515.

579 Gehrz, R.D., Becklin, E.E., de Pater, I., Lester, D.F., Roellig, T.L., and Woodward, C.E. (2009) A new window on the cosmos: the Stratospheric Observatory for Infrared Astronomy (SOFIA). *Advances in Space Research*, **44**, 413–432.

580 Ishiguro, M., Morita, K., Kasuga, T., Kanzawa, T., Iwashita, H., Chikada, Y., Inatani, J., Suzuki, H., Handa, K., Takahashi, T., Tanaka, H., Kobayashi, H., and Kawabe, R. (1984) *The Nobeyama Millimeter-Wave Interferometer.* Proceedings of the International Symposium on Millimeter and Submillimeter Wave Radio Astronomy (International Union of Radio Science), p. 75.

581 Guilloteau, S., Delannoy, J., Downes, D., Greve, A., Guelin, M., Lucas, R., Morris, D., Radford, S.J.E., Wink, J., Cernicharo, J., Forveille, T., Garcia-Burillo, S., Neri, R., Blondel, J., Perrigourad, A., Plathner, D., and Torres, M. (1992) The IRAM interferometer on Plateau de Bure. *Astronomy and Astrophysics*, **262**, 624–633.

582 Blundell, R. (2007) The Submillimeter Array. *Microwave Symposium, IEEE/MTT-S International*, pp. 1857–1860.

583 Beasley, A.J. and Vogel, S.N. (2003) CARMA: specifications and status. *Proceedings of the SPIE*, **4855**, 254–264.

584 Wootten, A. and Thompson, A.R. (2009) The Atacama Large Millimeter/Submillimeter Array. *Proceedings of the IEEE*, **97**, 1463–1471.

Index

Organic Matter in the Universe, First Edition. Sun Kwok.
© 2012 WILEY-VCH Verlag GmbH & Co. KGaA. Published 2012 by WILEY-VCH Verlag GmbH & Co. KGaA.